RAPESEED

RAPESEED

CULTIVATION, COMPOSITION, PROCESSING AND UTILIZATION

edited by

L.-Å. APPELQVIST

Division of Physiological Chemistry, Chemical Center, University of Lund, Lund, Sweden

and

R. OHLSON

AB Karlshamns Oljefabriker, Karlshamn, Sweden

ELSEVIER PUBLISHING COMPANY

AMSTERDAM LONDON NEW YORK

1972

ELSEVIER PUBLISHING COMPANY
335 JAN VAN GALENSTRAAT
P.O. BOX 211, AMSTERDAM, THE NETHERLANDS

AMERICAN ELSEVIER PUBLISHING COMPANY, INC.
52 VANDERBILT AVENUE
NEW YORK, NEW YORK 10017

LIBRARY OF CONGRESS CARD NUMBER: 75–135491
ISBN 0–444–40892–4
WITH 97 ILLUSTRATIONS AND 92 TABLES

PRINTED IN THE NETHERLANDS

List of contributors

E. Aaes-Jørgensen, Department of Biochemistry Royal Danish School of Pharmacy, Copenhagen (Denmark)

K. Anjou, AB Karlshamns Oljefabriker, Karlshamn (Sweden)

L.-Å. Appelqvist, Division of Physiological Chemistry, Chemical Center, University of Lund (Sweden)

L. Bengtsson, Oil Crop Division, Swedish Seed Association, Svalöf (Sweden)

A. von Hofsten, Institute of Physiological Botany, University of Uppsala (Sweden)

E. Honkanen, AB Karlshamns Oljefabriker, Karlshamn (Sweden)

E. Josefsson, Swedish Seed Association, Svalöf (Sweden)

B. Lööf, Oil Crop Division, Swedish Seed Association, Svalöf (Sweden)

R. Ohlson, AB Karlshamns Oljefabriker, Karlshamn (Sweden)

U. Persmark, AB Karlshamns Oljefabriker, Karlshamn (Sweden)

Ü. Riiner, AB Karlshamns Oljefabriker, Karlshamn (Sweden)

R. Wettström, AB Karlshamns Oljefabriker, Karlshamn (Sweden)

Preface

At the present time, world production of seed from rape and its relatives amounts to about six million tons per year, in fifth place among oilseed crops after soyabeans, cottonseed, groundnuts and sunflower seed. With an oil content of about 40 per cent and a protein content of about 25 per cent, rapeseed ranks fifth in the production of oilseed protein, but sixth in the production of vegetable oil, preceded by the aforementioned species and by coconut. Nevertheless its nutritional potential is often overlooked in discussions of the world food situation. Part of this may be due to old misconceptions about rapeseed and its products, and to the fact that relatively little research has been done on rapeseed compared to that devoted to soyabeans, cottonseed and groundnuts. During the past five to ten years, increased interest has been shown in rapeseed oil and meal, and thus we felt that a monograph dealing with the many different aspects of this important agricultural commodity would indeed satisfy a real need, especially since no handbook on the subject has hitherto been compiled. It is our hope that this book will remove many of the prevailing misconceptions about rapeseed and promote interest in this valuable source of vegetable oil and protein.

India is at present the largest rapeseed grower in the world, and considerable amounts are also grown in China and Pakistan. In all of these countries large amounts of rapeseed have been produced for centuries. In Canada, now one of the major production areas, rapeseed was first grown and utilized during World War II. Although some rapeseed was grown in northeastern Europe at the beginning of the 20th century, the breakthrough in production and utilization did not come until after the War. Knowledge about the composition and proper utilization of rapeseed remained meager until about twenty years ago, but currently much attention is being given to improvements in the production, composition and utilization of rapeseed oil and meal. Thus, the greater part of the literature cited in the present monograph is from rather modern work, mainly carried out in Canada and Europe.

Our intention has been to have the various chapters written as uniformly as feasible in style and pattern, and also to prevent unnecessary overlap and information gaps. Nevertheless it is inevitable that a few chapters, in view of the nature of the information presented, must follow a slightly different style. Thus Chapters 4, 5 and 6 present the best information available to the authors on the cultivation, harvesting, storage and breeding of rapeseed without giving detailed references for each important statement. In these fields so many 'practical' experiments have been performed that it was considered inexpedient to present the results with detailed references. A similar style has been used for Chapter 2. The other chapters are however styled in the manner expected for a review on a scientific subject.

We wish to express our sincere thanks to all of the authors of this volume for their great patience in following our suggestions, sometimes involving very large changes in the first manuscripts. Much valuable information on rapeseed presented in this volume has been gathered by visits to and correspondence with colleagues in other countries, mainly Canada, France, Germany and Poland. It is a pleasure to extend our thanks to them for their generosity in supplying interesting data and stimulating ideas. The English of this book has been corrected by Dr Robert E. Carter, Assistant Professor of Organic Chemistry at the Lund Institute of Technology. His great interest and kind cooperation are gratefully acknowledged.

Lund and Karlshamn, March 1972   Lars-Åke Appelqvist
                                                  Ragnar Ohlson

Contents

CHAPTER 1

Historical Background

L.-Å. APPELQVIST

Division of Physiological Chemistry, Chemical Center, University of Lund (Sweden)

CONTENTS

1.1 INTRODUCTION

The cultivation of rape and mustard seed for oil production has a long history from which a few glimpses will be presented in this chapter. Since the second World War drastically changed the pattern of rapeseed production in Europe, which had previously imported essentially all edible oil needed for margarine production, and, furthermore, initiated rapeseed production in Canada, statistics comparing the situation before and after World War II will also be discussed in this chapter. Production from 1940 and onwards in various parts of the world is presented in detail in chapter 2.

1.2 EARLY HISTORY OF RAPESEED CULTIVATION IN EUROPE

Natural populations of turnip rape (*Brassica campestris* L.) are reported from several different areas of Europe and Asia, whereas rape (*Brassica napus* L.) seems to occur naturally in more restricted areas, mainly in Europe and North-western Africa. It has been suggested that the domestication of these two species occurred whenever the value of the seeds collected from weed plants in grain fields was appreciated. It is also assumed that the two species have been cultivated as oil crops for a very long time but only in those countries of Europe where the olive tree and the poppy were unknown. Thus it was never used by the Romans. Seeds of one or both of the two Brassicas have been found in old German settle-

Fig.1:1. The threshing of rapeseed in Skåne (Southern Sweden) in the nineteenth century.
Artist: J. W. Wallander.

Sixteen men are standing in a large bin, probably made of ordinary bed-sheets (or sail-cloth). To the left, the landlord and his assistant inspect the work, one man rests while beer is being offered to others. Two violinists obviously entertain the 'party'. To the far right, two women take care of the threshed seed while three men in the central background are engaged in stacking the straw. Including the man on the horse (function unknown) it thus appears that altogether 29 people were engaged in this undertaking. The picture was kindly supplied by Prof. N. A. Bringeus, Institute of Ethnology and Folklore, University of Lund, Sweden.

ments together with millstones. In view of the difficulty of distinguishing seeds of the two species from each other (*cf.* Chapter 3) it is understandable that, in general, nothing can be learned about the geographic distribution of cultivated rape *vs.* turnip rape. However, a finding near Zurich of about 35 seeds of the *B. campestris* type from the bronze age has been reported.

The cultivation of rape and turnip rape supposedly occurred in Europe in the thirteenth century. In the latter part of the Middle Ages it was the most important lamp oil in Europe, north of the Alps, until it was replaced by petroleum oils. Even earlier, however, seeds from the weeds were collected and used for illumination and soap making. Whereas one author reports no field size cropping until about 1700, another reports cultivation on a field scale as early as the thirteenth century.

It seems to have been in Belgium that the occurrence of rape or turnip rape as a weed was so annoying that the oil first began to be utilized. It is also stated that Holland was the only country to grow rapeseed at the end of the seventeenth century. From there on, it spread into Germany; it was *e.g.* introduced into Sachsen in 1781 (The information so far presented under 1:2 was taken from the text of Schiemann[1]). A Spaniard who visited Belgium in 1772, reported on rapeseed cultivation and utilization there and recommended its cultivation in Spain[1a]. From the reports of Linnaeus[2, 3] it is also known to have occurred on a very small scale in Sweden in the 1750's, which is discussed in detail below. In the nineteenth century, the cultivation of rapeseed declined, probably due to the appearance of mineral oil for illumination. On the other hand its cultivation extended into new areas — eastwards into Switzerland, Poland and Russia, northwards into Denmark and Sweden[1]. In a German textbook on farming from the early 1800's, the difficulties as well as the advantages of the production of industrial crops, *e.g.* rapeseed, are discussed. It is emphasized that rapeseed would be a good 'cash' crop and that it has given a good return for farmers in Holland and other parts of Europe[4].

The Danish acreage of rapeseed amounted to 15,500 hectares (about 35,000 acres) in 1866, which was probably the highest level attained during the nineteenth century[5]. The peak in rapeseed production in Sweden during the nineteenth century occurred in 1866 when about 3000 tons were produced, mainly in southern Sweden[6]. From 1866 on, the production declined rapidly; at the end of the century only about 2 tons were produced. The reasons for this decline were many, *e.g.* the price reduction due to the importation of other vegetable oils for margarine and soap production[6].

It is reported that during the first World War there was a short period of cultivation of rape and mustard in Sweden, but some years thereafter, in 1923, the production ceased completely[7]. According to other sources, no rapeseed was produced during World War I in Sweden, but brown mustard (*Brassica juncea*) and white mustard were grown[8].

In view of the world-wide recognition of the achievements of the great Swedish
scientist, Carl Linnaeus, a translation of the parts of his reports concerned
with rape and mustard seed is no doubt of considerable interest. This is presented
below, together with an abbreviated description of other observations. Carl
Linnaeus made three different 'expeditions' to various parts of Sweden in 1741,
1746 and 1749[2, 3, 9].

In the book covering his visits to the islands of Öland and Gotland[9], he
repeatedly remarks how field mustard and field turnips overgrow the barley or
the rye, thereby greatly reducing the yield of grain. On page 240 in that report he
writes:

'Oil seed plantings, so common in Flanders and Braband etc., and of great
benefit for the inhabitants, have not yet been established with us, although there
are very good reasons for doing this in view of the large abundance of wild-flax,
field-mustard and field-turnips in the grain fields. From these seeds, just as good
an oil could be pressed as from the kale-seed and rape-seed of the foreigners.
Field turnips (*Napus sylvestris*) grew in great abundance among the barley in
'Burs' although far from the sea. The root was thick as on kale or carrots, the
leaves were glabrous and embraced the stalk like with kale, the flower-cups
stretched out somewhat, the pods were large and the seeds considerable, which
would give plenty of oil if they were pressed. No herb can be more easily planted
than this one, which hardly can be eradicated from the fields, and thus none could
be planted to greater advantage for oil production.

The field mustard (*Rapistrum flore luteo* C.B.) was so plentiful here that we saw
a whole field all yellow from it; the barley was so suppressed that it had not grown
even to the height of about ten inches. We would have in fact believed that the
field-mustard had been sown here instead of barley, unless the farmers had told us
the reverse. We tried to convince the farmers to press oil from the seed, which
actually yields a nice oil, but a difficulty may be that not all pods mature at the
same time. One pod shatters the seeds while the other pod is producing new ones.
The root of field-mustard is thin, the leaves are irregular and divided, the flower-
cups are outstretched'.

Also on his journey to Västergötland, a province in western Sweden, in 1746,
he noted the abundance of field-mustard as a weed. More interesting however,
is that he found a planting of rapeseed outside the town of Alingsås, where
another famous Swede, Mr. Jonas Alströmer, who first introduced the potatoe
to Sweden, had planted rapeseed for processing at an oil mill. Linnaeus remarks
on page 132: 'He (Alströmer) did not succeed too well in his test planting; I suppose
that the 'Gottlandish' turnips would be better suited for oil production'[2].

In his report from the 'expedition' to Skåne, the southernmost province of

Sweden, Linnaeus describes the constructional details of an oil mill, built by a certain Johan Haeggardt in Malmö. Basically the process was as follows: the seeds were crushed in a 'pounding mill' drawn by horses, followed by heating in a copper pot. The crushed seeds were then transferred to a piece of hair-cloth and pressed in a wooden press. It is interesting to note than the cakes remaining from pressing linseed were used to feed the horses, whereas the rapeseed cakes were used as fuel under the copper pot (*cf.* Chapter 15).

After his description of the oil mill, Linnaeus continues on page 189 of his report:

'Rapeseed was sown annually on 4–5 acres by Johan Haeggardt near Malmö, starting in 1737. He gave me the following description of the planting: The soil should preferably be a good humus, although poorer soils can also be used when fertilized, well-cultivated and drained. The soil should be turned over 4 to 5 times a year, so that it becomes well loamy and free from weeds. Sowing is performed between the 25th of July and the 10th of August, using 6 'cups' (1 cup = 4.58 liters) of rapeseed per acre. The crop stands throughout the Winter and must therefore be protected by a fence all the year round; it grows an ell (about 0.6 meters) high before the Winter, yet all that is above ground rots, so that in the Spring only the small cotyledons are visible, which again grow so rapidly that blooming is finished by the end of May. In the beginning of July, somewhat earlier or later depending on the weather, mature pods are ready and the crop can then be cut for drying, which takes 6–8 days. Later the crop is threshed on specially manufactured, large sheets of fabric or sail-cloth, and the seed is cleaned. Four men can thresh 9–10 barrels a day. From four acres, seeded with 24 'cups' of seed, 65 barrels of seed are usually obtained, from which the same number of 'ankers' of oil may be pressed. Domestic seed has always yielded more oil per barrel of seed than foreign seed. As soon as the rapeseed is harvested, the soil is ploughed, further cultivated and planted with winter-rye; the following year barley is sown.

Whereas in the neighbourhood the yield of barley is usually 6-fold, he (Haeggardt) obtains a 16- to 20-fold yield and never less than 12-fold. Therefore, those who become interested in this crop have no reason to regret their toil when they in this manner can derive rich remuneration from a well-cultivated soil. This town owns 1320 acres for crop farming and cattle pasture, of which each house-owner has a share. Some of it, less than a thirteenth, is used for tobacco plantings and about 50 acres for linseed and 'naked' barley. All the rest, except for the 18-acre fenced plot for rapeseed, is only used for growing grains. Besides the aforementioned acreage, there is another 1020 acres near the town. Some of this is privately owned and some belongs to the town council and is used as a subsidy for the town officials. Therefore, supposing only 150 acres of these fields were planted with rapeseed each year, this would bring in 65,000 Swedish 'silver dollars'. The public would indeed profit if other towns in this country, which have suitable soil, would plant this crop. The tobacco plantings near Malmö, which are well kept and yielded 200 kilos yearly, would not at all suffer from it. All of this information was given to me by Mr. Haeggardt, who was so actively engaged in building this oil mill for the benefit of the country. Three 'skeppor' of rapeseed usually yield eight 'pots' of oil, each pot being 1/3 of a tankard. Rapeseed is *Napus sylvestris* (Fl. 547) *calyce femipatents* or the 'Gottlandish' field turnips (Fl. Gottl. 240), which is a weed in the eastern parts of 'Gottland', where it ruins the barley fields like the field-mustard does in this part of Sweden. Where else in the world would rapeseed grow more easily and be more prolific than where it is a weed? Indeed, the 'Gottländer' could be rich from the oil, that is now only wasted, as are the fields which are stifled by the weed. What can be more stupid than to let a plant stifle fields, that could pay off 10-fold the best year's crop!

Felices Agricolae, si sua bona norint'.

1.4 NOTES ON THE HISTORY OF RAPESEED CULTIVATION IN ASIA AND AMERICA

The earliest direct references to oilseed rape or sarson in India are found in ancient Sanskrit writings from 2000 to 1500 B.C.[10]. Singh considered the Indian *Brassica campestris* variety Yellow Sarson to be the oldest of the various rapes and mustards grown in India[10].

The history of rapeseed cultivation in China is obscure. Old Japanese literature indicates that rape was introduced to Japan 2000 years ago directly from China or via the Korean Peninsula[11].

In Canada, nowadays the largest exporter of rapeseed, the commercial growing of rapeseed started in 1942. The primary aim was to supply lubricants for marine engines[11]. In view of its short history the entire development of rapeseed cultivation in Canada is treated in Chapter 2.

Very little, if any, rapeseed is grown in the United States for oil production. In statistical tables on world rapeseed production, nothing (= less than one thousand metric tons) is reported for the U.S. in the years 1930–1954[12]. As a fodder crop it is however used in some states, but the discussion of *Brassica* fodder crops, such as fodder rape (see Chapter 3) is outside the scope of this book. Furthermore, it has been emphasized that none of the *Brassica* vegetables is native to the American continent[13].

Presently there is extensive production of rapeseed in Chile (see Chapter 2). The only South American country with a rapeseed production prior to World War II seems to be Argentina with an average production of 26,000 metric tons in 1935–1939[13].

1.5 WORLD PRODUCTION OF RAPESEED IN THE 1930's COMPARED WITH THE POST-WORLD WAR II PERIOD

As previously mentioned, the outbreak of World War II markedly changed the production pattern of rapeseed in several parts of the world. In Europe the total production in 1948–1952 was three to four times greater than that during the years 1934–1938 (Table 1:1). The increase mainly refers to France (from 13 to 154

TABLE 1:1

THE PRODUCTION OF RAPESEED IN SOME EUROPEAN COUNTRIES IN 1934–1938 COMPARED WITH 1948–1952 (according to ref. 14)

Country	Annual production, thousand metric tons	
	1934–1938	1948–1952
Austria	2	5
Belgium	—	5
Bulgaria	18	14
Czechoslovakia	7	25
Denmark	—	7
Finland	—	6
France	13	154
Germany,	73	193
Eastern	(40)	(110)
Western	(33)	(83)
Great Britain	—	—
Hungary	8	—
Italy	2	14
Netherlands	5	33
Poland	48	100
Romania	24	4
Sweden	—	146
Switzerland	—	5
Yugoslavia	11	5
Total	210	720

thousand metric tons), Germany (from 73 to 193), Poland (from 48 to 100) and Sweden (from 0 to 146). Several European countries had a small and irregular production both before and after World War II. The increased European production was a necessity owing to the shortage of edible oils, but once established it was found to be of great value for a balanced agriculture in northern Europe (*cf.* Chapter 4).

In contrast with the European rise in rapeseed production shortly after 1945, the increased production of this crop occurred much later in Canada (*cf.* Table 1:2 and Chapter 2). The Asian production of rapeseed displayed some fluctuations but little overall change during this time. Thus, there are considerable differences in

TABLE 1:2

Country	Annual production, thousand metric tons			
	1930–1934	1935–1939	1945–1949	1950–1954
America, total	—	26	37	(16)
Argentina	—	26	18	2
Canada	—	—	12	7
Mexico	—	—	7	(7)
U.S.	—	—	—	—
Asia, total	3,577	3,189	(4,121)	(4,207)
China	2,227	2,102	(3,100)	(2,854)
India ⟩	1,264	969	1,001	866
Pakistan ⟨				277
Japan	86	118	20	210
Europe, total	111	241	343	794

The figures within parenthesis are estimates only and those for India and Pakistan include mustard (mainly *B. juncea*) as well.

the length of experience with rapeseed production in different parts of the world. The changes in rapeseed production from 1940 to the present for all countries with appreciable cultivation of this crop are reported in detail in Chapter 2.

REFERENCES

1 E. SCHIEMANN, E. BAUER AND M. HARTMANN (Ed.) in *Handbuch der Vererbungswissenschaft*, Band III: L, 1932, pp. 271–279, Verlag von Gebrüder Borntrager, Berlin.

1a J. NAVARRO, *Memoria Sobre las Utilidades y Cultivo del Colzat, y Sobre el Modo de Extraher el Aceyte de su Semilla, que en la Junta del Dia 10 de Junio de 1772*, Francisco Suria, y Burgada, Barcelona, 1773.

2 C. LINNAEUS, *Västgöta Resa*, Stockholm, 1747, p. 132. (In Swedish).

3 C. LINNAEUS, *Skånska Resa*, Stockholm, 1751, p. 189. (In Swedish).

4 A. THAER, *Den Nationella Landthushållningens Grundsatser*, Stockholm, 1817, pp. 228–253 (original printed in German, Berlin, 1812).

5 H. BAGGE, *Nord. Jordbr. Forskn.*, 36 (1954) 136. (In Danish).

6 G. ANDERSSON AND I. GRANHALL, *Odling av olje- och spånadsväxter*, Stockholm, 1954. (In Swedish).

7 G. ANDERSSON, in *Sveriges Oljeväxtodlares Centralförening, 1943–1953*, Kristianstad, 1953. (In Swedish)

8 J. GOTTFRIES, in *Skånes Oljeväxtodlares Centralförening 1943–1953*, Kristianstad, 1953. (In Swedish)

9 C. LINNAEUS, *Öländska och Gothländska Resa*, Stockholm och Uppsala, 1745. (In Swedish)

10 D. SINGH, *Rape and Mustard*, Indian Central Oilseeds Committee, Bombay, 1958.

11 R. K. DOWNEY, in J. P. BOWLAND, D. R. CLANDIAN AND L. R. WETTER (Ed.), *Rapeseed Meal for Livestock and Poultry*, Can. Dept. Agr., Publ. 1257, Ottawa, 1965, p. 7.

12 Commonwealth Economic Committee, 1936–64. Her Majesty's Stationery Office, London.

13 V. R. BOSWELL, *Nat. Geogr. Mag.*, 96 (1949) 145.

14 G. ANDERSSON, *Proc. Intern. Assoc. of Seed Crushers*, 1961, p. 1.

CHAPTER 2

Production of and Trade in Rapeseed

R. OHLSON

AB Karlshamns Oljefabriker, Karlshamn (Sweden)

CONTENTS

2.1.1 Introduction

Today rapeseed ranks fifth among the major oilseeds of the world.

Cultivation of the plant for oilseed production is almost entirely confined to the temperate and warm temperate zones of Asia and Europe where both summer and winter varieties are common.

TABLE 2:1

WORLD PRODUCTION OF MAIN OILSEED CROPS (IN THOUSAND TONS)

	1960	1965	1967	1970
Soya beans	26,585	35,704	39,500	46,500
Cottonseed	18,542	21,131	19,125	22,000
Groundnuts	13,652	15,403	16,400	18,100
Sunflower seed	5,929	7,886	9,400	9,600
Rapeseed	3,741	4,918	5,450	6,500
Copra	3,246	3,389	3,050	2,600
Linseed	3,135	3,462	2,500	4,100
Sesame	1,329	1,577	1,645	1,800

Rapeseed thrives best in rich soil in a cool, moist climate. Principal producing areas are India, Mainland China, and Pakistan in Asia; Canada in the Western Hemisphere; and Poland, France, and Sweden in Europe. Table 2:2 shows the area under rapeseed in principal countries.

It is traditionally the most important oilseed crop of Western Europe, but the increasing use of mineral oils as lubricants caused a sharp decline in European cultivation after 1860. Since the Second World War cultivation has been established on a more permanent basis, with some measure of state assistance, in several European countries, and the crop is now particularly important in France, Germany, Sweden, Denmark, and Poland. In the United States, very little is grown as a forage and soiling crop and not primarily as an oilseed.

The oil content of rapeseed varies in seeds of different varieties and from different localities (cf. Chapters 6 and 7). Typically, rapeseed contains about 40 per cent oil and 58 per cent meal. The commercial extraction yield of oil is as much as 98 per cent in modern mills, as in Sweden, but in India the yield is much lower, averaging 80 per cent. The greater part of the oil produced is used as an edible oil in the manufacture of salad oils, margarine, and shortening (Chapter 11). For industrial purposes rapeseed oil is often further processed into blown and sulphated oils (Chapter 12). Rapeseed meal is used as a feed concentrate but must be used in limited quantities because of its glucosinolate content.

TABLE 2:2

AREA UNDER RAPESEED IN PRINCIPAL COUNTRIES (IN 1000 HECTARES)

Country	1948–52	1956	1957	1958	1959	1960	1961	1962	1963	1964	1965	1966	1967	1968	1969	1970
Argentina	—	1	1	1	—	1	1	2	3	1	1	1	—	—	—	—
Austria	4	6	6	6	4	4	4	4	4	6	6	6	7	6	—	—
Belgium	2	1	1	1	1	—	—	—	3	4	—	1	—	1	—	—
Bulgaria	—	2	—	3	1	6	21	9	—	—	—	1	—	—	—	—
Canada	10	142	250	253	87	309	287	150	193	320	581	615	690	425	815	2040
Chile	—	9	11	24	38	29	30	42	44	48	54	45	48	—	—	58
China*	1592	2165	2210	2500	2700	2800	2750	2800	2900	3000	3000	3080	3080	2970	—	—
Czechoslovakia	23	33	33	9	47	39	55	34	38	48	51	48	49	35	—	—
Denmark	5	2	1	4	5	8	11	25	16	25	27	21	20	15	—	—
Ethiopia	50	—	—	—	—	—	—	—	—	—	—	14	14	16	—	—
Finland	4	12	7	10	19	3	6	6	6	9	5	3	5	5	6	—
France	120	58	108	149	89	57	70	89	80	124	173	179	215	248	288	300
Germany, Eastern	80	119	136	134	130	118	123	104	107	118	112	114	117	120	—	—
Germany, Western	54	18	30	33	28	32	36	48	45	50	53	47	49	63	74	80
Hungary	—	3	1	2	2	3	9	3	5	6	6	7	—	—	—	—
India**	2030	2539	2411	2437	2894	2883	3168	3127	3046	2881	2883	2884	3006	3204	2992	—
Italy	15	7	8	8	7	8	5	7	6	5	5	6	3	3	3	20
Japan	114	252	258	227	190	193	195	174	141	120	85	66	54	40	30	—
Netherlands	17	10	6	4	2	3	4	4	4	4	5	5	5	7	6	—
Pakistan**	643	755	734	796	796	725	690	728	677	674	633	633	656	757	643	—
Poland	133	107	112	89	89	108	165	250	197	238	274	272	315	361	—	—
Roumania	—	22	12	19	13	18	10	—	—	—	—	—	—	—	—	—
Sweden	101	20	7	81	81	35	50	62	54	79	85	48	93	104	100	—
Switzerland	2	4	4	5	5	4	6	6	6	6	7	7	8	8	8	—
Taiwan	—	3	3	4	4	5	8	9	10	19	19	8	5	(5)	—	—
Turkey	—	3	1	2	3	3	6	7	7	8	8	6	7	—	—	—
United States	—	1	3	1	1	—	—	—	—	—	3	—	—	—	—	—
Yugoslavia	10	8	4	10	7	7	8	2	3	3	3	3	4	5	4	—
Total	5009	6302	6438	6871	7213	7434	7720	7684	7589	7831	8114	8177	8482	8512	—	—

* Estimated   ** Rape and mustard seed

11

TABLE 2:3

PRODUCTION OF RAPESEED IN DIFFERENT COUNTRIES (IN 1000 TONS)

Country	1934–38	1948–52	1953	1954	1955	1956	1957	1958	1959	1960	1961	1962	1963	1964	1965	1966	1967	1968	1969	1970*
Argentina	50	—	—	—	—	—	—	—	—	—	1	1	4	2	2	1	—	—	—	—
Austria	2	5	10	6	9	9	9	9	6	6	9	9	7	11	12	14	15	12	12	—
Belgium	—	5	2	1	1	2	2	2	2	—	—	—	1	2	1	4	1	2	2	—
Bulgaria	18	14	—	—	—	1	2	3	—	5	15	6	—	—	—	—	—	—	—	—
Canada	—	9	12	13	35	136	196	176	81	252	254	133	190	300	517	585	561	424	757	1617
Chile	—	—	—	—	—	—	11	18	41	36	36	29	49	56	75	65	55	60	—	—
China Mainland	1800	782	350	450	1050	925	890	1100	950	900	850	965	1036	1120	1120	1120	1100	1050	900	975
Czechoslovakia	7	25	—	21	43	48	39	48	73	55	84	48	41	46	74	78	85	73	—	24
Denmark	—	9	20	11	3	2	2	8	11	13	27	52	26	52	50	33	39	30	20	—
Ethiopia	—	20	20	20	20	20	19	19	21	23	5	5	5	5	5	6	6	6	—	—
Finland	—	6	23	13	18	10	5	13	25	4	6	8	7	9	7	3	9	13	9	—
France	13	154	95	88	107	81	160	196	131	83	107	160	135	247	335	317	433	449	520	560
Germany East	33	110	—	120	214	164	176	128	189	182	173	165	128	176	214	210	270	274	190	215
Germany West	40	83	32	15	21	39	67	58	59	68	74	115	96	109	107	99	125	170	156	185
Hungary	8	4	—	1	3	2	1	2	2	3	10	4	5	8	8	7	8	8	—	—
India	745	815	839	1035	860	1042	938	1041	1064	1356	1347	1346	1294	903	1466	1276	1228	1480	1570	1500
Italy	2	14	11	7	10	6	10	11	10	10	9	10	8	8	9	10	5	5	4	2
Japan	120	129	289	220	270	320	286	267	262	264	274	247	109	135	126	95	79	66	50	30
Mexico	1	6	5	6	7	8	6	8	6	8	8	7	7	7	7	6	7	—	—	—
Netherlands	5	33	10	17	19	26	16	9	8	8	10	10	10	10	12	13	15	18	12	19
Norway	—	—	—	—	—	—	—	—	—	—	—	3	3	4	5	8	14	19	20	—
Pakistan	232	270	276	327	326	318	298	337	323	313	310	363	302	307	300	278	306	390	350	380
Poland	48	100	—	100	152	80	101	80	131	147	257	361	227	267	504	448	650	710	200	570
Roumania	24	4	—	7	10	8	6	6	7	11	6	1	1	1	1	—	—	—	—	—
Sweden	—	146	80	169	132	26	195	132	181	61	118	146	103	181	210	95	245	228	184	173
Switzerland	—	4	5	5	6	6	4	10	10	10	9	12	12	13	14	14	18	19	17	—
Taiwan	—	—	1	1	1	1	1	2	2	4	8	10	13	27	23	8	5	4	30	—
Turkey	4	4	3	2	2	2	1	2	3	4	5	4	5	7	8	7	7	7	—	—
United Kingdom	—	—	—	—	—	—	4	1	1	—	2	2	3	3	3	6	15	13	12	8
U.S.A.	—	1	—	—	2	1	—	—	—	—	1	1	1	1	1	—	—	—	—	—
U.S.S.R.	88	63	—	—	—	—	—	30	28	28	25	25	25	25	26	25	—	—	—	—
Yugoslavia	11	5	12	3	8	5	3	7	7	7	9	2	4	4	3	4	5	7	—	—
World Total	3890	2828	4700	5300	5900	3300	3500	3700	3600	3800	4048	4190	3905	4031	5241	4733	5254	5565	5240	6600

* Preliminary figures.

Intensive research is being carried out to eliminate this toxic element. The protein content of rapeseed meal is about 33 per cent in expeller pressed meal, and 35–40 per cent in solvent extracted meal. Very little rapeseed meal enters world trade.

2.1.2 Production

In the seven years between the 1959–60 and 1966–67 seasons world production of rapeseed went up by one third, from 3.5 million tons to 4.7 million tons, the only decreases registered being those for the 1963–64 and 1966–67 seasons, mainly due to bad weather. The annual average compound rate of increase was 4.3 per cent, and in those seven years the only oilseeds to show greater rises were soya and sunflower. Statistics indicate that there was a striking recovery in world rapeseed output in the 1967–68 season, total production attaining a record 5.6 million tons. Figures for 1969 indicate that the production was 5.4 million tons (3 per cent less than in the year before mainly due to smaller crops in Canada and India). Preliminary figures for 1970 show a record production of 6.6 million tons.

The production of rapeseed in different countries is shown in Table 2:3.

The expansion in production in recent years has taken place mainly in five areas, *viz*. Canada, Western Europe, Poland, India and China, but in the decade ending with the 1968–69 season the relative shares of world production provided by Canada, Poland and Western Europe increased, that of China declined, while the Indian proportion was virtually unchanged. The cultivation of rapeseed has expanded in Canada as a profitable alternative to wheat, and the general trend in plantings and production has been upwards. Most of the crop is exported, although domestic usage has been increasing. In most Western European countries rapeseed enjoys some measure of state assistance and production has tended to grow, especially in France and West Germany, encouraged by the establishment of new increased European Community prices for rapeseed in the 1967–68 season, above those previously paid by member states. The Polish authorities appear to have given a considerable stimulus to rapeseed cultivation, and by 1965 and 1966 production of the crop, which is the principal oilseed grown, had reached a level at which substantial exports could be made. Over the passed years Indian crops have become larger with the expansion in areas and generally favourable weather conditions, but the increments in harvests have all gone towards meeting increases in domestic requirements. Little is known of production in China in recent years, although the country is apparently the world's third largest producer; it is believed that the crop has been of approximately the same size in the last few seasons. Japanese production has shown a steep decline in recent years, continuing in both 1970 and 1971. Rapeseed is also an important oilseed crop in Chile, and both the area cultivated and the production have risen in recent years.

Associated with the rise in world production of rapeseed in recent years has been the expansion of world trade in rapeseed and rapeseed oil. Since so much of the growth in world output has been in exporting countries, net exports of rapeseed (Table 2:4) and oil (Table 2:5) from primary producing countries expanded by almost four and a half times between 1961 and 1966, rising from 73,000 tons of oil equivalent to 320,000 tons, an increase proportionately much larger than that in world production.

TABLE 2:4

EXPORTS OF RAPESEED (IN 1000 TONS)

Country	1961	1962	1963	1964	1965	1966	1967	1968	1969
Austria	3	—	—	7	7	8	11	3	8
Bulgaria	1	1	2	3	2	2	2	—	—
Canada	121	192	140	82	240	312	331	323	500
China, Mainland	1	—	1	—	6	33	30	10	2
Denmark	27	25	43	48	38	32	20	7	5
Ethiopia	2	2	2	1	2	1	4	—	—
France	31	81	70	118	124	129	102	119	185
Germany, Eastern	—	—	—	—	1	20	17	60	24
Germany, Western	1	—	—	3	5	4	9	7	21
Nepal	6	5	3	3	(3)	(3)	—	—	4
Netherlands	12	4	13	12	5	10	8	20	12
Poland	2	21	4	—	59	87	105	170	50
Sweden	7	15	30	48	73	14	21	58	101
Total	211	343	308	320	559	634	628	776	800

TABLE 2:5

EXPORTS OF RAPESEED OIL (IN 1000 TONS)

Country	1961	1962	1963	1964	1965	1966	1967	1968	1969
China, Mainland	2	—	—	—	4	35	19	18	15
France	11	10	7	12	32	43	35	30	12
Germany, Western	4	9	13	14	24	25	30	53	30
Hungary	2	..	..	2	2	5	2	1	6
Japan	—	—	3	2	4	11	10	6	8
Netherlands	—	1	1	—	1	3	3	7	5
Poland	—	—	4	—	8	21	65	80	10
Sweden	5	18	15	9	16	14	22	15	15
Total	25	39	44	40	91	152	190	200	172

The major exporters of rapeseed and rapeseed oil are Canada, France and Sweden. In certain years China has been important, for example in 1966 and 1967, and Poland has lately begun to ship large quantities of seed and oil. The very heavy world rapeseed production from the 1964–65 season onwards occasioned a large increase in net exports, which more than doubled between 1964 and 1966. The greater part of the Western European crop tends to be exported in the year in which it is harvested, but since the Canadian crop is somewhat later in the year most of the exports take place in the calendar year following harvest. Despite the lower levels of output in 1966–67 in most of the major exporting countries, world exports of rapeseed and oil were well maintained in 1967, totalling 326,000 tons of oil equivalent which was slightly greater than in 1966.

Prices of rapeseed oil were consistently below quotations for soyabean oil during 1967, and prices of both seed and oil were well below those for groundnuts and groundnut oil, thus making rapeseed and rapeseed oil more attractive to the importer. In 1967 rapeseed oil shared the position of the cheapest edible oil along with sunflower oil.

The growth in exports is associated with a corresponding rise in imports of rapeseed (Table 2:6) and rapeseed oil (Table 2:7). The larger part of the increase in supplies was taken by the principal importers, *i.e.* Italy, West Germany, Algeria

TABLE 2:6

IMPORTS OF RAPESEED (IN 1000 TONS)

Country	1961	1962	1963	1964	1965	1966	1967	1968	1969	1970
Algeria	(65)	(52)	73	65	58	53	55	55	35	
Belgium	2	4	1	2	4	3	4	3	3	2
Czechoslovakia	(2)	4	14	17	36	(10)	1	3	24	
Finland	11	6	0	4	4	8	5	10	6	
France	33	15	11	7	4	8	5	18	37	50
Germany, Western	24	31	45	32	100	91	63	108	134	60
India*	6	5	3	2	5	3	6	5	4	
Italy	64	102	90	64	132	210	221	150	166	213
Japan	20	36	86	75	101	211	215	250	250	330
Netherlands	3	13	5	3	(21)	(18)	25	26	20	24
Norway	—	2	3	3	3	3	4			
Pakistan	—	—	—		11	20	18	4	1	
Poland	—	—	4	7	17	3	..			
Soviet Union	—	—	—	—	—	(—)	..			
Taiwan	—	—	—	6	—	—	6	47	20	
United Kingdom	5	7	8	12	32	42	40	70	70	50
Total	235	277	343	310	508	646	632	739	816	728

* Includes mustard seed.

IMPORTS OF RAPESEED OIL (IN 1000 TONS)

Country	1961	1962	1963	1964	1965	1966	1967	1968	1970
Algeria	—	—	—	—	17	15	13	10	
Australia	0	0	0	2	4	6	5	1	
Austria	2	3	3	5	8	12	8	12	
Belgium	0	0	0	0	1	2	2	10	3
Cyprus	1	2	1	1	2	3	1	1	
Czechoslovakia	1	1	1	2	3	(3)	(3)		
Denmark	0	—	1	0	1	1	0	1	
France	—	0	5	1	1	2	1	7	4
Germany, Western	5	5	3	5	8	22	27	27	15
Hong Kong	1	0	1	—	4	22	20	20	21
Italy	0	2	1	0	1	1	19	30	22
Netherlands	3	5	6	4	11	14	10	16	6
Ryukyu Islands	—	0	2	2	1	(1)	3	3	
Switzerland	—	—	2	1	2	4	4	3	
United Kingdom	..	(2)	4	0	0	—	—	10	15
United States	2	2	1	3	2	3	4	4	
Total	24	28	39	33	58	113	107	163	85

and Japan, but the United Kingdom and Hong Kong greatly enlarged their imports of seed and oil respectively in 1966. In 1967 most of the growth in imports of rapeseed oil was accounted for by Italy, but purchases by other leading importers generally remained substantial.

The European Communities Commission has submitted guide-lines for an international agreement on fats and oils to the European Council, which will affect the future trade.

2.1.4 Prices

World prices of rapeseed tend to move with those of soya beans, while those of rapeseed oil are influenced by prices of competing edible oils, including soya and sunflower. Thus the rising quotations for soya beans in the first eight months of 1966 were associated with higher prices for rapeseed, while the subsequent drop in soya bean values led to a fall in rapeseed prices. Since the price of soyabean oil did not generally follow that of beans in 1966, owing to plentiful world supplies of soft oils, quotations for rapeseed oil tended to ease during the year. In 1967 the reduction in soyabean quotations in the latter months of the year also brought rapeseed prices down. The pressure from various edible oils, including soya and sunflower, as well as unusually cheap marine oils, resulted in considerably lower prices for rapeseed oil during 1967. Quotations for both rapeseed and rapeseed oil weakened a good deal more in the first half of 1968. In 1969 however prices increased steadily but fluctuated a little more during 1970–71.

2.1.5 Projections

The estimated world production of fats and oils has increased at an average rate of nearly 900,000 tons annually since 1950 and reached well over 37 million tons in 1968, corresponding to about 80 million tons of oilseeds. The rate of increase in total production exceeded the rate of increase in world population and world per capita supplies of fats and oils for all purposes have increased from approximately 9 kilograms in 1950 to 10.5 in 1965.

The biggest increases in production have occurred in the developed countries which now account for nearly 50 per cent of total world production. The most striking increase has been in the production of soyabean oil but production of sunflowerseed and rapeseed oil has also shown a strong upward trend.

The FAO projection of production of oilseeds is 102 million tons in 1975 and 144 million tons in 1985.

It seems more than likely that world rapeseed production will continue to grow, although the upper limits of Western European crops may be reached in the next few years. A rather good guess for the world production in 1975 would be at least 7 million tons of rapeseed and 12 million tons in 1985. The main reason for this is the promising preliminary results to eliminate the toxic components in the seed. The use of the meal would increase dramatically and provide stiff competition for soyabean meal in feed stuffs, and also for soya proteins in human consumption.

2.2 EUROPEAN ECONOMIC COMMUNITY

Rapeseed production in the European Economic Community (EEC) reached 630,000 tons in 1968 (*cf.* Table 2:3). France and West Germany are the major producers of rapeseed within the Community. They are also the main producers of rapeseed oil together with Italy, which crushes mainly imported seed.

2.2.1 EEC regulations governing rapeseed

The Common Agricultural Policy (CAP) for Fats and Oils, which became effective on July 1, 1967, provides for a pricing and marketing system for rapeseed. In accordance with CAP regulations, a single 'target price', 'basic intervention price', and 'derived intervention prices' are established each year for rapeseed produced in the Community. The target price is established at a level which will encourage the volume of production by assuring the producer a favourable return. It is also intended to maintain a balance between the acreage under rapeseed and that under other crops. The basic intervention price, which guarantees the producer the sale of his rapeseed is fixed as close as possible to the target price. The spread between the target and intervention prices allows for free price fluctuations. Derived intervention prices are determined according to the location of the regional intervention centers, costs of transportation, and handling charges.

The EEC Council fixed the target and intervention prices for the rapeseed marketing year beginning July 1, 1967 as follows:

Target price	202.50 US $/ton
Basic intervention price	196.50 US $/ton
Lowest derived intervention price	180.50 US $/ton

The prices apply to rapeseed of EEC standard quality, *i.e.*, seed containing 42 per cent oil, 10 per cent moisture, and 2 per cent impurities, in bulk, on a wholesale basis.

The basic intervention price as officially stated is applicable in Ravenna, Italy, the principal deficit area in the Community. The lowest derived intervention price, while not officially noted, applies in Chateauroux, and Bourgis, France. Thirty principal intervention centers were established at the beginning of the marketing year, and the derived intervention prices were only slightly below the basic intervention price, with the exception of the price in Florence, which was at the same level as in Ravenna. The Commission authorized 68 additional intervention centers on August 1, 1967. The prices applicable at these centers were fixed on the basis of the intervention prices in the principal centers and the lowest freight rate of these centers.

Aid to rapeseed producers is provided in the form of a subsidy or deficiency payment which is given only for rapeseed grown and processed into oil in the Community. The right to the aid is acquired at the time the seed is processed at the oil mill. Member states may, however, pay the subsidy at the time the seed is submitted provided that a guarantee is given that it will be processed. In order to assure that subsidies intended for EEC rapeseed producers are not paid on rapeseed imported from third countries, certificates are issued certifying the eligibility of EEC-grown rapeseed and regulating the amount of the subsidy which may be fixed up to two months in advance of the submission of the rapeseed to government control. Certificates are not issued for rapeseed for export nor for denatured seed, seed for planting, or seed which has been processed into meal or feed.

The subsidy is equal to the difference between the target price and the world market price as determined by the Commission. In accordance with EEC regulations, the Commission calculates the world market price at least once a week from current representative offers of at least 500 tons, bulk, c.i.f. Rotterdam, for rapeseed of EEC standard quality. If the qualities offered do not correspond to the EEC standard quality, coefficients based on the country of origin are applied to the prices offered in order to adjust them to the EEC quality standards. In case no representative offers are made, the price is assessed on the value of the oil and meal produced from crushing 100 kilograms of seed. The Commission also determines each week the subsidy—the difference between the target and calculated world price—and notifies each Member State of the prevailing rate of producer aid. The subsidy may be modified, if necessary, for the stabilization of the Community market.

Beginning September 1, 1967, a premium of US $ 0.18 per 100 kilograms was added to the target and intervention prices. The premium was to be added each month for seven consecutive months during the 1967–68 marketing year. Aid to producers was therefore increased to this extent. In the EEC, aid is also given to the processers of rapeseed of EEC origin. In September 1970 this was US $68.00 per ton.

Although the EEC imports rapeseed on a large scale, market regulations provide for export subsidies on rapeseed exported to third countries. The subsidy may be granted by a Member State upon the request of a Community exporter.

Oil consumption in the EEC is expected to increase in line mainly with population growth and, to a lesser extent, with income growth, over the next ten years. But the demand for oilseed cake and meal for livestock feed is expected to increase even more rapidly. Thus the EEC's import demand will strengthen the market for oilseeds, with a high yield of cake or meal.

2.2.2 France

France is the largest country in Western Europe, with a population of approximately 50 million. The population density is however the lowest in the EEC. The most important agricultural regions are the northern plains area, especially the Paris Basin and Brittany, and the Mediterranean area. Winters are relatively mild and summers warm. The best soils for farming are in the alluvial plains of the Paris Basin, there where rapeseed is grown.

France's agricultural policy is greatly influenced by the relative importance of its agricultural production within the EEC, as well as in Western Europe at a whole, and by its large volume of agricultural trade.

Domestic oilseed production is now equal to 25 per cent of domestic requirements. Oilseed output is limited to small quantities of sunflower seed and flax seed and the expanding rapeseed production. The 1960–64 average rapeseed area of 84,200 hectares increased to 288,000 hectares in 1969 (Table 2:2). Production during the same period climbed from 146,000 tons to 520,000 tons—an increase of more than 250 per cent (Table 2:3). A further increase in production is anticipated for 1970. Winter rapeseed *(B. napus)* is the dominating variety. Both acreage and production have responded to the high prices guaranteed to producers (*cf.* Table 2:8). The intervention prices for rapeseed will be unchanged in French francs for 1970.

TABLE 2:8

Year	Producer price Francs/100 kg		Rapeseed:wheat price ratio
1953	74.00		2.1
1954	62.30		1.8
1955	62.30		1.8
1956	62.78		1.8
1957	62.10		1.9
1958	65.03		1.8
1959	69.45		1.8
1960	75.00		1.9
1961	80.00		1.7
1962	80.00		1.8
1963	80.00		1.8
1964	80.00		
1965	75.00		
1966	75.00		
	min	max	
1967	89.11	93.80	
1968	89.21	93.95	
1969	89.61	93.95	

TABLE 2:9

FRENCH EXPORTS OF RAPESEED BY COUNTRY OF DESTINATION (IN TONS)

Destination	1962	1963	1964	1965	1966
Belgium-Luxembourg	75	—	—	1,689	678
Germany West	1,981	1,561	447	11,081	12,323
Italy	36,656	6,416	44,754	36,660	47,934
Netherland	—	1,534	190	7,734	—
Total EEC	38,712	9,511	45,391	57,164	60,935
Algeria	38,536	61,416	66,477	59,400	54,038
Germany East	127	—	4,413	923	—
Morocco	—	—	—	—	5,013
United Kingdom	1,175	—	3,070	6,531	11,034
Others	3,372	185	53	2,347	315
Total	81,922	71,112	119,404	126,365	131,335

Domestic crushing of rapeseed, which amounted to only 61,700 tons in 1963 and reached 175,000 tons in 1966, was even higher in 1967 and 1968 due to increased domestic consumption and export demand for oil. In 1968 320,000 tons of rapeseed were used for internal consumption and France was expected to use even more in 1969. Rapeseed oil is now sold in France also as 'table oil'. Consumption increased from 13,000 tons in 1964 to an estimated 42,000 tons in 1967.

Exports markets have been a main outlet for French rapeseed, with Algeria, Italy and West Germany as the principal buyers (Table 2:9). France is the second largest exporter of rapeseed in the world (*cf.* Table 2:4) and a leading world supplier of rapeseed oil (*cf.* Table 2:5). Rapeseed oil has been exported mainly to Algeria and the Netherlands. About 35 per cent of the 1967 crop was exported as seed and about 25 per cent as oil, while the remainder was consumed domestically, mostly as oil. Due to the devaluation in 1969, exports of rapeseed would be cheaper in terms of other currencies, and imports more expensive. Therefore a tax will be imposed on exports to prevent French exports from undercutting the market intervention system in other EEC countries and a subsidy on imports will prevent diversion of trade away from community suppliers.

The consumption of vegetable oils in France in 1970 will be about 700,000 tons and the projection for 1975 is above 800,000 tons. The demand for oil cake and meal will be about 14 million tons in 1970 and 16.6 million tons in 1975. The main part of it will be imported.

2.2.3 West Germany

West Germany has had the highest price level for agricultural products of any

TABLE 2:10

AVERAGE DOMESTIC PRODUCER AND IMPORT (C.I.F. EUROPE) PRICES FOR RAPESEED, SELECTED YEARS 1950–63, IN WESTERN GERMANY (IN U.S. DOLLARS PER METRIC TON)

Year	Producer price	Import price	Difference between domestic and import price
1950	160	148	12
1951	205	205	0
1955	188	148	40
1956	188	160	28
1961	165	118	47
1962	165	110	55
1963	165	121	44

TABLE 2:11

AREA PLANTED, PRODUCTION AND YIELD OF RAPESEED VARIETIES IN WESTERN GERMANY

	Area planted in 1000 hectares			Production in 1000 tons			Yield in deciton/hectares		
	B. napus		B. campestris	B. napus		B. campestris	B. napus		B. campestris
	winter	summer		winter	summer		winter	summer	
1935/38	19		6	36		8		18.6	13.7
1950/55		29			49				
1956	11	5	2	27	9	3	23.5	18.8	17.6
1957	20	9	2	47	17	4	23.8	19.5	18.1
1958	24	7	2	44	11	2	18.3	15.3	14.2
1959	20	5	2	47	9	3	23.3	15.9	17.3
1960	25	6	1	57	11	2	22.8	17.3	17.2
1961	27	8	1	57	15	2	21.6	17.8	17.5
1962	39	8	1	98	16	2	25.4	18.7	18.9
1963	35	9	1	77	17	2	22.1	18.2	17.5
1964	41	8	1	92	15	2	22.4	18.2	16.8
1965	46	5	2	95	9	3	20.5	16.9	15.9
1966	40	6	1	86	10	2	21.5	17.7	16.0
1967	41	7	1	110	13	2	26.8	19.7	18.2
1968	56	6	1	156	12	2	27.6	20.6	19.3
1969				145				22	
1970				179					

EEC country and much of the last years' policy has been directed towards adjusting the existing price structure to CAP.

West Germany encourages production of rapeseed by maintaining producer prices equal to those in France (Table 2:10). Although the rapeseed area is indirectly controlled through the farmer price support program, acreage increased 15 per cent in 1967 from the 1960–64 average of 48,600 hectares. Production, however, rose from the 5-year average (1960–64) of 92,500 tons to 170,000 tons in 1968—an increase of 84 per cent. Of the total rapeseed crop in West Germany, about 70 per cent is estimated to be harvested in Schleswig-Holstein and Bavaria. An increase of almost 15 per cent in acreage seeded to winter rapeseed in the autumn of 1969 is indicated.

West Germany is a big importer of rapeseed (*cf.* Table 2:6).

In the crop year 1968–69 West Germany imported a total of 122,000 tons of rapeseed, re-exported about 7,000 tons, but was also a net exporter of some 53,000 tons of rapeseed oil. The bulk of the import came from Sweden, followed by Poland and France. Total consumption of vegetable fats and oils in 1970 is about 750,000 tons and will increase to about 780,000 tons in 1975.

The total consumption of oilseed meals and cakes in 1962/63 was 1.9 million tons and is expected to reach 3.3 million tons in 1970 and 3.6 millions tons in 1975. Of this 80,000 tons will be domestically produced rapeseed meals in 1970 compared to 100,000 tons in 1975. There will still be a large import of oilseed meals, of which rapeseed is only a minor part. In 1964 it was 41,000 tons, in 1965 43,000 tons and in 1966 50,000 tons.

2.3 EUROPEAN FREE TRADE ASSOCIATION

Since May 1960 Austria, Denmark, Norway, Portugal, Sweden, Switzerland, and the United Kingdom form the European Free Trade Association (EFTA). Finland became an associate member in 1961 and Iceland from March 1970. The purpose of EFTA is to eliminate barriers to industrial trade and to encourage trade in agricultural and fishery products.

A striking dissimilarity between EFTA and EEC is that the EFTA members retain their individual agricultural policies. Agricultural goods were exempted from EFTA's dismantling of industrial tariffs which was completed by January 1, 1967. EFTA regulations provide for special bilateral agreements concerning agricultural trade. However, in EEC and EFTA there are similar objectives for agriculture which apply to most members countries. They are: To raise agricultural income to the level of income in other sectors, to encourage price stability, and to improve agricultural efficiency by increasing farm size wherever possible.

Of the EFTA countries only Sweden has thus far begun to modify traditional agricultural policy. The change involves a lowering of self-sufficiency targets for farm products, which should result in some increase in agricultural imports. Sweden has already adjusted prices of certain products to EEC levels.

The EFTA countries started negotiations with EEC countries in 1971.

The proposals for Nordic cooperation (NORDEK) in agriculture are in an embryonic stage, and intensive studies and discussions will have to be undertaken concerning all their aspects.

2.3.1 Denmark

There is reported an important cultivation of rapeseed in Denmark in 1760. It was grown in 'marshgrounds' and on some of the best 'clay soil'.

Towards the end of World War II the area of white mustard was considerable and this was the most important oil crop in Denmark. The acreage and production of rapeseed in Denmark during the last years can be seen in Tables 2:2 and 2:3, respectively.

In 1958 there was an agreement between the Danish margarine industry and the government that the industry should buy Danish rapeseed at a fixed price. The amount of seed bought by the industry should correspond to 10 per cent of the fat used in the production of margarine in Denmark. In 1961 this principle was established in the form of a new law. The industry was also obliged to pay a certain fee (42 Danish öre/kg) to a fund in the department of agriculture on top of the price of the seed (80 Danish öre/kg). This fund is used to encourage the cultivation. It was also decided that the farmers had the right and obligation of buying the meal resulting from the rapeseed extraction in Denmark. The law is still valid (February 1970).

The export of rapeseed from Denmark is shown in Table 2:4.

2.3.2 Sweden

As already mentioned in Chapter 1 rapeseed was grown in Sweden during the 18th century. There was a considerable crop during World War I but then it was discontinued until World War II. In the early 1940's most of the area under oil crops was confined to southernmost Sweden, and many different species were grown, e.g. winter rape, white mustard, and linseed.

Over the years the area under oil crops has been extended northwards and experiments have shown that the winter types were well adapted to the plains of Central Sweden. Winter turnip rape, which had been grown only to a small extent in the 1940's, proved to be well suited to northwestern Götaland and Svealand.

The harvested acreage of different oil crops in Sweden from 1940 is shown in Table 2:12.

The production of seed is shown in Table 2:3. The average yield in kilograms/hectare for the last twenty years is shown in Chapter 4.

TABLE 2:12

HARVESTED AREA OF OIL CROPS IN SWEDEN (IN 1000 HECTARES)

	1940	1942	1944	1946	1948	1950	1951	1952	1954	1955	1957	1958
Winter rape	0.06	1.4	11.9	5.4	14.3	60.2	85.5	64.5	57.4	57.7	52.5	49.3
Winter turnip rape	—	—	—	0.4	1.0	5.8	14.4	20.4	17.3	16.8	19.7	14.7
Summer rape	—	0.2	8.7	6.9	23.6	50.2	32.6	15.8	4.6	2.4	2.1	5.1
Summer turnip rape	—	—	—	—	—	7.0	10.8	3.9	1.5	0.7	0.5	1.9
White mustard	—	14.0	7.7	5.2	18.0	6.6	11.1	17.2	8.2	6.0	4.9	4.5

	1959	1960	1961	1962	1963	1964	1965	1966	1967	1968	1969
Winter rape	46.4	21.2	36.6	46.4	34.8	42.7	54.7	22.9	53.9	56.2	60.7
Winter turnip rape	21.2	6.0	11.6	13.0	9.7	16.4	15.7	8.1	19.4	15.9	13.2
Summer rape	2.7	2.3	4.2	4.9	6.2	24.5	12.1	12.9	16.1	20.0	14.5
Summer turnip rape	0.4	0.6	0.4	0.6	1.3	1.8	2.3	3.8	5.8	6.0	6.2
White mustard	3.3	2.0	15.0	18.2	18.7	20.4	7.2	6.0	3.4	2.6	1.5

A government control on fats in Sweden originated during the crisis of the 1930's. The Parliament of 1933 decided that the margarine industry should help support butter sales by means of an excise tax on margarine. From 1940 to 1956 imports of raw materials for margarine were relegated to a central association for the import of fats. With the introduction of large-scale cultivation of oil crops on a more permanent basis in 1944 imports were restricted.

In early years the prices of different oilseed crops were fixed by the Government (*cf.* Table 2:13). In 1956 this practice was abandoned and prices were

TABLE 2:13

GUARANTEE PRICES FOR OIL SEEDS, ÖRE PER KILOGRAM

Year	Winter type rapeseed	Summer type rapeseed	Mustardseed
1941	—	—	108.5
1942	—	100	80
1943	97	85	70
1944	80	75	65
1945	69	75	65
1946	69	75	65
1947	75	85	75
1948	90	90	78
1949	90	85	70
1950	75	75	61
1951	70	70	75
1952	90	83	80
1953	85	75	65
1954	75	67	54
1955	73	67	54

regulated according to a special system which was influenced by world market prices. To guarantee farmers a market for their oil seeds the National Agricultural Marketing Board has been authorized by the government to compel the use of domestically produced oil in margarine. However, it has not been necessary to exercise these powers because in the so-called Rapeseed Agreement of 1956 the margarine industry pledged to accept oil seed in quantities corresponding to its margarine production. For 120,000 tons of margarine this involved an obligation to buy approximately 105,000 tons of oil seeds—an amount which, converted into oil represented about 40 per cent of the margarine industry's requirements of fats. According to the agreement, the rapeseed oil had either to be used in manufacturing margarine in Sweden or be exported.

On September 1, 1968 a new agricultural policy was introduced in Sweden with the aim of making agriculture self-supporting and internationally competitive to the greatest possible extent.

2.3.3 United Kingdom

During the last decades the rapeseed producing area in the United Kingdom has been less than 2,000 hectares, but in 1966/67 about 11,500 hectares were grown.

The revival of interest in growing oilseed rape in England and the relatively large increase in acreage since 1966 has been mainly due to the quest by many producers with large areas for an effective breakcrop. Rapeseed is an attractive crop for the farmers of the chalk lowlands of Hampshire and adjacent counties.

At the present time, rapeseed of any origin enters the United Kingdom market free of duty, but when it comes to the products the situation is somewhat different. Rapeseed oil has a 15 per cent preference from Commonwealth countries and the meal enjoys a 10 per cent preference over the full rate or the rate applied to EFTA countries. Home production has lagged behind the rise in demand, as imports of rapeseed rose steadily from 4,000 tons in 1960. Rapeseed imports to the United Kingdom in 1969 totalled some 76,000 tons. This compares with total import in the calendar year 1968 which was 78,600 tons of rapeseed, or nearly double that of 1967. Sweden was the leading supplier in 1969, while in 1968 East Germany and Poland led. In 1968 10,657 tons of rapeseed oil were also imported.

It would appear that the crushing capacity of operating plants in the United Kingdom after mid-1970 will fall considerably short of the market demand for oil and meal from rapeseed, so that the deficiency could result in a broadening of imports in these products. There would appear to be an opportunity for exporters of rapeseed oil and meal to capitalize on this new development.

Increases in United Kingdom net imports of oilcake and meal are indicated for 1970, 1975, and 1980. Declines in imports are projected for vegetable oils and oilseeds, but increases for other fats.

2.4 EAST EUROPEAN COUNTRIES

Agricultural policy in Eastern Europe has passed through several phases since the solidification of Communist control after the Second World War. Domination by the Soviet Union resulted in attempts to introduce policies similar to those adopted earlier in the Soviet Union.

2.4.1 Czechoslovakia

The area planted to rapeseed in Czechoslovakia during the last 15 years has varied from 30,000 to 50,000 hectares (Table 2:2). During the same period production has varied from 20,000 to 80,000 tons (Table 2:3).

The crushing capacity (for oilseeds) is about 250,000 tons/year in Czecho-slovakia, with four large solvent extraction mills.

2.4.2 East Germany

The acreage under rapeseed in East Germany has, during the last twenty years, been rather constantly around 120,000 hectares (Table 2:2). The production has varied mainly due to the weather conditions from 110,000 to 210,000 tons during the same period. In 1967 the production increased to 270,000 tons (Table 2:3).

2.4.3 Poland

Poland is the leading rapeseed producer in Europe and by far the largest producer in Eastern Europe. About 15.7 million hectares, or nearly half of Poland's total land area, is arable.

In 1934–38 the yearly production of seed from rape and turnip rape in Poland was about 48,000 tons, corresponding to a planted area of 50,000 hectares. After World War II acreage increased rapidly to 200,000 hectares in 1955, but the yield decreased to about 750 kg/hectare. The area under rapeseed and the production of rapeseed in Poland during the last 15 years can be seen from Tables 2:2 and 2:3, respectively. *Brassica napus*, winter type, is by far the dominating seed.

As can be seen from Fig. 2:1 the yield has increased steadily during the last 15 years. Poland's rapeseed crop in 1968 was about 750,000 tons compared with 650,000 tons in 1967, 448,000 tons in 1966 and the previous record of 505,000 tons in 1965. In 1969 the crop is estimated to be 30–50 per cent lower than in 1967, due to bad weather. In 1971 it will be of the order of 600,000 tons.

The crushing capacity for oilseeds in Poland is about 600,000 tons, with seven large mills. The rapeseed oil share of the total vegetable oil utilization in Poland increased from 31 per cent in 1957 to 84 per cent in 1966.

Exports from Poland have shown a marked increase since 1960 when 400 tons of rapeseed were exported, compared with 1966 exports of 68,300 tons of seed and 8,800 tons of oil. Shipments from the 1967 harvest have reportedly entered Western European markets and Japan at prices somewhat below the world market price. Practically all of Poland's trade in rapeseed and oil has been with European countries.

2.5 AMERICA

2.5.1 Canada

Rapeseed production began in Canada in 1942, when the so-called Argentine

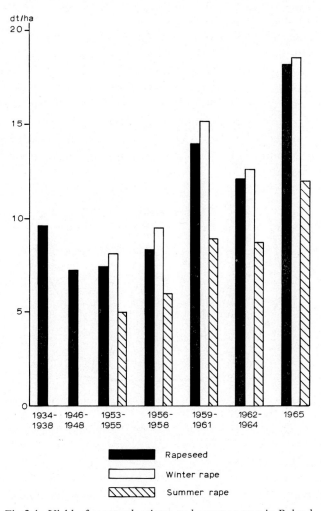

Fig.2:1. Yield of rapeseed, winter and summer rape in Poland.

type* of rapeseed (see Chapter 3) was grown commercially in Western Canada as a war measure to supply oil for lubrication of marine engines. The crop appeared to be well adapted to the northern parts of the Prairie Provinces, where it gave high yields. Following the war, private interests established markets abroad for rapeseed. Acreage and production were erratic until about 1956, when rapeseed was widely adopted as a crop by Western Canadian farmers. Production has expanded rapidly so that rapeseed is now the second most important grain crop for Canadian farmers and the fastest-growing seed crop in Canada. Rapeseed is the main oilseed

* The name Argentine type was adopted because the seeding stock came from Argentina.

crop grown and in recent years the annual value of rapeseed production has been over 50 million Canadian dollars.

The popularity of rapeseed has increased because it is a cash crop that gives returns equal to those from wheat and, in addition, the early maturing varieties allow extension of the busy seeding and harvest seasons. These advantages suggest that the rapeseed acreage in Western Canada could be more than doubled if economic conditions continue to favour its development.

Rape is grown mainly in north-central Saskatchewan and Alberta (52–54 °N) and in southern Manitoba. The map in Fig. 2:2 shows the geographic distribution of the 1968 rapeseed crop.

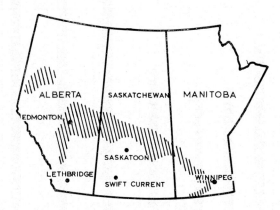

Fig.2:2. Areas in Western Canada (shaded) adapted to rapeseed production.

Table 2:14 shows rapeseed acreage, yield per acre, production, and average farm prices from 1943 to 1970. In 1971 rapeseed production is estimated to be 2,300,000 tons, nearly 50 per cent above the 1970 figure.

Two distinct species of annual rape are grown in Canada: Argentine *(Brassica napus)* and Polish *(Brassica campestris)*. A variety of *Brassica napus*—Oro— has been developed which contains no erucic acid and only traces of eicosenoic acid in the seed oil. The oil obtained from this zero-erucic rapeseed, known as Canbra, is a distinct type, and is particularly good as a salad oil. At present there is only a limited domestic market and in 1968 110,000 tons of this new strain weer grown under contract. Rapeseed varieties seeded in 1969 in per cent were: Echo *(B. campestris)* 52.8, Arlo *(B. campestris)* 17.2, Polish *(B. campestris)* 13.9, Target *(B. napus)* 11.9, Tanka *(B. napus)* 2.6, and others 1.6. In August 1970 the Canadian Department of Agriculture took steps to bring about the gradual change-over to the production of low- or zero-erucic acid varieties. The target is to have these available in substantial commercial quantities for planting in 1972.

The use of summer-fallow for the growing of rapeseed has increased during recent years to an average of 65 per cent in 1969 *vs.* 35 per cent for seeding on stubble. The yield per acre of rapeseed planted on summer-fallow land is much

RAPESEED ACREAGE, YIELD, PRODUCTION, AND AVERAGE FARM PRICE IN CANADA

Year	Acreage, hectares	Yield, kg/hectare	Production, 1000 tons	Average Farm Price, $ Can/bushel
1943	1,300	935	1,000	—
1944	4,360	760	2,780	—
1945	5,000	890	3,820	—
1946	9,500	740	5,900	—
1947	23,600	505	9,950	—
1948	32,400	1075	29,000	—
1949	8,100	1140	7,700	$2.50
1950	160	405	55	$1.90
1951	2,700	1250	2,720	$1.75
1952	7,500	1000	6,320	$1.70
1953	12,000	1120	11,150	$1.80
1954	16,200	970	13,200	$1.65
1955	56,000	760	35,300	$1.75
1956	142,000	1145	136,000	$1.75
1957	250,000	940	196,400	$1.60
1958	253,000	835	176,000	$1.25
1959	86,500	1120	80,800	$2.00
1960	308,000	980	252,400	$1.65
1961	287,000	1060	254,400	$1.80
1962	150,000	1060	134,000	$2.05
1963	194,000	1180	189,500	$2.50
1964	320,000	1120	300,000	$2.54
1965	580,000	1070	518,000	$2.33
1966	615,000	1140	585,000	$2.47
1967	700,000	1030	567,000	$1.92
1968	425,000	1240	441,000	$1.53
1969	813,000		757,000	n.a.
1970 proj.	2,040,000		1,617,000	

higher than when seeded on stubble. In 1969, for example, the average yields were 20.8 and 14.1 bushels per acre, respectively.

The first oil extracted for edible use in Canada was processed in 1956–57. Since 1963 the domestic market for rapeseed oil and meal has developed rapidly. In 1968 the crush was almost 1.1 million tons compared to 340,000 tons five years before. The ratio of crush to production has remained around 1/5 during those years. The domestic crush in 1969–70 is projected to be 170,000 tons.

Rapeseed oil is a new commodity in Canada, and the consumer is therefore still not aware of the quality properties of rapeseed oil. When the Canadian consumer becomes fully conscious of rapeseed oil as a household item, and if rapeseed oil could be used in the manufacture of food items instead of imported oils domestic crush and consumption would increase to 650,000 tons.

Canada is the world's largest exporter of rapeseed (see Table 2:4). Major markets have been Japan, Italy, the Netherlands, and Western Germany. Since the 1967 rapeseed production in Western Europe reached record levels, increased amounts of rapeseed in the future are expected to go to Japan rather than Europe. Projections for export in 1970–71 are in the area of 680,000 tons, of which 545,000 tons will go to Asia and 135,000 tons to Europe.

2.5.2 Chile

The acreage of rapeseed in Chile during 1956–1966 has steadily increased from about 10,000 to 50,000 hectares (Table 2:2). Estimates of the area planted to rapeseed in 1967 were down by 28 per cent. The dominating varieties in Chile are *B. napus* of Swedish origin.

The production of rapeseed has varied from 10,000 tons to 75,000 tons during the period of 1956–1966. The entire output in 1966 was crushed domestically for oil. The output of refined edible rapeseed oil was 25,000 tons in 1965, 32,340 tons in 1966 and 24,000 tons in 1967.

2.6 ASIA

More than half of the world's rapeseed crop is produced in Asia, principally in India, mainland China and Pakistan. The output of rapeseed in Japan and Taiwan shows a steady decline.

2.6.1 China

The area under rapeseed, primarily winter-grown, in China for different years is given in Table 2:2 and the production in Table 2:3.

The rapeseed crop in China for 1969 is estimated at 688,000 tons. Compared with 1968 the crop is about 12 per cent smaller.

China's export has increased during the last decade, from the 1960–64 average of 6,100 tons, oil basis, to an estimated 45,800 tons (as oil) in 1966 (*cf.* Tables 2:6 and 2:7).

2.6.2 India

Oilseed crops play a very important role in the agricultural economy of India in that they account for about 8 per cent of the value of the total agricultural production.

About one-tenth of the total land under cultivation in India is devoted to the growing of oilseed crops. Most of India's soils are exhausted by centuries of crop raising without replenishment of nutrients. It takes 25 years to rejuvenate a depleted soil. Consequently there is hardly room for dramatic expansion in oilseed acreage and increases will have to come through higher yields per acre. The output of oilseeds in India, like her other agricultural production, depends largely on the weather; only 6 per cent of the total area planted to oilseed was under irrigation in 1968.

In the past 15 years the average increase in demand for oilseed in India has been an estimated 5 per cent a year.

Rape and mustard seed are important oilseed crops in India, particularly in the northern States (Uttar Pradesh, Punjab, Rajasthan, Assam and Bihar), and India is one of the largest producers of rapeseed in the world. During 1951–56 the annual average area under rape and mustard seed in the world was 23.4 million acres, to which India contributed 25 per cent.

Agricultural progress has been rather slow and the yields per acre have remained almost static. The area under rapeseed during different years can be found in Table 2:2 and the production in Table 2:3.

1966–67 seedings of rapeseed and mustard seed were about 7,000,000 acres, less than the corresponding estimate the year before (7,144,000). The production of rapeseed and mustard seed in 1967 was 1.0 million tons, compared with 1.3 million tons harvested in 1966. The output of rapeseed and mustard seed is expected to be up slightly to 1.45 million tons in 1969 from 1.4 million in 1968.

The different types of *Brassica* grown in India are generally divided into three groups, *viz*. rai *(B. juncea)*, sarson *(B. campestris)* and toria *(B. campestris)*. There is also another oilseed crop belonging to the family Cruciferae, *viz*. taramira *(Eruca sativa)*.

Practically all of the rapeseed produced in India is consumed domestically to meet the demand for food oils. Rapeseed oil ranks number two after peanut oil in production. The oilcake is used as cattle feed as well as compost.

2.6.3 Japan

The Japanese farmer grows mainly winter rape *(Brassica napus)* and sows it either as first crop or as a second crop in a rice paddy immediately following the rice harvest. The output of rapeseed in Japan steadily decreased during recent years, as can be seen from Table 2:3. The acreage of rapeseed in Japan may be seen in Table 2:2. The domestic production of rapeseed was approximately 70,000 tons in 1967. A further decrease in plantings of rapeseed in Japan and unfavourable yields resulted in a 1969 harvest of only 48,000 tons, compared to 66,000 tons in 1968.

The imported volume of rapeseed was about 15,000 tons in 1967 and increased to 250,000 tons in 1968, whereas no rapeseed oil was imported. In 1969 the imported quantity is estimated to be about 300,000 tons of rapeseed, the main supplier of which is Canada. The expected increase results from the government's sharp increase in the rapeseed import quota for the Japanese fiscal year 1969 (April 1969–March 1970). A further step towards liberalization of rapeseed trade will be taken by the end of 1971. The attitude towards rapeseed liberalization is attributed not only to the decline in domestic production but also to the favourable profit margins on rapeseed crushing.

In 1964 660,000 tons of vegetable oils were used in Japan compared to 180,000 tons in 1954. Rapeseed oil holds a preferred position in Japan as an edible oil since it has been used traditionally and is the most popular oil for cooking purposes. In 1968 about 112,000 tons or rapeseed oil together with 1,800 tons of mustard seed oil was used in food products, mainly cooking oil.

In Japan rapeseed meal corresponding to about 160,000 tons of rapeseed is used as fertilizer for citrus fruits, tobacco plants, eggplants, cucumbers, and green vegetables in the greenhouses. Some meal is used for animal feeding. Both government and industry groups are intensely studying the use of rapeseed meal as a livestock feed component. About 12 million tons of feedstuffs will be used annually in Japan in the beginning of the 1970's. If only about one per cent of this would come from rapeseed, the demand for this crop would be something like 300,000 tons.

2.6.4 Pakistan

In Pakistan, as in India, the statistics for rapeseed and mustard seed crops are combined but rapeseed accounts for the larger part of the plantings. In the northern part of Pakistan the main type is *B. campestris* and in the southern part it is *B. juncea*.

The area under rapeseed has slightly decreased from nearly 800,000 hectares in 1958 and 1959 to about 650,000 hectares in 1966 (Table 2:2), but this trend is now broken. Acreage in rapeseed for harvest in 1968 was 5.5 per cent larger than in 1967. Prices have been attractive and growers have been induced by this and by government exhortations to produce more oilseeds.

The production of rapeseed was rather constantly around 300,000 tons during 1948–1967, but in 1968 and 1969 it was nearly 400,000 tons (Table 2:3).

As in India, all rapeseed is crushed domestically and the oil is used as food.

2.6.5 Taiwan

The production of rapeseed in Taiwan has been reduced considerably in recent years as a result of decreased plantings resulting from the unprofitability of the crop. The drop in the price of vegetable oils in the last two years has resulted in reduced rapeseed prices, and farmers who have suffered losses have shifted rapeseed acreage to other more profitable crops. This shift can be expected to continue.

34

In 1964 cultivation of this crop rose to 19,600 hectares producing 27,343 tons of rapeseed, but by 1967 only 4,500 hectares were planted and 4,931 tons harvested. Production in 1969 increased to 30,000 tons.

Rapeseed is usually planted as a fall crop following the second rice crop and is planted largely in the west central part of Taiwan.

Most of the rapeseed is crushed for oil, which is particularly desirable as a cooking and frying oil, while part of the rapeseed oil is used in paint. Canada has become a new supplier of rapeseed for Taiwan, and imports are now significant.

2.7 AUSTRALIA

Growers in New South Wales are experimenting with the growing of rapeseed. They plan to sow 1,300 hectares in 1970 but reports indicate that experiments to date have not been too successful. Climatic influences result in heavy scattering losses in the inland areas. Australia has been importing refined rapeseed oil from Canada since 1968.

REFERENCES

The tables are compiled from the following references:
Agricultural Commodities—Projections for 1975 and 1985, Vol. 1, Food and Agriculture Organization of the United Nations, Rome, 1966.
Monthly Bull. Agr. Econ. Statistics, FAO, Rome.
Foreign Agr. Circ., USDA FFO 4-68.
Rapeseed Digest, Vancouver, Canada.
Trop. Prod. Quart., Commonwealth Secretariat, London.
Vegetable Oils and Oilseeds, Commonwealth Secretariat, London.

CHAPTER 3

Botany of Rapeseed

LENA BENGTSSON*, ANGELICA VON HOFSTEN** and B. LÖÖF*

* *Oil Crop Division, Swedish Seed Association, Svalöf (Sweden)*;
** *Institute of Physiological Botany, University of Uppsala (Sweden)*

CONTENTS

3.1 DEFINITION OF RAPESEED

The oilseed denoted rapeseed on the world market does not derive from one species in the way that soya bean, linseed, peanut and other oils do, but may come from several species generally belonging to the genus *Brassica*. Sometimes rapeseed will be a mixture of two or more types or species. A survey of the species from which the commercial rapeseed may be derived is given in Table 3:1. These species are related and rather similar in appearance. They are also divided into subspecies, formae and varieties or cultivars. Turnip rape *(Brassica campestris)* seems to be the most variable and was originally the most widespread of the various *Brassica* species. Winter and summer types of rape *(Brassica napus)* as well as turnip rape occur, which are not very different in morphological characters but physiologically very different as the winter type does not come into the generative phase (forming of seed) if the crop has not been exposed to subzero temperatures for a certain period[1, 3].

Rapeseed from Europe generally consists of winter rape and rapeseed from Canada is generally summer turnip rape. Rapeseed from the Far East may be

36

TABLE 3:1

DEFINITION OF RAPESEED

Botanical (Latin) name	Correct English	Synonyms	Canadian	French	German
Brassica napus ssp.* oleifera forma biennis forma annua	Winter rape Summer rape	Oil rape, rapeseed swede rape, oilseed rape	Argentine	Colza d'hiver Colza de printemps Colza d'été	Winterraps Sommerraps
Brassica campestris ssp.* oleifera. forma biennis forma annua	Winter turnip rape Summer turnip rape	Rapeseed, oil turnip	Polish	Navette d'hiver Navette de printemps Navette d'été	Winterrübsen Sommerrübsen
forma annua var. chinensis	Summer turnip rape	Chinese mustard		Moutarde chinoise Pak-choi	Chinasenf
var. pekinensis	Summer turnip rape	Celery cabbage Chinese kale		Chou chinois Pet-sai	Chinakohl
var. dichotoma	Summer turnip rape	Toria		Toria	Toria
var. trilocularis	Summer turnip rape	Sarson		Sarson	Sarson
Brassica juncea	Brown mustard	Leaf mustard Indian mustard	Oriental mustard	Moutarde brune	Brauner Senf Sarepta Senf

* subspecies.

either summer turnip rape, brown mustard *(Brassica juncea)* or a mixture of these. Both species are grown together in many districts and rather often the crop is not pure but a mixture of *Brassica campestris*, *Brassica juncea* and sometimes also *Eruca sativa* (Rocket salad). The names Toria and Sarson used for rapeseed grown in India and Pakistan do not define any special type or species, since Toria comprehends several types of early sown *Brassica* and Sarson late sown *Brassica*. Sarson is *Brassica campestris* in some regions, *Brassica juncea* in other regions.

3.1.1 Systematic and genetic relation of Brassica species

The relationships of the rapeseed-yielding species belonging to the genus *Brassica* are illustrated by Fig.3:1, called the triangle of U (this relation was first illustrated

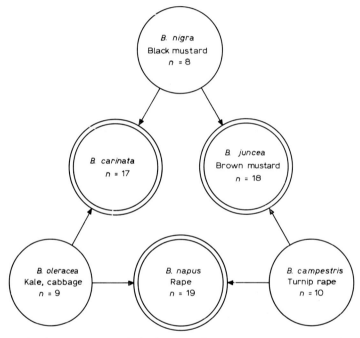

Fig.3:1. Triangle of U. The 'basic' diploid species form the corners with the 'derived' amfidiploid or allotetraploid species on the sides of the triangle.

by the Japanese scientist U). There are three basic species, *Brassica nigra*, *Brassica oleracea* and *Brassica campestris*. The chromosome numbers of the gametes (pollen and ovary cells) are denoted by $n = 8$, $n = 9$ and $n = 10$. By hybridization three new species, *Brassica carinata*, *Brassica juncea* and *Brassica napus*, are obtained. This hybridization can be made artificially but was done by Nature very long ago. The hybrid species obtain the added number of chromosomes in their gametes, *viz.* the sum of the numbers of the basic species[1, 3].

Seed types (formae oleiferae)

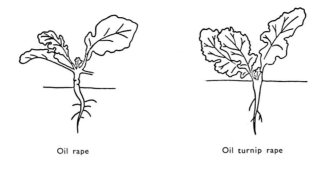

Oil rape   Oil turnip rape

Fodder (leaf) types (formae foliferae*)

Fodder (root) types (formae rapiferae)

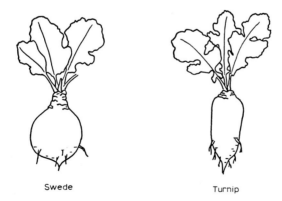

Swede   Turnip

*Name suggested by the authors.

Fig.3:2. Different types (formae) of rape *(Brassica napus)*, left, and turnip rape *(Brassica campestris)*, right, in the vegetative phase.

3.1.2 Different types within species

By natural selection and by breeding for perhaps several thousand years, different types and forms of most *Brassica* species have been developed. *Brassica oleracea* for instance has, by selection, been divided into many very different types such as kale, cabbage, cauliflower, brussels sprouts and several other types. Of *Brassica napus* and *Brassica campestris* there are, as can be seen from Fig.3:2, one seed type (forma *oleifera*) and two fodder types, one of which yields bulby roots, called swede *(Brassica napus)* and turnip *(Brassica campestris)*. The other fodder type has been obtained by selection for high yield of leaves and stalks.

Besides the species included in Fig.3:1, there are some other *Brassicas*. The relations of these are not yet completely clear. These species have the same chromosome numbers as the basic species in the triangle of U and are as follows: *Brassica tournefortii, Brassica barrelieri, Brassica oxhyrrina* ($n = 10$), *Brassica arvensis** ($n = 9$) and *Brassica fruticulosa* ($n = 8$). *Brassica tournefortii* sometimes is included with the rapeseed-yielding species.

3.2 MORPHOLOGY

3.2.1 How to distinguish the rapeseed-yielding species

Most examination keys use the inflorescence to distinguish the species. However, this way is to some extent uncertain as far as the *Brassica* species in the triangle of U are concerned. Systematization according to the colour of the petals is unreliable, as there are not less than five different colours[10]. The shape of the inflorescence can generally provide indications to distinguish species. *Brassica oleracea* has the buds at a higher level than the flowers just opened. This character is dominant in the hybrids *Brassica napus* and *Brassica carinata*, whereas in *Brassica campestris* the buds are at a lower level than the flowers just opened (Fig.3:3). However, there are exceptions to this rule, and in *Brassica napus* varieties with the same bud position as in *Brassica campestris* may be found. The most reliable character used for distinguishing the *Brassica* species in the generative phase is the shape of the upper leaves (Fig.3:4). Exceptions in this character are almost never found. In the basic species *Brassica campestris* the lower part of the blade (lamina) grasps the stalk completely, whereas in *Brassica oleracea* the blade just reaches the stalk. The hybrid *Brassica napus* is intermediate and the blade is half-grasping the stalk. In *Brassica nigra* the lamina ends far from the stalk and the hybrid *Brassica juncea* has obtained much of this character, but the petiole is

---

* Also referred to as *Sinapis arvensis*.

40

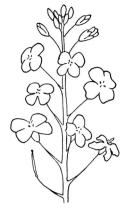 

Fig.3:3. How to distinguish the rapeseed-yielding species. The inflorescence.
Left: Flowering rape *(Brassica napus)*. The buds are generally at a higher level than the flowers just opened.
Right: Flowering turnip rape *(Brassica campestris)*. The buds are at a lower level than the flowers just opened.

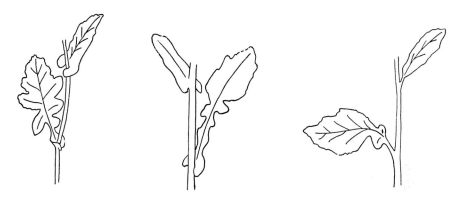

Fig.3:4. How to distinguish the rapeseed-yielding species. Generative phase: The upper leaves.
Left: Turnip rape *(Brassica campestris)*. The leaves of the inflorescence grasp the stalk completely.
Centre: Rape *(Brassica napus)*. The leaves of the inflorescence only half-grasp the stalk.
Right: Brown mustard *(Brassica juncea)*. The laminae (blades) of the upper leaves do not reach the stalk.

rather short (Fig.3:4). Distinguishing the *Brassica* species by their seed is also possible, as will be mentioned below.

3.2.2 The Brassica pod

The family Cruciferae, to which the genus *Brassica* belongs, has a quite special type of fruit called pod (siliqua). The pod is made up of two carpels, which are separated by a false septum, thus providing two chambers (Fig.3:5). The number of seeds in each pod is rather different in different species and there is also variation

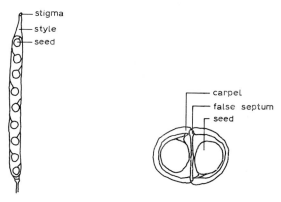

Fig.3:5. The *Brassica* pod.
Left: Longitudinal section.
Right: Cross section.

between types and varieties within species. Modern varieties of rape generally contain 15–40 seeds per pod. At maturity the two carpels are easily split from the false septum. Thereby the seeds are shed to the ground. The resistance to shedding is very different in different species and varieties, and for the plant breeder it is quite important to improve resistance to shedding (see Chapter 6).

3.3 THE SEED

Besides large differences in morphology, tolerance to different growing conditions and other agronomical properties, the various species, subspecies and types of *Brassica* give seeds, which are different in size, content of oil and other substances, colour of seed coat and so on.

3.3.1 Seed colour

The seed from turnip rape is generally somewhat lighter coloured than that of rape. Seed colours from light yellow to brown and black are found in brown mustard. In turnip rape there are also yellow-coloured types grown in India and Pakistan (*viz.* yellow-seeded Sarson).

3.3.2 Seed size

The seeds of *Brassica napus* are, as shown in Table 3:2, generally larger than those of *Brassica campestris* and *Brassica juncea* and the winter types give larger seeds than the summer types[1].

TABLE 3:2

NORMAL VARIATIONS IN SEED SIZE OF DIFFERENT TYPES OF BRASSICA OILSEED

	Weight grammes/1000 seeds
Winter rape	4.5–5.5
Winter turnip rape	3.0–4.0
Summer rape	3.5–4.5
Summer turnip rape	2.0–3.0
Brown mustard	2.0–3.0

3.3.3 Seed coat

The structure of different layers of the seed coat of *Brassica* oilseeds shows great variation between species (Figs. 3:6, 3:7 and 3:8). This character has been used

Fig.3:6. Coat structure of some *Brassica* seeds (after Berggren[2]).
1 a–d Summer rape *(Brassica napus)*
2 a–d Swede *(Brassica napus)*
3 a–d Siberian kale *(Brassica napus)*
4 a–d Winter rape *(Brassica napus)*
(a 70 ×, b–d 9 ×)

Fig.3:7. Coat structure of defatted rapeseed *(Brassica campestris)* (220 ×)

to distinguish seeds from different species and varieties[2, 3, 4, 8]. The seed coat has a rather low oil content and a very high content of fiber[5, 9]. The seed coat in per cent of the total seed is different in different species and varieties. Yellow-coloured seeds have lower percentage seed coat than dark-coloured ones[13].

3.3.4 The embryo

Contrary to many other important seed plants, *e.g.* cereals, the *Brassica* species have very little endosperm and when the seed coat, holding 12–20% of the seed weight, is removed, the interior contains the embryo (Fig.3:8). Most of the embryo consists of the cotyledons, which are double (conduplicate) and contain about 50 per cent oil and protein-rich grains similar to those in the aleurone cells lying just under the seed coat. The central part of the mature rape seed is occupied by meristematic tissue (ET of Fig.3:8) from which the radicle (root), hypocotyle (stem) and epicotyle (bud) develop[5, 6].

44

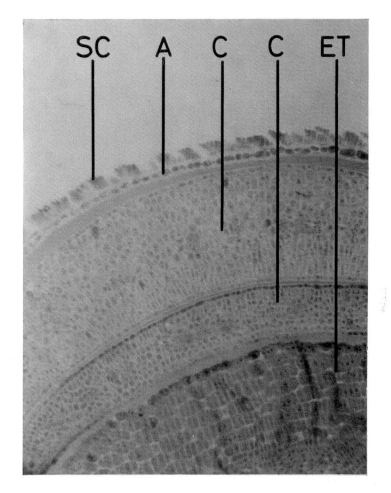

Fig.3:8. Section through a rape seed *(Brassica napus)* (200 ×)
SC   Seed Coat   C   Cotyledon (double)
A   Aleurone Layer   ET   Embryonal Tissue

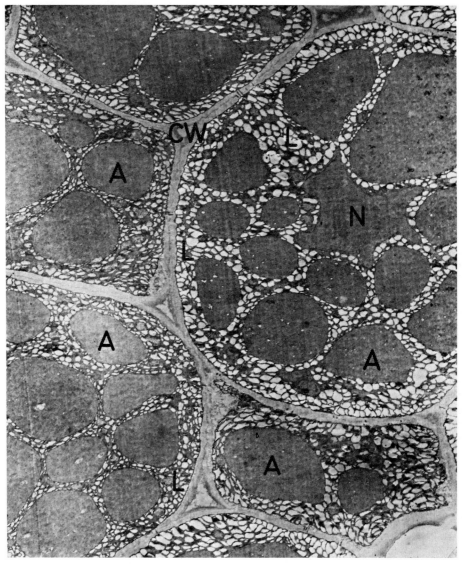

Fig.3:9. Electron micrograph of oilseed rape *(Brassica napus)* (8,000 ×)

CW	Cell Wall	L	Lipid droplets
A	Aleurone grains	N	Nucleus

Fig.3:9 shows an electron micrograph of an ultrathin section through cells of the meristematic tissue of a *Brassica napus* seed. Each cell is surrounded by a rather thin cell wall (CW) mainly composed of cellulose. Lipid droplets (L) are scattered through the cytoplasm and the nucleus (N) is seen in the middle of the cell. The aleurone grains (A) are electron dense and measure 2–10 μ in diameter.

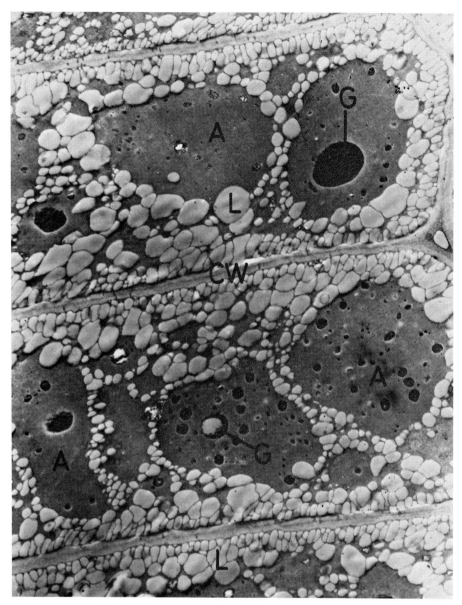

Fig.3:10. Electron micrograph of embryonic cells of turnip rape *(Brassica campestris)* (8,000 ×)

CW	Cell Wall	L	Lipid droplets
A	Aleurone grains	G	Globoids

The protein in these aleurone grains is considered to be storage products, and has been termed ergastic protein[7, 14].

The embryonal cells of *Brassica campestris* are also rich in lipids, as shown in Fig.3:10. The aleurone grains of this specimen contain large numbers of very electron dense bodies, which have been called globoids (G). They are assumed to consist of calcium and magnesium salts of phytic acid and occur both in the cotyledons and the central embryonic tissue[4, 11, 12]. However, marked variations in their abundance and distribution in different varieties of rapeseed have been observed.

REFERENCES

1 G. ANDERSSON and G. OLSSON, in H. KAPPERT and W. RUDORF (Eds.), *Handbuch der Pflanzenzüchtung*, Vol. 5, Paul Parey, Berlin, 1961, p. 1.
2 G. BERGGREN, *Meddelanden från Statens Centrala Frökontrollanstalt*, No. 35 (1960) 28. (In Swedish)
3 G. BERGGREN, *Svensk Botanisk Tidskrift*, 56 (1962) 65. (In English)
4 I. M. DECHARY and A. M. ALTSCHUL, in R. F. GOULD, *World Protein Resources, Am. chem. Soc.*, 1966, p. 148.
5 R. K. DOWNEY, in J. P. BOWLAND, D. R. CLANDININ and L. R. WETTER (Eds.), *Rapeseed meal for livestock and poultry*, Queen's Printer, Ottawa, 1965.
6 K. ESAU, *Anatomy of seed plants*, Wiley, New York, 1960.
7 E. M. ENGLEMAN, *Am. J. Bot.*, 53 (1966) 231.
8 J. A. OMS and A. PÉREZ TORT, *Identificación histologica de algunas Brassicas en subproductos oleaginosos*, Casa Jacobo Peuser, S.A.C.E.I., Buenos Aires, 1959. (In Spanish)
9 B. LÖÖF, *Sveriges Utsädesförenings Tidskrift*, 73 (1963) 101. (In Swedish with English summary)
10 J. MORICE, *Annales de l'amélioration des plantes*, I.N.R.A., (II, 1960) 155. (In French with English summary)
11 A. NIETHAMMER and N. TIETZ, *Samen und Früchte des Handels und der Industrie*, Cramer Verlag, Weinheim, 1961.
12 D. C. PEASE, *Histological techniques for electron microscopy*, Academic Press, New York. 1964.
13 L. BENGTSSON, *Sveriges Utsädesfören. Tidskr.*, 80 (1970) 186. (In Swedish with English summary)
14 I. N. SVESHNIKOVA, J. P. BOLYAKINA and N. P. BUZULUKOVA, *Int. Congr. Electron Microscopy*, 3 (1970) 455.

Cultivation of Rapeseed

B. LÖÖF

Oil Crop Division, Swedish Seed Association, Svalöf (Sweden)

CONTENTS

4.1 IMPORTANT PRODUCTION AREAS

As mentioned in Chapter 2, India and Canada are the most important production areas of rapeseed. The most important producer of summer turnip rape *(Brassica campestris)* is Canada, while winter rape *(Brassica napus)* is almost exclusively produced in Poland, France, Germany, Sweden and some other European countries as well as in some regions of South America, mainly Chile. Winter turnip rape *(Brassica campestris)* holds a very small area in some European countries, e.g. Sweden has about 20,000 hectares. In Finland winter turnip rape is the most important oil crop, but the area has not surpassed 20,000 hectares[1].

Among the more important rapeseed producing countries, Sweden seems to have the highest yields. Averages for the years 1961–1968 are as follows:

	kg/ha
Winter rape	2700
Winter turnip rape	1900
Summer rape	1700
Summer turnip rape	1300

The average yield of winter rape is 2000–2500 kg/ha for other European countries, except Poland which has an average somewhat under 2000 kg. The average yield of summer turnip rape in Canada is about 900 kg/ha, whereas the summer types of *Brassica campestris* grown in India and Pakistan yield only about 400–500 kg/ha. The yields of winter rape in the main rape growing districts of South America are similar to those of Europe[2, 3].

4.3 OIL CONTENT OF THE SEED

The oil contents of rapeseed given in the literature are somewhat difficult to compare, as they sometimes refer to a dry matter basis and sometimes to the oil content of a seed with varying water content. The content of dockage in the seed also influences the oil percentage calculated on total sample. Swedish oil crops (pure seed, dry matter basis) have the following average contents of crude fat:

	% of dry matter
Winter rape	47
Winter turnip rape	45
Summer rape	43
Summer turnip rape	43

There is a rather great variation between fields, regions and years, for example in winter rape there has been a variation between 40 and 53%[1, 4, 5]. (Oil quality and oil content are further treated in following chapters.)

4.4 EDAPHIC AND CLIMATIC REQUIREMENTS

All the types and varieties of *Brassica* yielding what is commercially named rapeseed are completely or at least to a considerable extent cross-pollinated. They are consequently rather variable and are thus to be considered as populations. This fact, in connection with the great number of different subspecies and types, results in a very good adaptability to different environmental conditions. In

Sweden and Finland hardy varieties of winter rape and winter turnip rape can survive very low temperatures and snow cover for long periods of time. Summer turnip rape can be grown above the polar circle with 24 hours of light a day, as well as in Pakistan with a day length which is shorter than 10 hours in winter. In some districts there is almost no rainfall during the growing season, whereas in other districts the winter and spring are very humid (Holland, France). These facts speak for rapeseed having a great future, if only the quality of the oil and the meal can be improved[4, 6, 7].

4.4.1 Soil

According to some old literature, rape and turnip rape have high requirements upon the soil, *i.e.*, the pH should be around 7. However, according to modern views and new experience rape and turnip rape can tolerate a wide range of pH and also be grown on marginal soils, not good enough for wheat and sugar beets. Also on light, sandy soils rape and especially turnip rape can do well in Europe, since they start growing early in spring and the root system is very deep. Winter rape and winter turnip rape suffer less from drought in summer than cereals (shallow root system) and sugar beets (spring sown crop). When rape and turnip rape start growing, they place rather great demands on the soil structure, but once established they can fairly well tolerate packed soil. The small seeds need shallow sowing (1–3 cm) and enough moisture in the surface layers for germination. Furthermore the soil surface should not be too fine, as heavy rain followed by drought may result in the formation of a crust, too hard for the small germs to penetrate[1, 8, 9].

4.4.2 Salinity and water

In arid climates salinity poisons certain crops. Rape and turnip rape seem to be fairly tolerant to salinity. In humid climates too much water causes problems. Rape and turnip rape prefer well-aerated soils and are liable to suffer from water-logging. Too much water may favour attacks by fungi on the roots and decrease resistance to low temperatures during winter. Thus it is important that surplus surface water is drained away. On irrigated soils excessive flooding may damage the crop, especially in the early stages of development. Lack of water during the generative phase may heavily influence the formation of seeds, seed size and oil content[3, 9, 10, 11].

4.4.3 Light and temperature

Rape and turnip rape may be grown under very variable light conditions. Since the vegetation period is very short, it is evident that these crops can make use of long days and strong light in the northernmost regions. Winter rape may become less frost hardy from too feeble light in connection with high temperature in winter. In comparison to most other oil crops rape and turnip rape prefer relatively low temperatures up to flowering. In the generative phase these crops are more tolerant to high temperatures, but heat in connection with drought may cause a reduction in seed size and oil content[9–13].

4.5 CULTIVATION TECHNIQUE

4.5.1 Preparation of seed and soil

In the technically advanced countries farmers generally buy seeds from seed firms. These are generally subjected to state control and for each class of seed there are strict regulations for purity, germination capacity and origin. The seed is ordinarily treated with some fungicide or a combined fungicide–insecticide. In less developed countries farmers use their own seed, sometimes resulting in impure, poor crops.

Proper preparation of the soil before sowing is most decisive for subsequent development and for the yield. Since the *Brassica* oil seeds are very small, the seed bed should not be too coarse. On the other hand the soil should not be laboured too much, since a heavy layer of fine particles will predispose hilly fields for erosion and lack of air if the soil is saturated with water in clay soils. The seeds should not be planted deeper than 1–3 cm. It is important to think of preserving sufficient soil moisture to secure germination even if the weather after sowing should be dry and warm[1, 8, 9].

4.5.2 Fertilizers

In the countries where there are enough fertilizers on the market and the farmers have means to buy sufficient amounts, rather many fertilizer trials have been carried out. Generally these trials have shown that rape and turnip rape have about the same need for potassium and phosphorus as other field crops such as cereals and sugar beets. The need for nitrogen is, however, higher than for most other crops.

Most of the fertilizer is taken up during the period of rapid growth and up to late flowering. For a seed yield of 4,000 kg/ha of winter rape the following quantities were used: nitrogen 135 kg, phosphoric acid 55 kg and potash 70 kg.

The time at which the application of fertilizer takes place is rather important. Potassium and phosphorus should be given before sowing. If the precrop contains much straw, which is plowed into the soil, it is important to give 50–60 kg of N before sowing winter rape or turnip rape. An additional 120–180 kg of N

Fig.4:1 Winter rape in early spring. The temperature is still low (nightly frosts) but strong light during the day favours rapid growth.

Fig.4:2. Field trials with winter rape. Full flowering.

Fig.4:3. Winter rape at full maturity. The field was combine-harvested and the yield was near 4,000 kg per hectare.

Fig.4:4. In windy regions the rape is windrowed to avoid shedding. It is very important to cut the crop at the right moment: Yellow pods, seeds almost black.

should be applied when the plants start growing in the spring. Nitrogen should be given spring-sown crops before sowing or dressed as nitrate shortly after emergence. Calcium nitrate generally is more efficient than other types of nitrogen fertilizers[1, 6-9, 14, 15].

4.5.3 The sowing of rapeseed

The date of sowing is generally rather important for the subsequent development of the crop. Autumn-sown crops should be sown early enough to insure sufficient development before winter. Too early sowing may result in too vigorous growth and less resistance to cold during winter. Late sowing may give too small plants, which are killed in winter owing to root damage and dehydration. The sowing time should be adapted to prevailing weather conditions. Too hot and dry weather speaks for the postponement of autumn-sowing, since germination may be poor and the stand thin and afterwards damaged by flea beetles, aphids or other insects. Sowing in spring should take place as early as possible since rape and turnip rape are favoured by rather cold and wet weather in the first stages of development. The risk for insect damage is also smaller the more developed the stand is before warm weather comes.

The seed rate and row distance may be varied within rather great limits without significant effect upon the yield. There is always an interaction between density of stand and other factors. In winter rape a thin stand is favourable for frost hardiness and overwintering capacity. A thin stand cannot compete with weeds as well as a rather dense, uniform stand. The ripening occurs earlier and is more uniform in a dense stand than in a thin one. Too high a seed rate may cause stalks to break or to lie down already at flowering, leading to reduction in yield.

The distance between rows must be accommodated to the type of hoeing machine used. If the soil structure is excellent and there are few weeds, hoeing is not necessary.

The sowing should always be carried out with machines planting the seed in rows. Broadcasting the seed followed by harrowing does not secure good germination and an even stand.

Depending upon seed size and type of machine the seed rate should be between 2 and 15 kg/ha. The lowest seed rates are used for small-seeded varieties, sown in autumn with precision machines and row distances of about 40-50 cm. The highest seed rates should be employed for large-seeded varieties of summer types of oilseed sown with narrow row distances (10-12 cm)[1, 6-9, 11, 15, 16].

4.5.4 Plant protection

In some countries rather much attention has been given to the place of oilseed in crop rotation. According to modern experience, rapeseed crops may be placed anywhere in the rotation, if the soil is laboured in order to insure a suitable structure and the pre-crops do not determine the frequency of weeds, insects or fungus diseases.

(a) Weeds

In modern mechanized agriculture weeds are no serious problem for oil crops. Most weeds cannot be controlled by herbicides, as in the case of cereals, since the rapeseed crops are more sensitive. Mechanized hoeing is, however, rather efficient, and new vigorous varieties can compete very well with most weeds. Weeds belonging to the grass family can be controlled by TCA (trichloroacetic acid) and Dalapone, especially in rape *(Brassica napus)*. Turnip rape is more sensitive to TCA. It is especially important to treat *Agropyron* and volunteers of cereals before sowing (5–25 kg of TCA depending upon soil type, grass frequency, soil moisture etc.).

Quite a number of trials with herbicides other than TCA to control different types of weeds have been carried out, but none of these has hitherto proved to be quite safe and efficient.

(b) Insects

During all stages of development, rape and turnip rape are attacked by several insects, the most important of which are the following:

Psylliodes chrysocephala and other flea beetles. Several kinds of flea beetles attack rape and turnip rape. Some of them feed on the young cotyledons shortly after emergence. Most important is the rather big, black rape beetle *Psylliodes chrysocephala*. The adults feed on the young plants and the larvae penetrate into the stem and roots of plants during autumn, winter and early spring. This attack can be fatal, especially in dense stands (slender plants). Since the introduction of seed treatment with Lindane this insect has been almost eliminated.

Ceutorrhynchus napi (quadridens), the stem weevil. This insect attacks the winter types in spring and summer types in summer. The larvae feed on the marrow of the stem, and the attack sometimes causes a heavy decrease in yield. This weevil is difficult to control.

Meligethes aeneus, the pollen (blossom) beetle. The pollen beetle attacks young buds and its larvae feed on the top flowers of the inflorescence. This insect contributes to the pollination and does not cause so much damage if the attack comes during flowering. If there are about five beetles on each plant before flowering, treatment with some insecticide is considered to be economic.

Ceutorrhynchus assimilis, the pod weevil. The pod weevil itself does not cause much harm to the rapeseed crop, but the punctures it makes in the pods are used by another insect, namely *Dasyneura brassicae*, the pod midge. This small insect introduces quite a number of eggs into the pods in which the larvae then feed from the interior of the walls, causing necrosis which leads to premature opening

54

and loss of the seeds. This insect is one of the most harmful to rapeseed crops. *Brassica napus* as well as *Brassica campestris* are attacked but not *Brassica juncea*.

Athalia colibri, the turnip sawfly. This rather big fly lays its eggs under the leaves of several Cruciferae. The young larvae are of the same green colour as the host plant and consequently difficult to observe, but with time they become darker and easier to detect. If not killed they can destroy young plants, especially on the edges of the field.

Brevicoryne brassicae, cabbage aphids. Aphids are not very dangerous for winter types and in a cool humid climate, but they are the most dangerous of all insects for the spring-sown types in dry, hot climates. Aphids have made the growing of summer rape almost impossible in Poland, and they are to a great extent responsible for the very low yields of rapeseed in India and Pakistan. However, if treatment is applied before the number of aphids is too great they are easily mastered by several systemic insecticides. There are several other insects that attack *Brassica* oilseed crops, but those mentioned are the most frequent ones.

(c) Other animals

Limacidae, snails. These animals can be very dangerous if the structure of the soil is rather coarse and the rapeseed is sown after ley. Attacks from snails are also to be expected on grassland edges of fields. In large numbers these animals may be quite dangerous to small plants. They can, however, be mastered by several chemicals, e.g. hydrate chalk and special snail poisons, mainly based upon metaldehyde.

Since rapeseed crops, especially in the vegetative stage, are very tasty to several large animals, fields are often attacked by for instance pheasants, pidgeons, rabbits, hares and deer. If there is a large number of animals feeding on the rapeseed crops the reduction in yield may be considerable. Control by hunting seems to be the only way of mastering this problem[1, 6-9, 17].

(d) Fungi

Fungus pests are generally not serious in *Brassica* oilcrops if these do not come again too often in the rotation. If *Brassica* species are grown more often than every fourth or fifth year some pests like *Plasmodiophora brassicae* (clubroot, attacking roots) and *Phoma lingam* (attacking hypocotyle and stem) may be very serious and cause complete destruction or quite low yield. Plant breeders have, however, succeeded in finding new varieties, now ready for marketing, which are fairly resistant to the races of these fungi now present in some important growing districts. (*Phoma* is very serious in parts of France; *Plasmodiophora* occurs, *e.g.*, in Sweden, Germany and England.)

Sclerotinia species (attacking stem and branches) probably become more severe if cultivation is too intensive, but this fungus is more dependent upon quite special weather conditions than the above-mentioned. Too little research has been done on *Sclerotinia* to map out its life cycle and influence upon the yield. If two or more types of *Brassica* oilcrops with different dates of maturity are grown in the same district (causing a stepwise increase of spores), *Alternaria* species may be quite harmful. These fungi, visible as black spots upon pods, stem and branches, do not cause complete destruction, and the attack is generally limited to two stages in plant development, *viz.* young seedlings and almost 'yellow-mature' plants, causing premature, smaller seeds and increased shedding as the pod walls become very brittle. Severe attacks are said to cause 15–20% reduction of the yield. The fungi mentioned are more frequent in humid climates than in dry ones. Other fungus pests of importance in *Brassica* oilseeds are *Peronospora brassicae*, most severe on young cotyledons, delaying the development before winter comes; *Botrytis cinerea*, generally secondary after primary damage by insects and *Alternaria* species; and *Verticillium* and *Typhula* (in regions with long-time snow cover). Viruses, sometimes spread by aphids, are most important in hot, dry climates[1, 8, 11, 18].

4.5.5 The harvest of rapeseed

The date and method of harvesting considerably influence the yield as well as the seed quality. Too early harvest may cause damage (dead seed), high chlorophyll content, free fatty acids (if artificial drying is not undertaken immediately) and heavy drying costs. As rapeseed crops are not very resistant to shedding of the seeds, the harvest should not be postponed for a long time, since strong winds or storms may cause considerable losses of seed. To avoid this, swathing is applied in windy regions, mainly along the coasts of Western Europe. Swathing is also undertaken (*e.g.* in Canada) to speed up and even out the maturation of the seeds.

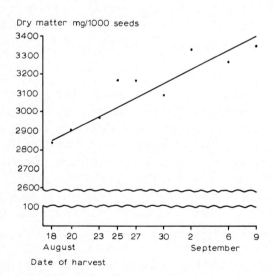

Fig.4:5. The relation between harvest date and seed yield (after Lööf[24], revised).

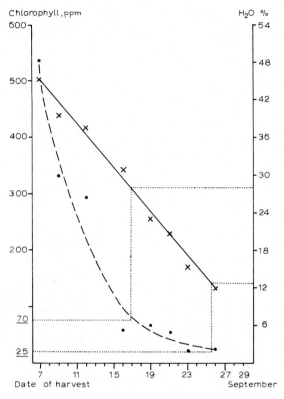

Fig.4:6. The relation of harvest date to moisture and chlorophyll content. Summer rape.

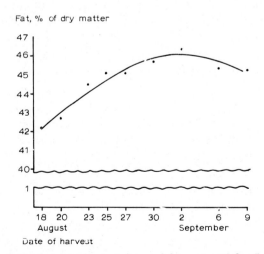

Fig.4:7. The relation between harvest date and fat content (after Lööf[24], revised).

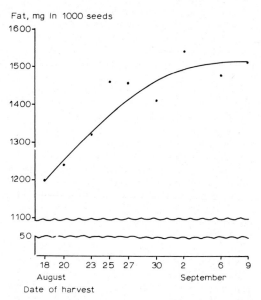

Fig.4:8. The relation between harvest date and fat yield (after Lööf[24], revised).

Premature cutting when swathing, however, causes respiration losses, resulting in smaller seeds and lower oil content than if the crop had been left to full maturity on root. Special studies of the influence of harvest date and treatment after cutting upon yield and seed quality of rape have shown that the moisture content of the seed is an adequate measure of the stage of maturity down to about 20% water, provided there is no influence by rain, dew or humid air, and the crop is normally developed. When the crop is uneven and the ripening delayed, the chlorophyll content is a better measure as it shows directly if the seed oil can be used for food. The chlorophyll content should not be higher than 25 ppm of the oil in satisfactory seed.

The oil content increases in winter rape up to the 'yellow pod' stage and shows a tendency to decrease after full maturity. In summer rape there is a constant increase up to full maturity. The decrease in oil content is more pronounced the earlier the cutting takes place. The chlorophyll content is already low in even, normal stands some time before threshing is possible. In uneven stands, damaged by winter, drought, insects and fungus pests, the chlorophyll content may be high, even if threshing is postponed until most plants seem to be ripe. Threshing should normally be carried out when the average moisture content of the seed is 12–20%. It is difficult to avoid crushing and damage of seed if the crop is too dry (6–12% moisture) at the harvest.

Germination capacity (vitality) is important if the seeds are to be stored for some time (several months). A high percentage of dead seeds generally causes a rise in the content of free fatty acids and oxidation products, causing off-flavours of the oil. Dead seeds are caused by mechanical damage during threshing or during handling and transport. When artificial drying is applied, too high temperatures (about 40 °C) may result in killing of the seeds. These are more sensitive the higher the moisture content (further treated in Chapter 5).

To be stored for more than a few days rapeseed should have a moisture content between 6–8%. Seeds dryer than 6% moisture content are very brittle and more easily damaged during handling and transport. In many rape producing areas facilities for artificial drying are not available. If available, however, equipment for artificial drying considerably improves the possibilities for the growers to avoid shedding by wind, damage during threshing, respiration losses and fat deterioration (free fatty acids)[1, 3, 6-10, 18-24].

REFERENCES

1 B. Lööf, *Field Crop Abstracts*, 13:1 (1960) 1.
2 R. K. Downey, S. H. Pawlowski and J. McAnsh, *Rapeseed Canadas 'Cinderella' Crop*, Rapeseed Ass. Can. Publ. No. 8, (1970).
3 B. Lööf, The winter oil crops. Recommendations for an accelerated winter oil seed improvement program in West Pakistan, *Report No. 10 Accelerated Crop Improvement Series*, (1969), Distr. by Planning Cell, Agriculture Department, Government of West Pakistan, Islamabad.
4 G. Andersson and G. Olsson, in H. Kappert and W. Rudolf (Eds.), *Handbuch der Pflanzenzüchtung*, 5 (1961) 1.
5 W. Schuster, *Fette, Seifen, Anstrichmittel*, 69 (1967) 831. (In German)
6 *Skånes Oljeväxtodlareförening 1943–1968*, Malmö, 1968, p.139. (In Swedish)
7 *Sveriges Oljeväxtodlares Centralförening 1943–1968*, Malmö, 1968, p.199. (In Swedish)
8 CETIOM, *Bulletins d'information* 1960–1969, 174 Avenue Victor-Hugo, Paris 16ᵉ. (In French)
9 R. Delhaye, *Rev. Agr.*, 5 (1967) 653. (In French)
10 F. Dembinski, E. Matusieyicz, E. Kopanska, A. Dukas and I. Wiatroszak, *Rocz. Wyz. Szkoly Roln. Poznań*, 4 (1958) 3. (In Polish with English summary)
11 B. Lööf, *Sveriges Utsädesfören. Tidskr.*, 73:1–2 (1963) 25. (In Swedish with English summary)
12 B. Lööf and G. Andersson, in E. Åkerberg and A. Hagberg (Eds.), *Recent Plant Breeding Research. Svalöf 1946–1961*, (1643) 246, Almquist och Wiksell, Stockholm.
13 B. Lööf and R. Jönsson, *Sveriges Utsädesfören. Tidskr.*, 75:1–2 (1965) 70. (In Swedish with English summary)
14 G. Andersson, R. Olered and G. Olsson, *Z. Acker-Pflanzenbau*, 107:2 (1958) 171. (In German with English summary)
15 H. Rüther, *Z. Landwirtsch. Versuchs-Untersuchungs. Rostock*, 2:4 (1957) 338.
16 A. Vez, *Agriculture Romande*, 6:5 Série A (1967) 59. (In French)
17 J. Mühlow and E. Sylvén, *Oljeväxternas skadedjur*, Natur och Kultur, Stockholm, 1953, p.163. (In Swedish)
18 B. Lööf, *Zeitschrift für Pflanzenzüchtung*, 46:4 (1961) 405. (In German with English summary)
19 L.-Å. Appelqvist, *Sveriges Utsädesfören. Tidskr.*, 71 (1961) 74. (In Swedish with English summary)
20 L.-Å. Appelqvist, *J. Am. Oil Chem. Soc.*, 44 (1967) 206.
21 B. Lööf and S.-Å. Johansson, *Sveriges Utsädesfören. Tidskr.*, 76:5–6 (1966) 344. (In Swedish with English summary)
22 B. Lööf and S.-Å. Johansson, *Sveriges Utsädesfören. Tidskr.*, 79:1–2 (1969) 16. (In Swedish with English summary)
23 K. Babuchowski, *Wyzszey Szkoly Rolniczej w Olsztynic*, 18 (1964) 191. (In Polish)
24 B. Lööf, *Sveriges Utsädesfören. Tidskr.*, 76 (1966) 31. (In Swedish with English summary)

CHAPTER 5

Postharvest Handling and Storage of Rapeseed

L.-Å. APPELQVIST* and B. LÖÖF**

* *Division of Physiological Chemistry, Chemical Center, University of Lund (Sweden);*

** *Oil Crop Division, Swedish Seed Association, Svalöf (Sweden)*

CONTENTS

In discussions of the world food situation, it is often emphasized that considerable amounts of food produced in technically less developed countries are spoiled due to improper post-harvest handling and storage[1]. It has been estimated that one third or more of the food raw materials thus disappears. This general spoilage, depending on the lack of suitable drying and storage facilities, hits such large rapeseed-producing areas as India, Pakistan and China. In Canada, there is generally little need for artificial drying, since the air is dry during the rapeseed harvest. In western Europe, on the other hand, with frequently occurring moist weather during the rapeseed harvest, especially during the harvest of spring-sown crops, there is a definite demand for rapid artificial drying. In Sweden the system for drying of rapeseed is well organized and has sufficient capacity, possibly except for extremely wet autumns. Some of the other European rapeseed producing countries, such as Poland and France, do not have quite sufficient capacity for immediate artificial drying of the entire crop. As a consequence of these differences in climate in the major rapeseed producing countries, and differences in capacity for artificial drying, there is considerable variation in the regulations on allowable moisture levels. Since the cultivation and trade of rapeseed in Sweden are well organized, and the wet and cool autumns in Sweden have necessitated the construction of a large number of high capacity drying plants, most examples in this chapter will be from the Swedish system.

5.2 EQUIPMENT FOR MOISTURE CONTENT DETERMINATION, SEED CLEANING AND ARTIFICIAL DRYING

In Chapter 4, the cultivation of rapeseed was discussed up to the point of threshing of the crop. This chapter is a direct continuation in the sense that it will cover the subsequent handling of rapeseed, up to the point where the seed is delivered to the oil mill. Depending on the weather conditions during the harvesting and also upon the method of harvesting, the threshed seed will vary in moisture content quite considerably from field to field and even within a given field. Before starting the artificial drying of the seed (if necessary) the farmer therefore has to know its moisture content.

5.2.1 Determination of moisture content and estimation of acceptable delay in starting the drying operation

Equipment for instantaneous determination of the moisture content of seeds and grains are generally based on one of two principles:

(1) The electric conductivity of the seed sample or

(2) The dielectric constant of the seed sample.

Instruments based on the first principle do not require a weighing operation but use a very small sample, generally a few grams, which means that considerable sampling errors can occur. Instruments based on differences in the dielectric constant of the sample are generally more reliable and are also more favourable in eliminating sampling error, since a considerably larger sample is used, generally ca 200 grams. A known weight of the sample must, however, be used in these instruments.

Of many different types of equipment tested at a leading grain laboratory in Sweden, the one shown in Fig.5:1 was found to be a good example of a highly

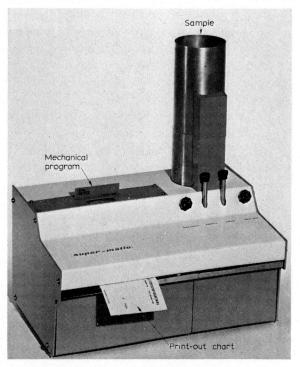

Fig.5:1. Instrument for instantaneous determination of moisture content of seeds and grains. See text for details. (Courtesy of A/S N. Foss Electric, Hilleröd, Denmark)

reliable instrument. Different mechanical programs are used for different seeds or grains, and the result is printed out on a chart. The standard scale ranges from 6 to 35 per cent moisture; the precision is reported by the manufacturer to be $\pm$ 0.25 per cent (standard deviation). Efforts to estimate the moisture content of the harvested seed must consider the aforementioned variations in moisture content between different portions of a large seed quantity. Proper methods for sampling of rapeseed and variations obtained in moisture, oil and chlorophyll content[2-4] between such subsamples from a large seed quantity have been discussed recently. The proper sampling technique is of importance in two connections: (1) To obtain a good approximation of the true value as a basis for payment between farmer and seed company and (2) To provide a basis for a decision on how long the seed can be stored before artificial drying and also under what conditions it should be dried to ensure a proper end product. There are many different types of automatic samplers for grains and seeds available on the market. As pointed out by Masson[2], it is of key importance that the sample is taken from the entire cross-section of a transporting pipe or line.

As previously mentioned the seeds harvested in certain countries are rather often so dry that they can be safely stored without additional artificial drying, especially if the air temperature soon after the harvest drops below zero degrees centigrade, as is often the case in Canada. According to French and Swedish experience, storage of rapeseed in these countries over a long period, such as a year, requires a max. moisture content of ca 7 per cent (see further page 84). Such a low moisture content seldom occurs naturally in rapeseed harvested in Europe. At a somewhat higher moisture content, the seeds can be stored temporarily if they are ventilated (aerated). Results from investigations carried out in France[5] are presented in Table 5:1.

TABLE 5:1

ALTERNATIVE TREATMENTS OF MOIST RAPESEED TO PREVENT QUALITY DETERIORATION

Moisture %	Temporary storage	Ventilation		Storage under		Drying with heated air
		To prevent deterioration	To dry	CO_2	Cold	
9–11	×	×	×	×	×	×
11–16		×	×	×	×	×
16–19				×	×	×
> 19						×

From *Bull. Cetiom No.* 37 (1968) (ref. 6).

Obviously temporary storage (in French 'Transilage') can be undertaken only if the moisture content is lower than 11 per cent. In this connection it is again necessary to emphasize that a seed lot can be a mixture of very wet seeds and dry seeds. Wet seeds in a sample with a rather low average moisture level can be detrimental for the whole seed lot. If the moisture content is higher than 11 per cent action must be taken shortly after threshing in order to prevent deterioration of the seed. At a moisture content between 11 and 16 per cent spoilage may be prevented by ventilation. Ten to twenty m³ of air must be blown through each m³ of seed per hour. According to the French experiments[7] seed lots can be kept at a temperature of 20 °C for 60 days at 10 per cent moisture, for 20 days at 12 per cent moisture and for only 6 days at 15 per cent moisture provided the seed is ventilated continuously (Fig.5:2).

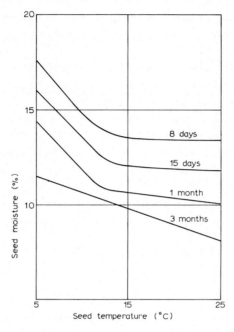

Fig.5:2. Relationship between the moisture and temperature of the seed and maximal time of safe storage at continuous ventilation. (From Bulletin CETIOM, 7)

By ventilation the rapeseed can also be dried to some extent. In this case considerable amounts of air must be used, about 40–100 thousand m³ of air for each m³ of grain. The ventilation should take place only during those hours of the day when the relative humidity is at its lowest. If the relative humidity even then is not under 70 per cent, the air must be heated a few degrees. The process of ventilation just described is useful for farmers who do not have the equipment for drying to a low moisture content and who may not immediately deliver their seed to a dryer-stocker company for complete drying and storage. The changes in quality of the seed by such short-time storage at rather high moisture levels with ventilation will be discussed further in connection with the effects of other processes on the quality of the dry end products. Although cold storage or storage under carbon dioxide has been studied in the laboratory[8], these methods do not seem to have found practical application. The data obtained in these investigations, as presented in Table 5:2, may however explain why higher moisture levels can

STORAGE TIME LIMITS FOR RAPESEED AS A FUNCTION OF ORIGINAL MOISTURE CONTENT AND
TEMPERATURE (MAINTAINED BY INTERMITTENT COOLING WITH COLD AIR)

Moisture content	Temperature of the seed			
of the seed %	0°	5°	10°	15°
19	5 weeks	3 weeks	1 week	—
17	2 months	5 weeks	3 weeks	1 week
15	3 months	2 months	5 weeks	3 weeks
13	more than	3.5 months	2 months	1.5 month
11	5 months	more than	4 months	3 months
9		5 months	more than 5 months	

From Bull. Cetiom (1963) (ref. 7).

be tolerated in the Canadian prairie provinces with frost soon after the rapeseed harvest and thereafter a persistently cold climate. The only method for bringing the seed to safe storage conditions in Europe is therefore to dry it artificially down to a moisture level of ca 7 per cent.

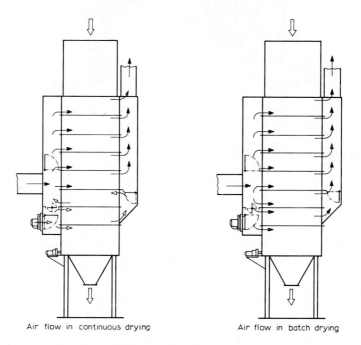

Air flow in continuous drying   Air flow in batch drying

Fig.5:3. The basic design of a drying silo. Drying air (warm) is indicated by filled arrowheads, cooling air by open arrowheads. (Courtesy of AB Svenska Fläktfabriken, Stockholm, Sweden)

Most equipment used for drying on a large scale at dryer-stocker companies is of the silo-type. The principal components of a silo dryer are shown in Fig.5:3.

The overall arrangements for a small drying silo usable at larger farms or minor seed companies for batch or continuous drying are presented in Fig.5:4.

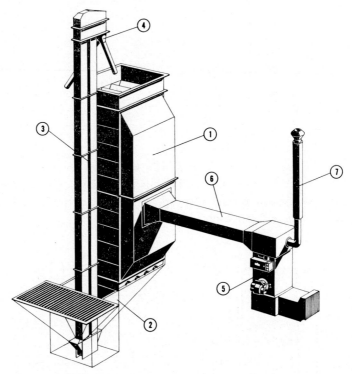

Fig.5:4. The overall arrangement of a small drying silo, usable for batch or continuous operation. 1, Drying unit; 2, Reception pit; 3, Feed elevator; 4, Discharge hopper; 5, Oil burner; 6, Hot air duct; 7, Chimney. (Courtesy of AB Svenska Fläktfabriken, Stockholm, Sweden)

A major part of the Swedish rapeseed harvest is dried on equipment of this type. The drying silo shown in Fig.5:3 comprises feed, drying, cooling and discharge zones. The drying and cooling zones consist of a variable number of 5-meter high sections depending upon the desired capacity.

(a) Continuous drying
The material to be dried enters at the top of the silo and descends between the transverse air ducts in the drying and cooling zones. The ducts are connected alternately to the main warm air and moist air ducts. On leaving the cooling zone the material enters the discharge zone where there are rollers that feed the

material into the discharge tables and down into an elevator hopper or into a special hopper. Drying is performed by means of warm air supplied by a direct fired furnace connected to the silo warm air duct.

Cooling air is drawn from the building in which the dryer is installed and forced by the cooling fan through the dried material at the lower part of the cooling zone. It then passes through the dried material at the upper part of the cooling zone and to the warm air duct, where it mixes with warm air from the furnace.

Under normal capacity conditions, the cooling air cools the material which has been heated in the drying zone to about five degrees centigrade above the temperature of the cooling air.

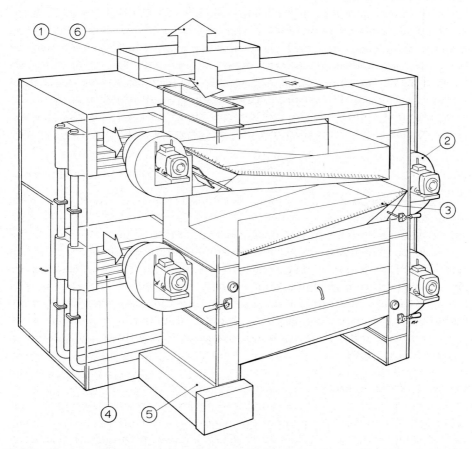

Fig.5:5. The basic design of a fluidised bed dryer. 1, Grain feed; 2, Fans for drying air; 3, Perforated bottom plate; 4, Heating units, 5, Discharge; 6, Moist air exhaust. On this drawing, the air inlet is not seen; it is situated under the heating units (4). (Courtesy of AB Svenska Fläktfabriken, Stockholm, Sweden)

References pp. 98–100

(b) Batch drying

In batch working, the cooling zone is also used for drying, and the lower parts of the warm air and moist air ducts form part of the drying zone. The cooling fan is not used and its outlet is covered (see Fig.5:3).

The warm air passes from the warm air duct, or a part of this duct, through the material and to the moist air duct, after which it is discharged from the silo. Cooling is achieved by air supplied by the furnace, after the furnace has been shut down and allowed to cool. The cooling air follows the same route through the silo as the drying air.

The drying silo can be used for both continuous and static batch drying. When a batch is dried, the silo is discharged by lowering the discharge tables.

(c) Fluidised bed drying

For very careful drying or when a large number of smaller lots is to be dried, the fluidised bed drier is preferable. This type is safer for preservation of the viability of the seed, since no local overheating can occur. As demonstrated in Fig.5:5 the seed moves downward on sloping perforated sheets and each time the seed moves from one segment to another the material is turned over. Such a drier is easily opened for inspection and cleaning but the disadvantages are that it requires more powerful fans and involves higher energy costs. Also the cost of the equipment is higher than that of a silo-dryer, calculated at equal capacity.

5.2.3 Equipment for seed cleaning

The rapeseed obtained from combine harvesting is contaminated by parts of the pods, straw, pebbles, sand etc. Before the seed is pure enough for processing into an edible oil product it is generally cleaned at two points in the processing: (1) before or after drying and (2) before crushing in the oil factory. The second seed cleaning operation is discussed in detail in chapter 9. Seed cleaning can involve use of three different 'components'. In all-purpose commercial seed cleaners, the first step generally involves the removal of larger straw and pod fragments by passing the sample over a so-called scalping reel or skimming sieve. Seeds and other material pass through the coarse wire cylinder or net and are then transferred to the second stage. To some extent very fine material is removed from the seed stream by aspiration when the seed passes from the first to the second step in the cleaning process. The screening unit can consist of one or more sets of screens inclined at a variable angle and shaken at an adjustable speed. The upper screen removes larger impurities while the lower one separates off sand and other minor impurities. The third section consists of vertical air channels in which the material is graded according to weight (aspirated). Light weight impurities, such as seed fragments, dust, and light, low quality seeds are removed from the main stream of high quality seeds. Fig.5:6 illustrates the basic construc-

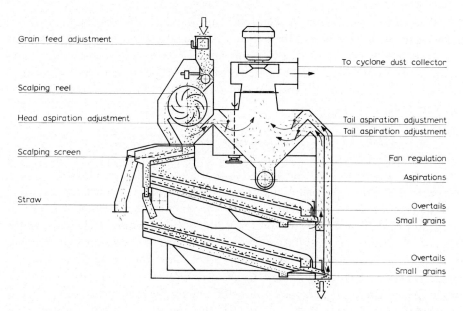

Grain feed adjustment

Scalping reel

Head aspiration adjustment

Scalping screen

Straw

To cyclone dust collector

Tail aspiration adjustment
Tail aspiration adjustment

Fan regulation

Aspirations

Overtails
Small grains

Overtails
Small grains

Fig.5:6. The principal components of a seed cleaner. (Courtesy of Kamas, Kvarnmaskiner AB, Malmö, Sweden)

tion of such a machine. It is possible to use only the first section, the first and the second in combination or all three consecutively. Such different combinations are available commercially from various companies.

The usual practice among Swedish farmers is to clean the seed after drying, whereas the dominant habit at the dryer-stocker companies is to pass the seed over a skimming sieve and through an aspirator before drying. Cleaning of the seed after drying is more efficient than cleaning of wet seed. Consequently cleaning should be carried out before the dried seed is put into storage or delivered to the crusher, even if some cleaning was undertaken upon delivery to the dryer-stocker company.

5.3 THE EFFECT OF VARIOUS OPERATING PARAMETERS DURING THE TIME FROM THRESHING TO THE END OF THE DRYING ON THE SEED QUALITY

Before entering into the details of this part of the chapter, it may be of interest to emphasize that the 'earmarkings' on the seed quality can differ from country to country. Quite obviously the moisture and oil contents form the basis of the quality evaluation since they directly indicate the yield of oil in industrial processing. Also it seems to be generally accepted that a high content of free fatty acids or a high 'acid value' is an indication of quality deterioration. Other important quality factors, such as the peroxide value of the seed oil are never or very seldom discussed as a basis of payment between seed seller and buyer[10]. Most probably, it is only in Sweden that the rapeseed crop is extensively analysed for chlorophyll content.

It is however necessary in this connection to also discuss those qualitative factors, that are infrequently or never considered in the present grading systems, but which are known to be of importance from laboratory investigations. Four different points will be discussed.

5.3.1 The effect of delay in drying and of ventilation of the seed

As previously mentioned, a delay in the drying of very wet seeds is detrimental to the seed quality. The following is a brief description of what can occur. At high moisture levels (20 per cent or more) the rapeseed is respiring although it is removed from the pod. This respiration causes a rise in temperature and a further increase in moisture content due to the respired water. If the seed is stored in large containers the dissipation of heat is reduced, so that at the center of the container the temperature increases to high levels. Under very unfavourable conditions this can cause fires, which actually occurred during the early phase of modern oil seed growing in Sweden. Provided the original moisture content is not too high and the size of the container is not too large an equilibrium will be established at a temperature above ambient and at such a high relative humidity that a rapid

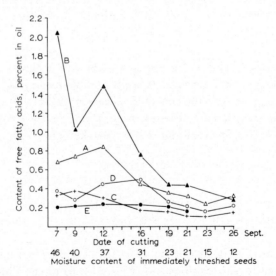

Fig.5:7. Rise in free fatty acid content during short time storage of wet seeds of summer rape
(From Lööf and Johansson[19])
    A   Analysed immediately after sampling.
    B   Kept for 5 days in a closed vessel before analysis.
    C   Immediately after sampling dried with hot air to 7.5 per cent moisture.
    D   Dried slowly at ambient temperature.
    E   Swathed ca. 10 days before threshing and drying to 7.5 per cent moisture.

70

microbial, mainly fungal, growth occurs which in turn contributes to the respiration heat. If the process is interrupted and the seed is dried, it will be characterized by an elevated level of free fatty acids. Such mold-infested seeds can easily have 5 per cent free fatty acid or more in the oil and are generally considered unsuitable for processing into an edible oil product. If the wet seed is ventilated, there will most probably occur a small rise in the level of free fatty acids (FFA), but if the seed lot contains a considerable portion of seeds with a high chlorophyll level the slight decrease in quality, judged by the increased FFA level, is balanced by a simultaneous decrease in chlorophyll content which is beneficial[11]. Fig.5:7 demonstrates the rise in free fatty acid content of wet rapeseed when stored in small glass containers in the laboratory ('B' in Fig.5:7).

It must be emphasized that these data are not directly applicable to practical conditions. It is however very difficult to undertake reproducible 'field' experiments to investigate the risk of spoilage unless large amounts of high quality

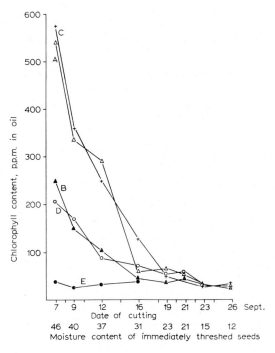

Fig.5:8. The effect of slow *versus* rapid desiccation of moist summer rape on the chlorophyll levels in the dried seeds.

A   Analysed immediately after sampling.
B   Kept for 5 days in a closed vessel before analysis.
C   Immediately after sampling dried with hot air to 7.5 per cent moisture.
D   Dried slowly at ambient temperature.
E   Swathed *ca.* 10 days before threshing and drying to 7.5 per cent moisture.

(From Lööf and Johansson[19])

rapeseed are deliberately spoiled, and such experiments are generally not carried out. Fig.5:8 depicts an experiment demonstrating the reduction in chlorophyll level of highly immature rapeseed when allowed to dry out slowly (ventilated) compared to drying out immediately by hot air ('D' versus 'C' in Fig.5:8). It may be seen that a considerable drop in chlorophyll content can be achieved in immature rapeseed by ventilation during a delay in drying. A Polish report presents data on the storage of moist rapeseed under laboratory as well as industrial conditions[11a].

5.3.2 The effect of drying temperature on the quality of the dry seeds

It is well known that too high a temperature during drying will kill the seed, and therefore drying at certain rather low temperatures is applied, when the dried seeds are to be used for seeding. Whereas it has rather long been known that it is essential to dry wheat in a similar careful way when it is to be used for baking (see *e.g.* ref. 12), it does not seem to be generally accepted that such cautious drying would also be necessary with oilseeds to be used for oil production. On the contrary, the authors of this chapter have often met the viewpoint that the drying temperature is of little importance if the oilseeds are to be used in the food industry. In view of the correlation reported in the literature between free fatty acid content and seed viability[13-16] it seemed however to be of importance to preserve the viability of the oil seed to prevent quality deterioration. The maximum

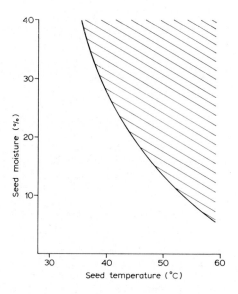

Fig.5:9. Relationship between moisture and highest possible temperature during drying to prevent decrease in viability. The hatched area represents moisture/temperature relationships under which germination capacity is lost or seriously impaired. (From Appelqvist[17])

allowable temperatures at different moisture levels of rapeseed have been studied in laboratory experiments[17]. A summary of the results obtained is presented in Fig.5:9. It should be noted that the temperatures given are those of the seed itself and not of the drying air or of the heating elements. The data presented in Fig.5:9 can therefore only be used as a guide for practical investigations on commercial large scale drying equipment. Whereas a delay in the drying of very wet seed will cause immediately observable damage to the seed, overheating—if not too extensive—will cause very little if any immediate change in the seed quality. It has however been found that heat-damaged seeds give an oil that is more susceptible to oxidation.

TABLE 5:3

CONTENT OF PEROXIDES AND SUSCEPTIBILITY TO OXIDATION OF OILS FROM UNDAMAGED AND HEAT-DAMAGED SEEDS OF RAPE AND WHITE MUSTARD

| | Method of analysis* | | | |
| | Steel extraction tubes | | Butt-type extractors | |
	A	B	C	D
Rape, undamaged	0.00 ± 0.00	5.85 ± 0.48	0.00 ± 0.00**	4.54 ± 0.21
Rape, heat-damaged	0.54 ± 0.02	8.48 ± 0.38	0.58 ± 0.06	6.95 ± 0.25
White mustard, undamaged	0.00 ± 0.00	1.91 ± 0.80	0.00 ± 0.00	1.26 ± 0.06
White mustard, heat-damaged	0.63 ± 0.07	4.68 ± 0.78	0.77 ± 0.08	4.56 ± 0.12

* All values reported represent the mean of three single determinations ± standard error except the sample marked ** in which case only two samples were analyzed. Results are given in mequiv./kg oil.
From Appelqvist (1966) (ref. 18).

Table 5:3 demonstrates the differences in peroxide values of undamaged and heat-damaged rape and white mustard seed analysed several months after harvest[18]. The oils were extracted from the seeds in different ways. Method 'A' involved use of a special type of stainless steel tubes and steel balls for disintegration and extraction with hexane[19]. A large number of precautions were taken to prevent any contamination of the sample with oxidized lipids and to prevent oxidation during the extraction and evaporation steps preceding the peroxide number determination. Thus all the equipment used was washed with hexane and air dried after conventional laboratory washing. The filtration of the seed slurry was carried out in a dark room, and after evaporation of the solvent at a maximum temperature of 30 °C in a vacuum oven, nitrogen was introduced upon the release of the vacuum. Weighing and addition of the solvent used in the peroxide determination were also performed in a dark room. Method 'B' involved the use of the same equipment but without any protection and with solvent evaporation on a hot plate as in a typical, gravimetric method for the determination of oil content[20-23].

Method 'C' involved an extraction as careful as that of method 'A' but with Butt-type extractors such as the ones used in official methods for the determination of the oil content of oil seed. Method 'D', finally, involves seed pretreatment and extraction according to the standard methods for oil content determination[20-23].

TABLE 5:4

SOME COMPOSITIONAL DATA FOR DIFFERENT SUBFRACTIONS OF FARMERS' 'DELIVERY-SAMPLES' OF THE 1966 AND 1967 SWEDISH CROP

| Sample | Oil content, % of dry matter | | Chlorophyll content p.p.m. | | | | Content of free fatty acids, % | |
| | | | in oil | | in dry matter | | in oil | |
	1966	1967	1966	1967	1966	1967	1966	1967
Winter rape								
High quality seeds	48.0	48.7	14	10	7	5	0.5	0.4
Cracked seeds and larger seed fragments	46.2	50.6	64	30	29	15	6.7	3.5
Germinated seeds	43.6	47.0	155	87	68	41	5.4	3.3
Weed seeds	6.7	10.6	1,388	496	93	52	10.4	7.2
Other waste matter	5.0	11.9	2,070	552	103	66	45.4	—
Winter turnip rape								
High quality seeds	44.9	47.8	16	7	7	4	0.5	0.3
Cracked seeds and larger seed fragments	44.7	48.9	66	29	30	14	7.1	4.8
Germinated seeds	42.6	45.1	134	83	57	38	4.4	3.4
Weed seeds	12.8	20.0	646	319	83	64	8.4	4.0
Other waste matter	8.6	11.6	2,793	1,064	241	123	35.9	—
Summer rape								
High quality seeds	44.9	42.9	33	33	15	14	0.3	0.5
Cracked seeds and larger seed fragments	45.6	43.4	108	89	49	39	4.6	6.3
Germinated seeds	44.8	40.5	104	157	46	63	2.5	3.7
Weed seeds	11.5	15.3	525	573	60	87	0.2	4.1
Other waste matter	9.6	14.1	2,184	2,812	210	396	17.8	—
Summer turnip rape								
High quality seeds	45.0	44.3	15	12	7	5	0.1	0.3
Cracked seeds and larger seed fragments	45.7	45.8	64	44	29	20	1.8	5.3
Germinated seeds	47.8	41.8	73	63	35	26	2.3	4.5
Weed seeds	11.9	16.4	301	324	36	53	0.1	3.4
Other waste matter	12.8	17.9	2,429	2,216	311	396	15.2	—

Appelqvist, Johansson and Lööf, unpublished results.

Three conclusions can be drawn from the data obtained. (1) The oil from un-damaged rape and white mustard seed had no detectable peroxide if the oil was extracted with all possible precautions, whereas the heat-damaged seeds had low, but easily detectable peroxide values. (2) The oil from heat-damaged rape and white mustard seed was more susceptible to oxidation since the increase in peroxide value during conventional solvent evaporation and other conditions initiating an oxidation of the oil will be significantly higher in the oils from the non-viable seeds compared to the oil from the undamaged seeds. (3) Oil from white mustard seed was less susceptible to oxidation than oil from rapeseed.

5.3.3 The importance of an optimal moisture content at the termination of the drying

As previously pointed out, rapeseed must be dried to about 7 per cent moisture for safe long-time storage. Whereas this has been known and practiced in Sweden and several other countries[25] for many years, it was not known until recently that drying to a too low moisture content could be dangerous, even if the drying was carried out carefully so that the viability is retained. In view of the very high levels of free fatty acids found in seed fragments (see Table 5:4), data reported on the relationship between seed moisture content and fragility[24] are of consider-able practical importance. It may be seen from Fig.5:10 that the fragility of the seeds increases only slightly with decreasing moisture content from about 6.5 to about 5.0 per cent. From the latter moisture level downwards there is a much more rapid increase in fragility with decrease in moisture content. Therefore, it

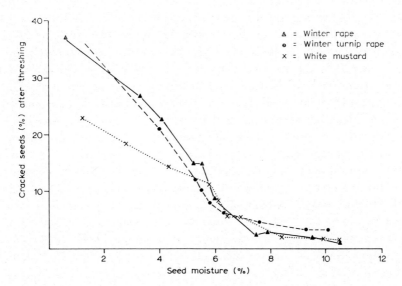

Fig.5:10. Relationship between seed moisture content and fragility. (From Appelqvist and Johansson[24])

would be highly desirable not to dry the seed below 6 per cent moisture. Similar to the heat-damaged seeds, the cracked ones show little change in quality immediately after cracking. As demonstrated in Table 5:5, on page 83, there is however a more rapid increase in free fatty acid content and microflora of the cracked seeds and seed fragments compared to the intact ones during storage.

5.3.4 The effect of seed cleaning on the quality of the oil and meal

The importance of removing any foreign material, such as straw, sand, stones etc., is easily understood, since their presence may be detrimental for the processing equipment (see Chapter 9) and would furthermore lower the feed value of the meal. It is however also of great importance to remove the 'fines' and small weed seeds, since their oil quality is very low. Besides having a high level of free fatty acids, they very often contain oil with large amounts of chlorophyll and high peroxide values. Table 5:4 presents data from investigations on the chemical composition of separate fractions of commercial seed samples. The fractions were separated by hand after visual inspection of about 50 to 100 individual 'farmer's delivery' samples of each seed type of both years' crop studied. Before analysis, the individual subsamples were pooled to yield enough material for chemical analysis. From the data of Table 5:4, the following points may be noted: (1) The chlorophyll content of high quality seeds was highest in summer rape in both years. (2) The cracked seeds and larger seed fragments in all cases had a higher chlorophyll content compared to the high quality seeds. More characteristic for this fraction is however the high free fatty acids levels, about ten times as high as those in the high quality seeds. (3) High free fatty acid levels can also be seen in the germinated seed and in this case the chlorophyll levels are still higher compared to the cracked seeds. (4) These two groups of inferior quality seeds have approximately the same oil content as that of high quality seeds and may therefore from the seller's viewpoint represent valuable material. The high chlorophyll and free fatty acid levels are however indicative of poor quality seeds and, if possible, such seeds and seed fragments should be removed from the sample as early as possible in the production chain. (5) The weed seeds and other waste matter, which generally represent a rather small percentage calculated on total sample weight, are characterized by their rather low oil content—only about a third to a tenth of that of high quality seeds—and, especially, by the very high chlorophyll levels. The levels of free fatty acids can vary from very high to rather moderate for 'other waste matter' which sometimes contains considerable amounts of very small seed fragments. It is quite evident that the two latter groups of material must be removed before processing of the seed sample into an edible oil product.

The low fat quality of the seed fragments also indicates the importance of a very careful mechanical treatment of the seed. Thus it is essential that the mechanical parts that come in contact with the seed, especially when moving at high speeds, do not have any sharp edges. To prevent the appearance of cracked seeds it is also of great importance that the seeds are not 'overdried' as discussed above, since very dry seeds crack more easily.

5.4 METHODS FOR STORAGE OF THE DRIED SEED AND TRANSPORT TO THE OIL MILL

Rapeseed is stored at the farms, at the dryer-stocker companies, and at the oil mill. Obviously this generally involves quite different ranges in the amounts stored and therefore will involve storage facilities of different types. In Canada and northern Europe the transport of the seed from the farm to the dryer-stocker company usually takes place rather shortly after the harvest and from thereon at different times during the winter season to the oil mill. The transport can be performed with trucks, railroad cars or boats. The basic facts about storage and transport of seeds, especially the economic aspects of these matters, are treated in other monographs (see *e.g.* ref. 26), and will not be repeated here.

In all industrially well-developed areas, the long-time storage of rapeseed as well as other seeds is almost exclusively done in silos of various sizes and constructions. In areas with less developed industrial capacity and rather hot and dry climate during the harvesting season the seeds can be stored in sacks or open bins during the entire storage period.

5.4.1 In sacks

Storage of rapeseed in sacks is practiced in many areas. Within Europe it occurs *e.g.* in southern France where the weather is normally so dry during threshing that the seed can be stored without artificial drying. The sacks typically contain about one hundred kilos of rapeseed each and are arranged in piles to a height of about four meters. Although this method can give good results in areas with a fairly dry climate, it is of course a more primitive method and involves more labour compared to storage in silos. Important disadvantages of storage in sacks are the following:
(a) The stock must be regularly inspected.
(b) Sampling for control of seed quality is laborious.
(c) Certain spoilage by attacks from rats, mice and insects occurs.
On the other hand storage in sacks involves less investment compared to storage in silos.

Since seeds are hygroscopic, it is of great importance for changes in quality if the seeds are stored in large silos made of concrete, with negligible access to environmental moisture, or if they are stored in sacks, bins or small open silos, under which conditions the seed moisture fluctuates with changes in atmospheric

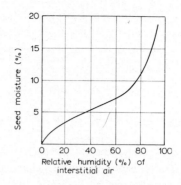

Fig.5:11. Relationship between the relative humidity of the air and the seed moisture content after an equilibrium has been established, at 20 °C. (From Bulletin CETIOM, 7)

moisture. Fig.5:11 demonstrates the relationship between seed moisture content and the relative humidity of the atmosphere.

A significant disadvantage of 'open' storage (sacks, bins, small silos), besides losses due to rats and insects, is the risk for growth of fungi during periods of high environmental moisture levels. According to Geddes[27] most fungi require 80 per cent relative moisture, corresponding to about 9 per cent moisture in the rapeseed[28]. However, some *Aspergilli* were reported to grow even at 65 per cent moisture, corresponding to ca 7 per cent seed moisture. Investigations in France[29] and Poland[25] have shown that the major part of the fungi present on rapeseed directly after harvest were *Cladosporium* and *Alternaria*. The total number of spores varied between 10 and 600 thousand per gram of seed. During the storage the microflora changes so that later *Penicillium* and *Aspergillus* dominated. Swedish investigations have shown that the most frequent fungus spores found on the 1963 crop was *Penicillium*, followed by *Aspergillus* and *Rhizopus*[30]. Among fungi parasitic on plants, *Alternaria* was the dominating genus.

Polish studies on rapeseed have shown that cracked seeds or seed fragments are more frequently attacked by fungi than undamaged seeds[31].

5.4.2 In open bins

When rape seed is produced on small farms with a rather small production of seed, it can be temporarily stored in open bins, with the same disadvantages as with storage in sacks. In this connection it should be mentioned that attacks by insects on rapeseed during storage are not frequent. The only insect that has been found in large amounts is the acarid *(Acaris)*, which of course is favoured by storage in open bins.

An inverse relationship between the oil content of the seed and attack by acarids has been reported. These insects seem to attack the seeds preferably at seed moisture levels of 8.5–11 per cent. Above 11 per cent other organisms develop on the seeds making the environment unfavourable for the insects and below 8.5 per cent they will not have sufficient water to develop[5].

A suitable method for the disposal of the acarids is to heat (the dry) seeds to about 60 °C whereby the acarids dry out. A subsequent regular seed cleaning operation (aspiration) is effective in removing the dry acarids. Seeds attacked by acarids have an unpleasant odour but no detrimental effect on oil quality has so far been reported to result from the insect attack.

5.4.3 In silos

Silos are by definition containers of solid material protecting the stored seed

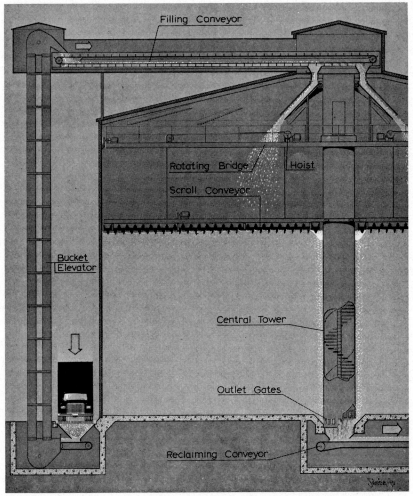

Fig.5:12. General layout of a very large silo for the storage of oil seeds and grains: Capacity about 60,000 tons. (Courtesy of Nils Weibull AB, Hässleholm, Sweden)

on all sides. The size varies from about one cubic meter up to 40,000 cubic meters. The material is usually concrete or metal, but the clay bins used in Asia may also be called silos.

The predominant type of grain and oil seed silo is about 10 meters in diameter and 30 meters high (with a capacity of about 2,000 tons) made of concrete and joined with several cells and with a central, 'service tower' to make a complex with a total storage capacity of 10,000 to 20,000 tons. Very large silos that can store up to 60,000 tons of seed in a single cell are of increasing importance. Fig. 5:12 shows the construction of such a silo, which according to the engineering company is designed to: (1) store the largest possible quantity of material within the minimum of surrounding surfaces, (2) keep the stored material in its original condition, (3) ensure both filling and complete emptying in a rapid manner, (4) minimize handling of stored material, (5) store material hygienically, and (6) reduce handling costs.

Because of the formation of water as a consequence of respiration, it is recommended to ventilate the seed at certain intervals[32] and thereby to use about 600 m^3 of air for each m^3 of seed or to transfer the seeds from one silo to another. For readers with no background in biology, it may be necessary to emphasize that high quality seeds (cf. p.72) are viable organisms that carry out respiration, although to an almost negligible extent in very dry seeds (6–8 per cent moisture). In conjunction with ventilation, sampling should be carried out in order to check the seed quality.

5.4.4 Storage under inert gas

In a previous section of this chapter (p.63) reference was made to French experiments with various treatments in order to prevent seed damage during temporary storage before artificial drying could take place. Experiments have been performed on the long-range storage of damp rapeseed in vacuo or under carbon dioxide or nitrogen[8, 9]. For this purpose specially constructed airtight silos of metal or concrete are being used, in which the interstitial air can be replaced by some gas, e.g. N_2, CO_2 or SO_2. Storage under carbon dioxide, which seems the most interesting alternative, can be undertaken with seeds with a max. moisture content of 19 per cent at a temperature range of 0–25 °C. The seeds should be brought into the CO_2 atmosphere before any damage has occurred due to the high moisture levels. The carbon dioxide is introduced from the bottom of the silo, which is already filled with the seed, and replaces the interstitial air.

5.5 CHANGES IN SEED QUALITY DURING STORAGE

The quality of the stored rapeseed depends, quite naturally, on the quality of the seed when put into storage and on the conditions of storage. Experimental data available in this field are generally subject to one of two significant limitations:

(1) Analysis of the composition of the seeds from large, commercial silos truly reflects the actual conditions under storage, but suffers from large sampling errors. This is because the quality of the seed in a given storage unit may vary quite considerably since it originates from many different farms (*cf.* refs. 30, 33, 34, 35). The storage conditions are also not uniform throughout the silo.

(2) Studies on the changes in composition of well-defined seed materials of known origin, stored in small quantities, may give small sampling errors and permit comparison of various types of specific damage, but cannot be directly 'translated' into large scale storage facilities, since *e.g.* the dissipation of heat and diffusion of gases (O_2 and CO_2) would be different in small and large storage units. Most of the examples of this chapter will be from laboratory experiments since data from commercial laboratories on changes occurring in large silos are generally not available.

Extensive experiments in this field have been performed in France for many years. The results are generally published as Bulletins from CETIOM (Centre Technique Interprofessionnel des Oléagineux Metropolitains) in mimeographed form. From the numerous data available the following can be mentioned from experiments with seeds stored in sacks or open bins (up to 20,000 lbs). Rapeseed with 8–10 per cent moisture was put into storage in early August, and samples were analyzed at regular intervals until March of the following year. The moisture level varied very little and was always between 8 and 10 per cent. The oil content was constant at 42.5–43.5 per cent for the different samples, which had acid values between 2.8 and 4.8 (corresponding to approximately 1.4 and 2.4 per cent F.F.A.). No increase could be observed in the acid values. The germination capacity (viability) was about 80 per cent during the entire storage period. From Polish studies[25] it is known that there is a parallel increase in FFA content and decrease in viability during storage of rapeseed at higher moisture levels.

From experiments with larger quantities similar experiences are reported[36] except that samples with a rather high moisture content (about 11 per cent) showed a decrease in moisture level during storage and that samples with low viability (about 40 per cent) when brought into storage reached still lower levels during storage.

Besides studying the effect of different conditions on the changes in seed quality, Masson also studied the correlation between various parameters[37]. He found no change in oil content during storage, essentially independent of the changes in quality (*e.g.* F.F.A.) provided the F.F.A. level did not rise above 3–4 per cent, when a drop in oil content could be observed.

A comparison was made between the influence of various methods of drying on the quality of the seed at various intervals. The best method from the quality point of view was drying with chilled air. The acid value of the dried seeds was 2.2 (about 1.1 per cent F.F.A.) and this low value remained unchanged during the storage period. Drying with warm air also resulted in seeds with low acid value (2.5) which remained low during storage. The two other methods of drying compared, namely 'transilage' and 'ventilation', yielded seeds with rather unfavourable properties. 'Ventilation', which means drying with air of ambient temperature, gave seeds with the highest acid value, namely 3.5–4. During the storage period, a considerable drop in viability and a further increase in acid value was observed. It should be pointed out that depending on the initial moisture and microflora of the seeds and the operating parameters of the process, the final results may vary considerably. A rather small increase in F.F.A. was noted for a similar process on the laboratory scale in a Swedish study[11]. Since the process leads to a marked drop in chlorophyll level, which might be essential to the overall seed economy, further experiments in this field seem warranted.

'Transilage' means that the wet seeds are transferred from one silo to another at various intervals. It has been found that this process is also detrimental to the seed quality, and therefore it should not be used except under special circumstances.

According to the French experiments, the viability of the seeds stored under CO_2 at high moisture levels decreases rapidly[8]. This naturally prevents the use of this method of storage for seeds to be used for sowing. Contrary to the case when non-viable seeds are stored under classical conditions, no major decrease in fat quality is observed for a storage of at least 6 months under these conditions[8, 38, 39]. For low quality seeds, storage under CO_2 seems even more favourable than storage in conventional silos[8, 38].

In Polish studies, it was observed that storage of rapeseed at 20 °C in closed containers initially with 13 per cent seed moisture was detrimental to the quality. Cracked seeds more quickly obtained very high F.F.A. levels than undamaged seeds[31]. There was also a considerable rise in moisture content during the 2–3 months in storage. Furthermore, it was reported that the peroxide value and the acid value of oils from immature rapeseed increased more rapidly during storage (at 18–20 °C and 60 per cent rel. moisture) compared to mature seeds[40]. Also mechanically damaged seeds demonstrated more rapid increase in F.F.A. and peroxide value compared to undamaged ones when stored under 'normal' conditions.

In another Polish investigation, it was found that the quality (F.F.A.-levels) of rapeseed directly after harvest was highest in the most mature seeds whereas the keeping quality was best with seeds harvested slightly earlier[41]. This might be due to the presence of field germinated seeds at later stages (*cf.* Table 5:5).

In laboratory experiments in Sweden the changes in free fatty acid content of three different species of cruciferous oil seeds, namely winter rape, winter turnip rape and white mustard, were studied from 5 months to 33 months in storage. With different damages imposed on high quality seeds, it was possible to study the differences in the increase of free fatty acids with different seed quality.

RISE IN CONTENT OF FREE FATTY ACIDS IN SEEDS OF RAPE, TURNIP RAPE AND WHITE MUSTARD DURING EXTENDED STORAGE AT LOW MOISTURE LEVELS (CA. 7 PER CENT)

Sample	Content of free fatty acids (per cent) after storage for					
	5 months	11 months	13 months	18 months	26 months	33 months
Winter rape						
High quality seeds	1.3	2.1	2.5	2.1	2.7	2.8
Heat-damaged, nonviable seeds	0.9	1.5	1.8	2.0	1.5	2.0
Germinated seeds	2.8	4.3	5.6	5.4	8.0	8.7
Mould infested seeds	9.3	12.8	12.2	12.6	18.4	15.5
Cracked seeds and seed fragments	11.0	28.1	32.9	37.8	63.6	—
Winter turnip rape						
High quality seeds	1.7	2.3	2.4	2.4	2.3	2.6
Heat-damaged, nonviable seeds	1.5	1.9	2.0	1.9	2.3	2.6
Germinated seeds	9.8	20.3	22.7	25.7	38.5	—
Mould infested seeds	7.8	8.1	9.0	8.7	9.8	10.6
Cracked seeds and seed fragments	7.0	28.3	32.0	37.3	61.2	—
White mustard						
High quality seeds	0.4	0.5	0.9	0.7	1.0	1.3
Heat-damaged, nonviable seeds	0.6	0.7	1.0	1.0	1.2	1.2
Germinated seeds	2.9	5.2	5.4	4.9	5.8	6.8
Mould infested seeds	3.4	3.5	3.8	2.2	2.6	2.5
Cracked seeds and seed fragments	2.8	7.6	9.0	7.9	14.4	14.7

Appelqvist and Johansson (unpublished).

Obviously the storage conditions in these trials were 'too good' since the total increase over a more than two-year period was only about one per cent absolute for the high quality seeds (Table 5:5). It should also be noted that the heat damaged non-viable seeds have levels of free fatty acid very similar to those of viable seeds. This is the case when the seed is stored in closed containers at constant temperature in the laboratory, and may be misleading in one respect, since it is tempting to assume that the fat quality of the heat damaged non-viable seeds is as good as that of high quality seeds. The deterioration in fat quality is however easily demonstrated by the rise in peroxide and the decrease in oxidative stability that is found for the heat damaged non-viable seeds (compare Table 5:3). The germinated seeds show a steady rise in free fatty acid content but, interestingly enough, there is a great difference between rape and turnip rape in that the latter accumulates much more free fatty acid in the germinated seeds. This is consistent with a finding of Wetter[42] that the increase in lipase activity in turnip rape is much greater than that in rape upon germination. The mould-infested seeds obtain a high free fatty acid content shortly after being put in storage and it does not increase very much during a period of two years. Especially noteworthy is the almost constant or slightly decreasing content of free fatty acids in mould-infested white mustard seeds. A difference between white mustard on the one hand and rape and turnip on the other has been noted in many connections. Similar to what was found earlier, the cracked seeds and seed fragments had high levels of free fatty acid and the levels rose considerably in storage, so that after about two years *ca.* 60 per cent of the total fat consisted of free fatty acids. Again white mustard behaves differently in that the levels of free fatty acid in cracked seeds and seed fragments of that species are much lower than the levels in rape and turnip rape.

Previous Swedish studies demonstrated that seeds which had been mechanically damaged while still wet (at *ca.* 40 per cent moisture) and then carefully dried to *ca.* 7 per cent moisture showed an inverse relationship between viability and rise in F.F.A. level during storage[16].

Based on the data presented above and other, unpublished, observations the following recommendations may be made, and should be followed in order to ensure a high-quality seed after some time (up to 12 months) of storage.

(1) The moisture content should not be higher than ca 7 per cent and not lower than 5 per cent. If higher, the percentage of free fatty acids will increase. At 9–12 per cent moisture a decrease in germination capacity from the original 90–100 per cent to about 40 per cent may occur. The recommended moisture level is applicable to maritime climates with rather mild winters. It is probable that slightly higher levels can be used with good results in continental climates with very cold weather from shortly after harvest on.

(2) The seeds should be dried at such a temperature, that they remain viable, since non-viable seeds will generally show a more rapid increase in F.F.A. level.

(3) Immature seeds and field-germinated seeds should not be stored for any length of time, since they rapidly deteriorate.

(4) Seed lots with a high percentage of cracked seeds and seed fragments should not be taken into storage since the F.F.A. level rises very rapidly.

5.6 NATIONAL GRADING SYSTEMS AND BASIS OF PAYMENT BETWEEN GROWER AND SEED COMPANY

Since rapeseed is grown and traded in countries with great differences in economic standard, as discussed in Chapter 2, it is understandable that the basis of payment between grower and seed company and the different systems of grading the seeds must vary considerably. The simplest way of estimating the amount of seed to be traded is to fill a certain volume (*e.g.* a bushel) with seed, reasonably dry and free from extraneous materials such as straw, stones etc. A somewhat more elaborate method is to base the payment on weight and weight per unit of volume. Furthermore, properties easily assessed by the eye or the nose, such as the number of damaged seeds or the odour (fresh versus a sour or rancid odour), can be included. Since the moisture content can be easily determined instantaneously by

TABLE 5:6

THE SYSTEM USED IN CANADA FOR GRADING RAPESEED

Canada grade rapeseed	Minimum pounds per bushel	Standard of quality Degree of soundness	Standard of cleanness (see note)
No. 1	52	Reasonably sound; cool and sweet; may contain not over 3 percent damaged seeds including not over 0.1 percent heated. Of good natural color.	May contain not more than 1 percent of other seeds that are conspicuous and that are not readily separable from rapeseed, to be assessed as dockage.
No. 2	50	Cool and sweet; may contain not over 10 percent damaged seeds, including not over 0.2 percent heated.	May contain not more than 1.5 percent of other seeds that are conspicuous and that are not readily separable from rapeseed, to be assessed as dockage.
No. 3	48	May contain not over 20 percent damaged seeds, including not over 0.5 percent heated. May have the natural odor associated with low quality seed, but shall not be distinctly sour, musty, rancid, nor have any odor that would indicate serious deterioration or contamination.	May contain not more than 2 percent of other seeds that are conspicuous and that are not readily separable from rapeseed, to be assessed as dockage.

NOTE: Assignment of rapeseed to the above grades shall not imply any guarantee of content of other seeds that blend with rapeseed.

The percentage of 'other seeds that are conspicuous and that are not readily separable' shall include weed seeds that do not blend with rapeseed and whole or broken kernels of other grains, when these are not removable by means of appropriate sieves and other cleaning devices.

Dockage shall be assessed on rapeseed for foreign material readily separated from the mass by ordinary mechanical cleaning methods, including any small whole or broken rapeseed removed with such foreign material, plus any other seeds, up to the limits established in the respective grades, that are conspicuous and remain in samples after ordinary mechanical cleaning methods have been applied, these to be added together and expressed as a percentage by weight of the whole; except that a reasonable allowance may be made for broken rapeseed not to be assessed as dockage in commercially clean rapeseed when this can be attributed to attrition in normal handling after cleaning.

From *Oil and Oilmeal from Canadian Rapeseed* (1963) (ref. 44).

the use of certain moisture meters (p.62), moisture content can also be determined and used as a basis for payment without too much extra effort.

A more elaborate but also more fair system (see Tables 5:10 and 5:11) makes use of chemical determinations of oil content[19, 43]. The chlorophyll content of the oil is also being used as a basis of payment in Sweden from the 1970 harvest.

5.6.1 Canada

The grading system for rapeseed in Canada takes into account mainly weight per unit of volume, cleanness and soundness (Table 5:6). The moisture content must not exceed 10.5 per cent for any of the grades. Obviously the Canadian system is relying on properties assessed by the eye and the nose. Whereas the price paid for the rapeseed to the farmer in most European countries is regulated by governmental agencies, the producers' price in Canada fluctuates with the bids for rapeseed of a certain grade at the Grain Exchange in Winnipeg. As a consequence of the rather tight regulations of the European oil seed market, a year with excessive Canadian production may result in very low prices[44]. The average farm price paid for rapeseed in Canada from 1950/51 to 1966/67 is shown in Fig.5:13. It is interesting to note that during a five year period (58/59 compared to 63/64) the price varied by 100 per cent whereas the greatest span noted for Sweden in a 23 year period is about 45 per cent (see Fig.5:14), both differences calculated on the lowest price paid. This serves to illustrate the situation in an unregulated versus a rather strongly regulated market. It may also be remarked

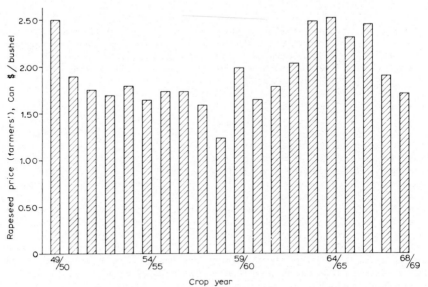

Fig.5:13. Average farm price paid for Canadian rapeseed from 1949/50 to 1968/69. (From Rapeseed, Canada's Cinderella crop[44])

86

that the only subsidy offered to Canadian rapeseed production is a slight reduction in freight tariffs for that commodity when transported by railway to the shipping ports.

5.6.2. *The EEC countries (France, Benelux, W. Germany and Italy)*

A common price regulation system has been planned for the EEC countries. The base price is calculated according to a so-called intervention system related to primary and secondary intervention centres (localities). The 0-locality is Ravenna, Italy (highest price). At other intervention localities the price is calculated for each year and for the months July–August (harvest). Then there is an increase in the base price from month to month up to March of the next year.

From the intervention base price there are reductions and increases in the price according to the contents of moisture, oil and dockage. As an example the French regulation system is given below according to 'Syndicat National du Commerce des Graines':

Base price: Moisture 10 per cent, crude fat 42 per cent, dockage 2 per cent.
Reductions or increases:
    For each per cent of moisture or dockage 0.296 F.
    For each per cent of crude fat 0.839 F.
Reductions for cleaning of the seed:
    less than 2.99 per cent dockage – no reduction
    3–7.49 per cent dockage   – 0.27 F
    7.50–9.99 per cent dockage   – 0.54 F
    more than 10 per cent dockage – rejected
Reduction for drying:
    up to 10.49 per cent moisture   – no reduction
    10.50–21 per cent moisture   – for each 0.5 per cent, 0.25 F
    more than 21 per cent moisture – rejected
In addition there are reductions for transports to the intervention locality and some fees and taxes.

5.6.3 *Poland*

Base price and general quality conditions: The present base price is 8 złoty per kilo for ripe and sane seed with uniform colour, without bad colour or fungi and not being 'self-heated', and containing not more than 4 per cent impurities, of which only 1 per cent minerals.

Moisture content: The base price is at 13 per cent moisture, and a price reduction of 1 per cent is levied for each per cent of moisture between 13 and 15 per cent and 1.5 per cent between 15 and 20 per cent moisture. For each per cent under 13 per cent moisture, the price is increased by one per cent.

Impurities (dockage): For each per cent of impurities between 4 and 10 per cent there is a price reduction of 1 per cent, between 10 and 20 per cent a price reduction of 2 per cent. For each per cent of impurities between 4 and 1 per cent there is a price increase of 1 per cent.

5.6.4 Sweden

The margarine factories of Sweden have formed an association, Svenska Extraktionsföreningen, SEF (The Swedish Extraction Association), with the object of extracting all of the rapeseed oil necessary for the domestic market. The oilseed growers of Sweden have formed an organization called Sveriges Oljeväxtodlares Centralförening, SOC (The Swedish Oil Seed Growers Association). These two parties together with the Government have formed an organization called Sveriges Oljeväxtintressenter, SOI (The Swedish Oil Seed Association), with the aim of handling the oilseed from the delivery by the grower to the reception at SEF, which implies that SOI purchases the oilseed from the growers and sells it to SEF. The price is regulated in relation to a base price (settled each year after negotiations between the interested parties; see ref. 45 for details) according to the following qualitative properties.

Moisture content: The base price is at 18 per cent moisture and a price reduction of 1.6 per cent is levied for each per cent of moisture between 18 and 25 per cent, and 1.7 per cent for higher contents up to 50 per cent, above which level the seed should be rejected. For each per cent of moisture under 18 per cent down to 7.5 per cent the price is increased by 1.5 per cent, and by 1 per cent between 7.5 and 6 per cent. Seed lots drier than 6 per cent are not rejected but there is no price increase.

Content of 'crude fat': The base price is at 47 per cent crude fat of dry matter. Increases are related to the base price for each 0.1 per cent above 47 per cent, and reduction is levied for each 0.1 per cent of crude fat below 47 per cent.

Dockage: Seed lots containing more than 10 per cent dockage should be rejected. In the near future, the per cent of pure, whole seed will be determined by machines in the laboratory and the seed quantity delivered reduced according to this analysis.

Chlorophyll: The base price is at 30 ppm chlorophyll in the oil. Seed lots containing over 70 ppm chlorophyll should be rejected. Scales for price reduction between 30 and 70 ppm are not yet settled. (This system is valid from 1970 onwards.)

General conditions: Seed obviously not sane, infested with fungi, having bad odour, being 'self-heated' etc., should be rejected.

Special fees: Special fees are levied for each kilo of oilseed delivered by the grower, to support the activity of SOC and to cover the costs for oilseeds in the Swedish Crop Failure Insurance system. The average prices paid to Swedish oilseed growers over many years are shown in Fig.5:14.

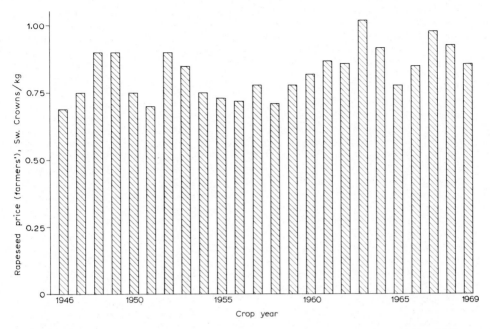

Fig.5:14. Average farm price paid for Swedish rapeseed (winter type) from 1946 to 1969.

5.7 SURVEYS OF THE QUALITY OF RAPESEED PRODUCED IN DIFFERENT COUNTRIES

As was shown above different grading systems are used in different countries for the evaluation of the quality of the rapeseed produced. Therefore, quite naturally no comparison may be made between the quality of rapeseed from different countries. Comparisons may however be made between different areas within a certain country and between the different years in any one country investigated.

The only systematic national survey of the quality of rapeseed during a period of many years is that from Canada, where each year a 'Crop Bulletin' is released in which the variation in oil content and protein content of the rapeseed received at various dryer-stocker companies is presented[46].

The Swedish crop has been investigated during a smaller number of years but in a somewhat more detailed manner. The Swedish investigations dealt not only with oil content but with content of free fatty acids and chlorophyll and also with the viability of the seed[30, 33–35]. An investigation of the quality of a few years' crop of rapeseed and white mustard seed harvested in Denmark has also been presented[47].

Besides these surveys little seems to be published on the variation in the quality of rapeseed produced in different countries. Most probably some data on the

FREQUENCY DISTRIBUTION OF OIL CONTENT AND PROTEIN CONTENT OF CANADIAN RAPESEED
DISCHARGED FROM DRYER-STOCKER COMPANIES IN 1968 AND 1969

Oil content*			Protein content**		
Range, %	No. of samples		Range, %	No. of samples	
	1969	1968		1969	1968
37.0–37.4	1	—	32.0–32.9	—	3
37.5–37.9	1	—	33.0–33.9	1	7
38.0–38.4	1	—	34.0–34.9	2	13
38.5–38.9	2	—	35.0–35.9	14	16
39.0–39.4	2	1	36.0–36.9	16	29
39.5–39.9	5	—	37.0–37.9	19	31
40.0–40.4	9	—	38.0–38.9	34	37
40.5–40.9	8	—	39.0–39.9	36	25
41.0–41.4	11	2	40.0–40.9	38	28
41.5–41.9	19	5	41.0–41.9	38	20
42.0–42.4	21	3	42.0–42.9	42	26
42.5–42.9	28	6	43.0–43.9	19	18
43.0–43.4	27	12	44.0–44.9	10	4
43.5–43.9	32	14	45.0–45.9	8	4
44.0–44.4	35	13	46.0–46.9	5	—
44.5–44.9	18	19	47.0–47.9	1	1
45.0–45.4	15	28			
45.5–45.9	11	17			
46.0–46.4	10	35			
46.5–46.9	8	27			
47.0–47.4	8	16			
47.5–47.9	4	19			
48.0–48.4	2	17			
48.5–48.9	2	9			
49.0–49.4	1	7			
49.5–49.9	—	5			
50.0–50.4	1	1			
50.5–50.9	—	2			
51.0–51.4	—	1			
51.5–51.9	—	1			
52.0–52.4	1	1			
52.5–52.9	—	—			
53.0–53.4	—	—			
53.5–53.9	—	—			
54.0–54.4	—	1			

* Moisture-free basis.
** Oil-free meal. Moisture-free basis. Kjeldahl, factor 6.25.

From *Flax and Rapeseed* (1969) (ref. 46).

variability are compiled in many other rapeseed producing countries, but like those from Poland[48], they are not published and therefore not easily accessible.

5.7.1 Canada

A survey of the quality of the rapeseed in Canada was started in 1956, when the acreage sown to rapeseed had risen to 357,000 acres with an estimated production of 305 million pounds. The results are published annually as a 'Bulletin from the Grain Research Laboratory of the Board of Grain Commissioners for Canada', Winnipeg, Manitoba. Table 5:7 presents the variation in oil and protein content of the new crop harvested in 1968 and 1969. The annual averages of oil and protein contents from 1956 to 1969 are compiled in Table 5:8. In inter-

TABLE 5:8

OIL CONTENT AND PROTEIN CONTENT OF WESTERN CANADIAN RAPESEED, HARVEST SURVEYS 1956–1969

Year	Oil content* %	Protein content** %
1956	45.0	39.4
1957	41.8	42.5
1958	41.4	42.8
1959	43.2	42.8
1960	41.3	43.2
1961	42.1	43.3
1962	43.0	41.4
1963	42.9	42.2
1964	43.8	42.0
1965	43.5	40.1
1966	44.8	39.3
1967	44.0	41.3
1968	46.0	39.0
1969	43.6	40.4

* Moisture-free basis.
** Oil-free meal. Moisture-free basis.

From *Flax and Rapeseed* (1969) (ref. 46).

preting the data, one should recall that only spring sown *Brassica* oilseed crops are grown in Canada, compared to predominantly winter rape in Europe (see Chapter 4). Also the proportion of summer rape, *B. napus*, to summer turnip rape, *B. campestris*, varies from time to time[49]. The samples have been taken at various so-called 'shipping points', namely when the seed is unloaded at the country elevators. This means that one sample represents the average of many consignments from different farmers. Although the most extreme samples have thus been 'lost' when mixed with other consignments, a range in oil content from about 40 per cent to 52 per cent may be noted for the 1967 crop. Since the price paid for rapeseed oil is generally 2 or 3 times that paid for the residue, the meal, trading of rapeseed without considering the oil content seems inadequate. Even the overall yearly figures show a great deal of variation, probably due mainly to climatic conditions. The protein content also shows considerable variability within and between years. Since rapeseed meal is being valued more and more, these data on protein content are of great interest.

5.7.2 Denmark

Results from analyses of selected samples of rape and white mustard seed received from the growers in Denmark in the four years 1959–1962 have been recorded[47]. The analyses included moisture content, content of dockage, oil content, acid value and iodine value. The average moisture content of rapeseed varied considerably from the low 10.0 per cent in the very dry year of 1959 to 15.0 per cent in 1962. On the average for the four years, 15 per cent of the samples had more than 18 per cent moisture. The fact that a considerable portion of the harvested crop has such a high moisture level is typical for the rapeseed crop from northwestern Europe and stresses the necessity for rapid artificial drying. A very large variation in oil content was also observed, from a minimum below 35 per cent to a maximum of 47 per cent, calculated at 10 per cent moisture. The magnitude of this variation has been used as an argument in Denmark in favour of basing the payment to the farmer on quantity of oil delivered, in the same way as practised in Sweden since 1956.

The iodine value of the oil extracted from the samples was found to vary comparatively little (cf. Wetter and Craig[50] and Appelqvist[51]). The high acid values reported may in part have resulted from improper handling of the very moist analysis samples before arrival at the laboratory.

5.7.3 Sweden

The crops harvested in 1962, 1963, 1964 and 1965 were studied in a manner somewhat similar to the Canadian samples but with more detailed chemical analyses[30, 33-35]. Thus, the major part of the samples was taken on release of dried seeds from the dryer-stocker companies. However, samples from smaller crop quantities, dried at the farms, were also analyzed to allow a comparison of the quality of the crop handled in the two ways.

The samples were analyzed for content of waste matter (dockage), oil content, content of free fatty acids, content of field-germinated seeds and viability in all four years and for chlorophyll content in the last three years. Furthermore,

TABLE 5:9

CHARACTERISTICS OF THE SWEDISH CROP OF RAPE, TURNIP RAPE AND WHITE MUSTARD SEED HARVESTED IN VARIOUS YEARS

Seed type	Crop Year	Artificially dried at: Dryer-stocker companies=D-S Farms=F	No of samples studied	Moisture content %	Oil content % in dry matter	Viability %	Free fatty acids %	Chlorophyll in oil ppm
Winter rape	1963	D-S	116	6.33	45.41	81	0.67	—
Winter rape	1963	F	37	5.75	44.97	94	0.56	—
Winter rape	1964	D-S	86	6.30	46.51	81	0.44	18.4[e]
Winter rape	1964	F	113	6.26	46.85	90	0.33	21.3[e]
Winter rape	1965	D-S	77	6.67[a]	46.87	79	0.97	23.9
Winter rape	1965	F	77	6.21[b]	46.58	90	0.91	29.9
Winter turnip rape	1964	D-S	59	6.38	46.05	90	0.32	10.5[e]
Winter turnip rape	1964	F	54	6.33	46.64	97	0.22	7.7[e]
Summer rape	1964	D-S	39	6.50	43.14	82	0.49	26.4[e]
Summer rape	1964	F	57	6.52	43.59	87	0.55	57.5[e]
Summer rape	1965	D-S	43	6.84[c]	40.43	69	0.96	49.3
Summer rape	1965	F	22	6.57[d]	39.89	80	0.93	62.8
White mustard	1964	D-S	51	6.55	32.27	81	0.28	6.8[e]
White mustard	1964	F	50	6.08	32.55	90	0.20	5.7[e]

[a] 2,587 samples analyzed for moisture content.
[b] 727 samples analyzed for moisture content.
[c] 509 samples analyzed for moisture content.
[d] 123 samples analyzed for moisture content.
[e] Calculated from original data (ref. 34) with correction factor given in (ref. 35).

From Appelqvist and Lööf (1964, 1965), Lööf and Johansson (1966) (refs. 30, 34, 35).

additional parameters were studied in various years, in efforts to clarify the effect of handling on the seed quality and to suggest improvements in the handling of the seed. Also quality differences between seeds from different districts and those handled by different companies were compared.

Of the numerous data thus obtained, a few mean values have been compiled in Table 5:9. It should be remarked that the content of free fatty acids was determined on samples from which the 'fines' had been removed by sifting (*cf.* Table 5:4). From Table 5:9 it is obvious that the average moisture content of the Swedish harvest of rapeseed after artificial drying is very low and is generally lower in lots dried at farms than in those dried at dryer-stocker companies. The average seed viability was consistently higher in lots dried at farms (*cf.* page 72). The average content of free fatty acids differed considerably between years for one crop (winter rape) and between different crops during a single year (1964). Finally, the large differences in chlorophyll levels between years and crop types is of major interest, and stimulated further investigations in Sweden on the 'chlorophyll problem' (*cf.* [52]).

Since the content of free fatty acids is used as the 'yard-stick' of oil quality, correlations between various degrees of physiological or physical damage of the

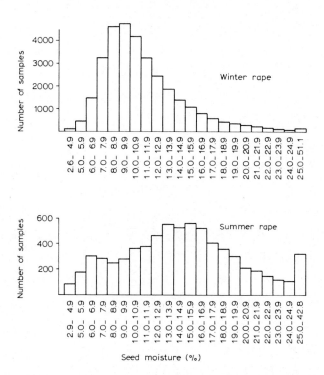

Fig.5:15. Variation in moisture content among the farmers 'delivery-samples' of the Swedish 1968 crop of winter and summer rape. (From Lööf[53])

94

seed and the level of free fatty acids were calculated to obtain insight into the degree to which the various types of damage contribute to the decrease in oil quality[34, 35].

From 1966 on, the information on the quality of the Swedish crop has been accumulated in a different fashion; utilizing the official analysis data of the Swedish Oil Seed Assoc., the data from the 'receiving' laboratory of the Swedish Extraction Association and supplementary analyses at the Swedish Seed Association[53-55]. A very large difference is observed in the moisture content of the farmers' 'delivery samples', a total of 27,699 samples of winter rape and 6,867 of summer rape in 1968 (Fig.5:15). Thus the moisture content of the winter rape varied between 2.6 and 51.1 per cent and that of the summer rape between 2.9 and 42.8 per cent. Since the recommended range of moisture in which the threshing should be undertaken is 12 to 18 per cent, it seems likely that a considerable portion of the winter rape and turnip rape was threshed at too low moisture levels, which might be disadvantageous for the seed quality (see Fig.5:10). The

TABLE 5:10

AVERAGE MOISTURE CONTENTS OF SAMPLES FROM FARMERS' DELIVERIES TO DRYER-STOCKER COMPANIES IN SWEDEN

Year	Crop				
	Winter rape	Winter turnip rape	Summer rape	Summer turnip rape	White mustard
1959	10.6	10.4	12.7	12.4	13.4
1960	13.0	13.6	18.3	16.7	17.8
1961	14.0	12.7	15.5	14.8	18.2
1962	18.7	14.2	21.7	17.1	23.1
1963	13.1	12.5	16.3	16.2	18.1
1964	14.6	12.1	17.1	16.3	15.3
1965	14.9	13.9	18.3	17.3	18.7
1966	15.3	12.1	16.0	15.5	17.8
1967	12.1	12.0	15.1	14.7	16.9

From Lööf (1969) (ref. 53).

average moisture levels of the harvested crop of the five oilseed types grown in Sweden were calculated for the years 1959 to 1968 (Table 5:10). Obviously there was a large variation in mean moisture levels, between crops as well as between years.

As seen from Fig.5:16 also the range of variation in oil content, calculated on dry matter of sample after passing a skimming sieve to remove major contaminants such as straw, pods etc., is remarkably large and certainly motivates the payment

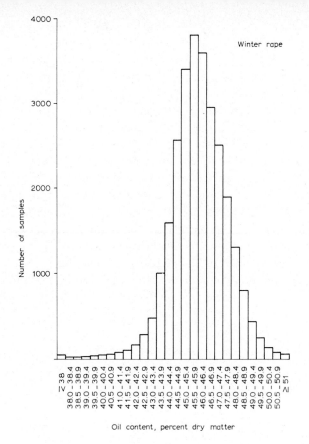

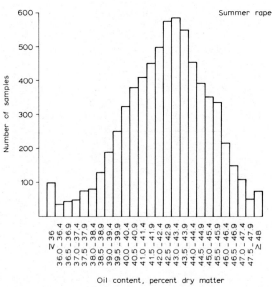

Fig.5:16. Variation in oil content among farmers 'delivery-samples' of the Swedish 1968 crop of winter and summer rape.

system in which the oil content is a major factor, as it has been in Sweden since 1956. It may be of interest to note that the price paid for the sample of winter rape with the lowest oil content (24.8 per cent) was 0.67 Sw. Crowns per kg dry weight, whereas that for the sample with the highest oil content (53.0 per cent) was 1.07 Sw. Crowns.

Because of the large variation in chlorophyll content noted in the survey samples[33-35], the entire 1969 crop was analyzed for chlorophyll content and this characteristic is to be included in the payment system for the 1970 crop in Sweden. Since the weather conditions prevailing during the 1969 oil crop season were atypical, Fig.5:17 presents the distribution with regard to chlorophyll content of samples of winter and summer rape from the 1968 survey studies ('unloading samples' from the dryer-stocker firms).

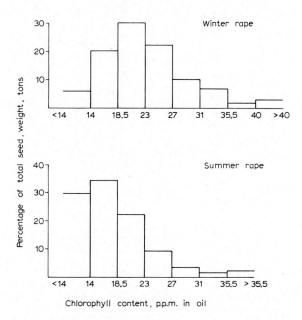

Fig.5:17. Distribution of the 1968 Swedish crop of winter and summer rapeseed with regard to chlorophyll level in the seed. (From Lööf[55])

A considerable variation in chlorophyll level is observed between the different survey samples. The level set as maximum for a top-quality seed is 30 ppm chlorophyll in the oil. By this standard a considerable portion of the 1968 crop is down-graded. From Table 5:11 it is obvious that the percentage of top-quality rapeseed, with regard to chlorophyll content, varies between crop types and years, turnip rape and white mustard obviously suffering very seldom from this type of quality deterioration.

TABLE 5:11

AMOUNTS OF OILSEED (PER CENT) WITH A CHLOROPHYLL CONTENT OF 30 PPM OR MORE HARVESTED IN SWEDEN DURING 1966–1968. SAMPLES FROM DRYER-STOCKER COMPANIES AND LARGE FARMS WHICH DRY THE SEED AND DELIVER IT DIRECTLY TO THE EXTRACTION PLANT

Sample	1966	Years 1967	1968
Winter rape			
Dryer-stocker companies	6	0	15
Large farms with drying equipment	13	2	20
Winter turnip rape			
Dryer-stocker companies	0	0	0
Large farms	2	0	2
Summer rape			
Dryer-stocker companies	39	2	4
Large farms	40	12	13
Summer turnip rape			
Dryer-stocker companies	0	0	2
Large farms	0	0	5
White mustard			
Dryer-stocker companies	0	0	0
Large farms	0	0	0

From Lööf (1969) (ref. 53).

REFERENCES

1 International action to avert the impending protein crisis, *UNESCO Rept. E/4343*, New York, (1968).
2 C. G. MASSON, *L'Echantillonnage des Graines de Colza*, (1968), C.E.T.I.O.M., 174 Avenue Victor-Hugo, Paris. (Duplic.)
3 B. LÖÖF AND S.-Å. JOHANSSON, *Svensk Frötidning*, 37:12 (1968) 179. (In Swedish)
4 B. LÖÖF AND S.-Å. JOHANSSON, *Svensk Frötidning*, 38:2 (1969) 17. (In Swedish)
5 C. G. MASSON, *La Conservation du Colza*, (1959), C.E.T.I.O.M., 174 Avenue Victor-Hugo, Paris. (Duplic.)
6 *Bulletin C.E.T.I.O.M.* 37 (1968), 174 Avenue Victor-Hugo, Paris. (In French)
7 *Le Colza de sa Recolté à sa Commercialisation*, (1963), C.E.T.I.O.M., 174 Avenue Victor-Hugo, Paris. (Duplic.)
8 C. G. MASSON, *La Conservation des Oleagineux en Atmosphère Confinée ou Gazeuse*, (1962), C.E.T.I.O.M., 174 Avenue Victor-Hugo, Paris. (Duplic.)
9 C. G. MASSON, in A. RUTKOWSKI (Ed.), *International Symposium for the Chemistry and Technology of Rapeseed oil and other Cruciferae Oils, Gdansk (1967)*, Warszawa, 1970, p.111.
10 H. PARDUN, *Deutsche Lebensmittel-Rundschau*, 62:1 (1966) 6.
11 B. LÖÖF AND S.-Å. JOHANSSON, *Sveriges Utsädesfören. Tidskr.*, 79:1–2 (1969) 16. (In Swedish with English summary)

11a K. RUHOWICZ, *The Storage of Oilseed Oils and Extracted Meals,* Report from the Institute of Fat Technology, Warsaw, 1963. (Duplic.)

12 J. E. LINDBERG AND I. SÖRENSSON, *Kungl. Skogs- och Lantbruksakad. Tidskr. Suppl. 1,* (1959) 73 pp. (In Swedish with German summary)

13 C. L. HOFFPAUIR, S. E. POE, L. U. WILES AND M. J. HICKS, *J. Am. Oil Chem. Soc.,* 27 (1950) 347.

14 K. TÄUFEL AND R. POHLOUDEK-FABINI, *Pharmazie,* 9 (1954) 511.

15 M. L. KINMAN AND R. E. IBERT, *J. Am. Oil Chem. Soc.,* 33 (1956) 637.

16 L.-Å. APPELQVIST, *Sveriges Utsädesfören. Tidskr.,* 71:1 (1961) 74. (In Swedish with English summary)

17 L.-Å. APPELQVIST, *Svensk Frötidning,* 34:9 (1965) 119. (In Swedish)

18 L.-Å. APPELQVIST, *J. Am. Oil Chem. Soc.,* 44 (1967) 206.

19 L.-Å. APPELQVIST, *J. Am. Oil Chem. Soc.,* 44 (1967) 209.

20 *Official and Tentative Methods,* American Oil Chemists Society, (1945–69).

21 *British Standard Specifications, Vegetable Oils,* British Standards Institution, 1961, London.

22 *DGF-Einheitsmetoden* (1950–69), Deutsche Gesellschaft für Fettwissenschaft, Wissenschaftliche Verlagsgesellschaft, Stuttgart.

23 *Standard Methods of the Oils and Fats Division of the I.U.P.A.C.,* International Union of Pure and Applied Chemistry, 1964, Butterworths, London.

24 L.-Å. APPELQVIST AND S.-Å. JOHANSSON, *Svensk Frötidning,* 32:6–7 (1963) 97. (In Swedish)

25 K. RUHOWICZ, personal communication.

26 J. KREYGER, *I.B.V.L. Wageningen, Holland, Procedium of the Int. Seed Test. Ass.,* 28:4 (1963) 827.

27 W. F. GEDDES, *Food Technology,* Nov. 1958, p.7.

28 J. KREYGER AND M. J. F. KOOPMAN, *Rapport van de Coördinatie-Commissie Onderzoek Landbouwzaaizaden van het centraal Orgaan, Wageningen* (1959), p.74. (In Dutch)

29 J. POISSON, A. GUILBOT AND R. DESVEAUX, *Bulletin C.E.T.I.O.M.* (1962), 174 Avenue Victor-Hugo, Paris. (Duplic.)

30 L.-Å. APPELQVIST AND B. LÖÖF, *Sveriges Utsädesfören. Tidskr.,* 74:2–3 (1964) 183. (In Swedish with English summary)

31 A. RUTKOWSKI, K. BABUCHOWSKI AND J. OLEJNIK, *Oléagineux,* 17 (1962) 91.

32 Anon., *Bulletin C.E.T.I.O.M.,* 33 (1967), 174 Avenue Victor-Hugo, Paris. (Duplic.)

33 L.-Å. APPELQVIST AND B. LÖÖF, *Sveriges Utsädesfören. Tidskr.,* 73:4–5 (1963) 310. (In Swedish with English summary)

34 L.-Å. APPELQVIST AND B. LÖÖF, *Sveriges Utsädesfören. Tidskr.,* 75:5 (1965) 318. (In Swedish with English summary)

35 B. LÖÖF AND S.-Å. JOHANSSON, *Sveriges Utsädesfören. Tidskr.,* 76:5–6 (1966) 344. (In Swedish with English summary)

36 C. G. MASSON, Deterioration des Graines de Colza, *Bulletin C.E.T.I.O.M.,* (1959), 174 Avenue Victor-Hugo, Paris. (Duplic.)

37 C. G. MASSON, La Conservation du Colza, *Bulletin C.E.T.I.O.M.,* 174 Avenue Victor-Hugo, Paris. (Duplic.)

38 C. G. MASSON, *Bulletin C.E.T.I.O.M.* (1969), 174 Avenue Victor-Hugo, Paris. (Duplic.)

39 J. POISSON, A. GUILBOT AND R. DESVEAUX, *Bulletin C.E.T.I.O.M.* (1962) 174, Avenue Victor-Hugo, Paris. (Duplic.)

40 A. RUTKOWSKI AND Z. MAKUS, *Fette, Seifen, Anstrichmittel,* 61 (1959) 532.

41 K. RUHOWICZ AND J. HORSKI, *Roczniki Nauk Rolniczych,* 72-A-3 (1956) 503.

42 L. R. WETTER, *J. Am. Oil Chem. Soc.,* 34 (1957) 66.

43 S. TROENG, *J. Am. Oil. Chem. Soc.,* 32 (1955) 124.

44 *Rapeseed, Canada's 'Cinderella' crop,* Publ. 8, Second Edition, Rapeseed Association of Canada (1970).

45 W. TINGNELL, *Paper presented at the International Conference for the Science, Technology and Marketing of Rapeseed,* St. Adèle, Canada (1970).

46 Flax and Rapeseed, *Can. Crop Bulletin,* 108 (1969) 1.

47 P. S. FREDERIKSEN, *Forskningsinstituttet for Handels- og Industriplanter Beretning,* Nr.39 (1964) 1. (In Danish with English summary)

99

48 W. Tczebny, personal communication.
49 R. K. Downey, personal communication.
50 L. R. Wetter and B. M. Craig, *Can. J. Plant Sci.*, 39 (1959) 437.
51 L.-Å. Appelqvist, *Acta Agric. Scand.*, 18 (1968) 3.
52 Sveriges Oljeväxtintresseter, *Svensk Frötidning*, 39:4 (1970) 49. (In Swedish)
53 B. Lööf, *Svensk Frötidning*, 38:6 (1969) 89. (In Swedish)
54 B. Lööf, *Svensk Frötidning*, 38:7–8 (1969) 108. (In Swedish)
55 B. Lööf, *Svensk Frötidning*, 38:9 (1969) 121. (In Swedish)

CHAPTER 6

Plant Breeding for Improved Yield and Quality

B. LÖÖF* and L.-Å. APPELQVIST**

* Oil Crop Division, Swedish Seed Association, Svalöf (Sweden);
** Division of Physiological Chemistry, Chemical Center, University of Lund (Sweden)

CONTENTS

6.1 THE AIM OF PLANT BREEDING

6.1.1 Higher productivity

There is very little difference in the production costs of *Brassica* oilseed whether or not the yield is high. The higher yield may cause higher transport and drying costs, but the costs for sowing, plant protection, fertilizers and harvesting are about the same. By means of selection (plant breeding) new higher yielding varieties can be obtained. These new varieties have a higher productivity owing to their genetic constitution. However, the seed does not contain only fat, and if the fat has a considerably higher value than the rest of the seed, the total production of fat can be increased by decreasing the content of other not so valuable substances in the seed[1].

6.1.2 Improved growing security

A high productivity under favourable conditions will not always result in a high average yield of the crop. On certain occasions during the growth period the crop may be set back by unfavourable circumstances (cold, drought, insects and fungus pests). If these conditions determine the average yield, varieties with high growing security, that is resistance to these factors, will give higher average yields[2, 3].

6.1.3 Improved quality

If there are large price differences depending upon the quality of the product or if the present quality of the oil or protein can not be accepted by a quality-conscious market, the economic value of the oilseed produced may be increased considerably by plant breeding for improved quality. In the developed countries consumers pay more and more attention to the quality, and consequently the producers of oilseed must increase their efforts to meet quality demands[4–7].

6.2 PLANT BREEDING IN RELATION TO OTHER MEANS OF IMPROVEMENT

There are of course other means of increasing the yield of the *Brassica* oilseeds, for instance a more adapted cultivation technique, including larger amounts of fertilizers. As mentioned in Chapter 4, the *Brassica* oilseeds give a very good response to increased nitrogen fertilization[6, 8, 9]. Increased amounts of fertilizers can, however, be utilized by the crop only if the root system is capable of taking up large amounts of ions, and if the stalk is strong enough to bear the increased bulk of green matter and higher weight of seed. Consequently there is always an interaction between genetic constitution and the cultivation technique applied that gives a result in the form of increased yield per unit area[6]. If research on proper cultivation techniques and plant breeding are carried out at the same time, a continuous increase of the average yield can be achieved, as illustrated by Fig.6:1.

When there is a demand for higher quality of the seed components or lower content of toxic substances, one has to consider whether this should be achieved by plant breeding or by innovation in the technological processes. However, new

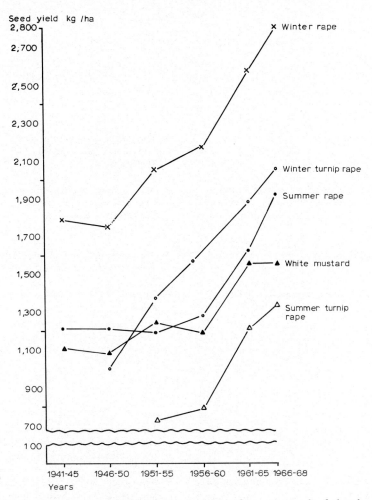

Fig.6:1. The increase in average yields of oil crops in Sweden as the result of plant breeding and improved cultivation techniques.

processes, high amounts of fertilizers etc. always convey higher production costs. On the other hand, a new higher level of productivity or better quality achieved by plant breeding conveys no further costs. Indeed, they may even be reduced: resistance to diseases and insect pests will reduce the costs for pest control, for example.

6.3 SPECIFIC OBJECTIVES OF PLANT BREEDING

Important objectives of plant breeding in rape are listed in Table 6:1, where reasons for introducing different objectives on the program are also given. Plant

OBJECTIVES OF PLANT BREEDING IN RAPE

Aim	*Reason for change*
A. Agronomic properties	
1. Higher yield of seed.	1. Decreases production costs per unit weight as these are rather independent of the yield.
2. Increased tolerance to low temperatures, light deficiency etc. during winter (winter types), water logging, salinity, heat, drought etc.	2. May increase yield, facilitate cultivation of the more productive winter types and extend production area.
3. Increased resistance to insects and fungus pests.	3. May increase yield and decrease production costs (pest control).
4. Adapted maturity, increased stalk strength, increased resistance to shedding.	4. Influences harvest costs, seed yield and quality.
B. Quality properties (of the seed)	
1. Increased seed size.	1. Larger seeds give higher technical output, are less brittle when dry and have lower percentage seed coat than small ones.
2. Lower percentage and less pigmentation of seed coat.	2. The seed coat has a low content of fat and protein and high content of fibre. Pigment gives undesirable dark colour of lecithin, meal and extracted protein.
3. Higher content of fat, protein and other valuable substances.	3. As production costs are the same per unit processed seed, a higher technical output will improve the economy.
4. Lower content of fibre, glucosinolates and other toxic or less valuable substances.	4. Components of no value or a 'negative' value increase production costs and decrease the quality of final products.
5. Suitable fatty acid composition.	5. Erucic acid is not desirable for nutritional as well as technical reasons. Linolenic acid causes rancidity and should be substituted by linoleic acid, preferred among the polyunsaturated fatty acids.

breeders are often forced to concentrate upon the yield, and other desired properties become of secondary importance. If, however, the seed price were regulated according to different quality properties such as oil content, protein content, colour of seed coat, content of fibre, content of glucosinolates, composition of fatty acids etc., it is evident that breeders and growers would make strong efforts to produce seed of a quality which gives the highest possible economic return. The reasons for a change of the properties listed in Table 6:1 as 'agronomic properties' can be rather easily understood. It seems necessary, however, to present some more detailed background information on the desire for changing the 'quality properties' of the seed, especially points 2, 4 and 5 in the table.

In the industrial processing of rapeseed (for details see Chapter 9) three products are obtained, namely oil, lecithin and meal. As pointed out earlier in this chapter, the oil is at present considered the most valuable constituent.

Lecithin is a complicated mixture of substances obtained by a special treatment of the oil (see Chapter 13). The yield of lecithin is only about 0.5% of the total oil yield, and therefore its value is not of too great importance for the overall economy of rapeseed processing.

A major part of the rapeseed oil is used as an edible oil, mainly in margarine (see Chapter 11). Rapeseed oil is characterized by containing variable but large amounts of erucic acid, generally *ca.* 45% in summer rape and slightly over 50% in winter rape, as discussed in detail in Chapter 7. The linoleic acid content of rapeseed oil is about 15%, and thus lower than that of the other major edible oils, and the linolenic acid content is 8–10%. The implications of this fatty acid pattern for problems related to food technology and human nutrition have been discussed at length in an earlier review[4] and are further discussed in Chapters 11 and 14 in this volume. Thus, the present chapter contains condensed information on the background for the desired changes in the fatty acid composition through plant breeding.

6.3.1 Seed yield

As mentioned previously, an increased seed yield cannot be attained if the plant does not have the appropriate constitution to produce it; that is to say the stalk must be stiff enough, and the root system efficient in taking up nutrients and water. These considerations are applicable under all conditions. In some districts other factors may have a great influence upon the yield. Where fungus pests reduce the yield, breeding for resistance will be a specific objective of high priority[10–12a], and in windy regions resistance to shedding may be very important[13,14].

6.3.2 Oil content

In countries where the seed is priced according to its content of oil, the breeders concentrate upon raising the yield of oil per unit area. It is easier to increase the yield of oil by raising the oil content of the seed than to raise the seed yield, since the oil content is less influenced by the environment. Furthermore, the oil content is controlled by a smaller number of genes than the seed yield[1, 15, 16].

6.3.3 Fatty acid composition

A lower erucic acid content is desirable because the present high erucic acid content of winter rape (the major seed type produced in northern Europe) prevents its sole use in several edible oil products such as salad oil and mayonnaise. The reason for this is that the triacyl glycerols containing erucic acid have a higher melting point than those containing C16 or C18 unsaturated fatty acids. A decrease in erucic acid content to a level below *ca.* 40% would be enough to permit its use in these products (see Chapter 11). The nutritional value of erucic acid-containing oils has, however, been disputed in the past and is presently the subject of extensive nutritional studies, as discussed in Chapter 14. From the nutritional point of view, the elimination of erucic acid in rapeseed oil by plant breeding seems to be a desirable goal. On the other hand, it has been found that the crystallization behaviour of erucic acid-free rapeseed oil is disadvantageous (see Chapter 11), and about 10% erucic acid would be necessary to prevent problems of polymorphic rearrangement. The opinion held by a majority of industrial chemists is that the reduction of the erucic acid content to very low levels, perhaps even total elimination, should be the goal of plant breeding.

The aim of decreasing the erucic acid content is appropriate for the predominant use of rapeseed oil, namely as an edible oil. If the rapeseed oil is to be used for the preparation of various industrial chemicals (see Chapter 12) the highest possible erucic acid content is desirable[17].

An increase in linoleic acid content is desirable for nutritional reasons. The linoleic acid content of rapeseed oil, about 15%, is considerably lower than that of the other four major edible oils, namely soya beans (about 55%), cottonseed (about 55%), sunflower seed (about 65%) and groundnuts (about 30%). Since the sole reason for increasing the linoleic acid content is to increase the nutritional value of the oil, the reader is referred to the material presented in Chapter 14.

Rapeseed oil and soyabean oil are the only two major edible oils which contain linolenic acid. This causes special food technological problems since a phenomenon called 'flavour reversion' is associated with these two oils (see Chapter 11). The alternative to a reduction in linolenic acid content by plant breeding, namely to hydrogenate this acid industrially, is presently not considered economically feasible, although it is technically possible to reduce one of the double bonds in linolenic acid to form a linoleic acid isomer, without touching much of the valuable linoleic acid. The products formed by hydrogenation of linolenic acid are not only the biologically valuable *cis,cis*-9,12-octadecadienoic acid (linoleic acid) but also other biologically inactive isomers. Therefore, plant breeding still seems to be the preferred alternative in efforts to reduce the linolenic acid content of rapeseed oil.

6.3.4 Lecithin

Major quality problems with rapeseed lecithin are its dark colour and unpleasant taste and odour. Since lecithin from white mustard seed with a yellow seed coat gives a much lighter coloured lecithin, it is understandable that a light coloured seed coat is a goal of interest for plant breeding. Whether or not the other qualitative factors of interest in rapeseed lecithin can be improved by plant breeding is not known at the present time.

6.3.5 Glucosinolates

Rapeseed meal is nowadays utilized as animal feed in many industrialized countries (see Chapter 15). It is, however, not considered a top quality feed ingredient although it has a high protein content and a well-balanced amino acid composition. This is mainly because of the presence of considerable amounts of glucosinolates, which can be split in the intestinal tract to toxic sulphur compounds (see Chapters 7 and 15 for details), and thus a reduction in the content of glucosinolates in rapeseed meal is of great importance.

6.3.6 Fibre content and pigments of seed coat

The rather high fibre content of rapeseed meal (see Table 1 in Chapter 7) contributes to the overall reduction of the feed value of rapeseed meal. When edible protein products are to be made from rapeseed meal the dark colour of the seed coat is a considerable problem, since it gives an unpleasant grey colour to most protein products made from rapeseed meal. Thus a reduction in seed coat colour is of interest from the point of view of protein quality as well as lecithin quality.

6.4 THE TECHNIQUE OF PLANT BREEDING

Plant breeding is based upon the laws of Mendel, worked out as early as the 1860's. Subsequently standard methods for breeding different types of plants were worked out, and there is a considerable amount of research in this field reported in the literature[1, 18-41]. Rather many anthologies and standard publications are available, and thus the fundamental principles and the details of practical plant breeding will not be treated here. This section will be concerned with some general views on the plant breeding of *Brassica* oilseeds, which may be of interest to specialists in other fields as well as to students of plant breeding.

6.4.1 *Type of plant and variation in characters of economic importance*

The first point the breeder has to consider is whether the plant is cross-pollinated or self-pollinated[42-45]. Which characteristics are of economic importance and is the variation sufficient? The most noticeable characteristics may be of no economic value, for instance the colour of the petals of the flowers. The height of the plant, the number of leaves, and the green bulk produced may be very important if the plant is to be used for fodder. If only the seeds are harvested the production of other plant parts is of no direct importance. However, the size of the leaves and the growth habit may be of importance if the crop has to compete with weeds or to cover the ground to protect it against heat and drought, and thus indirectly contribute to a higher yield of seed. In this case the breeder speaks of a correlation between morphological characters and seed yield. Some breeders pay very much attention to this type of correlation[31, 38]. This may be of great importance in case direct screening for yield for example is very expensive but screening for a character correlated to the yield is very cheap.

The mode of fertilization, that is whether the plant is cross-pollinated, self-pollinated or partly self-pollinated, is very important for the choice of selection methods. Self-pollinated plants become constant after some generations and

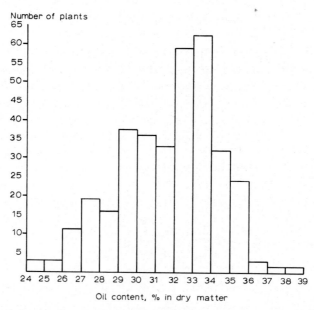

Fig.6:2. The variation in oil content between single plants from a population of white mustard *Sinapis alba* (a cross-pollinated plant).

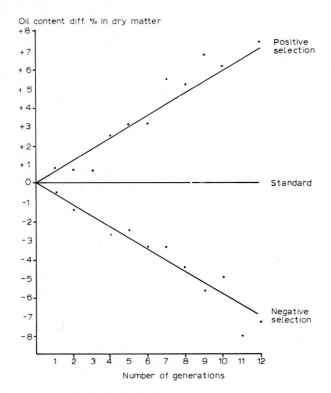

Fig.6:3. The effect of repeated selection for oil content in a cross-pollinated population of white mustard (*Sinapis alba*).

further selection will have no effect. Cross-pollinated plants on the other hand will recombine their genes for each new generation. Selection of single plants will have an effect upon the first generations of the plant progeny but by and by there will be an approach to the original population. Cross-pollinated plants can be forced to self-pollinate, which is called inbreeding. Inbreeding generally causes a great reduction in vigour, but the vigour can be restored by hybridization. Some plants are very difficult to inbreed because they are highly self-sterile, but the improvement can be achieved stepwise by selecting for certain characters generation after generation. Continuous selection may result in considerable improvement (Figs.6:2 and 6:3). This improvement is not lost as rapidly in the following generations[18] as is the case with hybrids. Continuous selection of marketed varieties of cross-pollinated crops is undertaken to maintain present properties or to improve them successively[46].

6.4.2 Means of widening the variation

In crop plants the variation has generally been sufficient to secure new important improvements by plant breeding. Even if none of the lines of the breeder's sortiment reaches the desired level in some property, he may eventually attain his goal by the recombination technique, which implies the crossing of lines with the desired properties but without exactly the same genes. This leads to recombination of genes, and in the progeny he can select new lines superior to the ones which were best in the previous generation. This is valid only for polygenic or quantitative characters. The fact that the progeny is better than both parents is called transgression. The process of continuous recombination, recrossing and selection may take a very long time, and therefore breeders always search for efficient means of widening the variation. By widening the variation the breeder may reach the desired level in one character but in other characters the line or population may be very inferior to the commercial material. The improvement of a standard variety in one or a few characters is, however, rather easy as it is achieved by repeated back-crossing. For the estimation of the possibilities of widening the variation the number of chromosomes and the ploidy level must be considered. Widening the variation by induced mutations is easier if the plant has a low number of chromosomes and is diploid. If the plant is an amfidiploid such as some of the *Brassica* species (see Chapter 3) the variation can be widened by re-synthesis of the amfidiploid using different types of the basic species[18, 47, 48].

6.4.3 The search for sources of resistance, special qualitative properties etc.

Most crop plants derive from wild types which generally are found in so-called gene centres, *i.e.* regions where the species probably was born very long ago. The variation between individual plants in such gene centres is generally much wider than in cultivated crop material. Numerous wild types of crop plants have been rescued, and are now found in so-called gene banks at universities or botanical gardens all over the world. As the plant breeding program tends to comprehend new items previously not considered, these gene banks can be of great value to the breeders. If, for instance, a fungus disease of minor importance suddenly increases and becomes serious, because there is no resistance in the varieties currently grown, the breeder will have to screen wild material or material from sortiments at the gene banks where genes for resistance may be found.

If resistance or other desired characters are dependent on one or a few genes, repeated back-crossing may rather soon result in a combination of high agronomic value and the desired resistance or quality properties. On the other hand if the breeder has to combine several desired characters, some of which are dependent upon polygenic factors, his work may be quite delicate[18].

6.4.4 Prerequisites for success in plant breeding

Generally four prerequisites are considered important for rapid success in any plant breeding project, *viz.*:

110

(1) Considerable genetic variation in the desired property between populations, cultivars, lines or plants.

(2) A rather limited environmental effect on the content of the desired or unwanted components.

(3) Access to rapid and sensitive analytical methods for the determination of the compound under discussion.

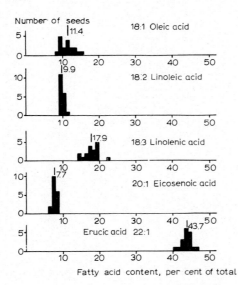

Fig.6:4. Variation in content of oleic, linoleic, linolenic, eicosenoic and erucic acids among 18 single seeds of *Brassica oxyrrhina*[58].

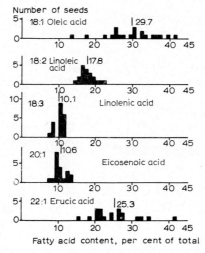

Fig.6:5. Variation in content of oleic, linoleic, linolenic, eicosenoic and erucic acids among 20 single seeds of *Brassica campestris*[58].

References pp. 121–122

111

(4) Sufficient area of even land for screening according to agronomic properties, as small trial errors as possible and general adaptation of the varieties selected.

Many papers have reported that there is greater variation in fatty acid composition between cultivars, than that caused by environmental conditions (see e.g.[55-57]). Most important, however, is the interseed variability in certain cross-fertilizing species, which is considerably greater than that in a comparable self-fertilizing species, as demonstrated in Figs.6:4 and 6:5 and discussed in detail elsewhere[58].

TABLE 6:2

CORRELATIONS IN THE GLUCOSINOLATES. CROSSINGS WITH BRONOWSKI

	Material number	Number of plants	r	t
Erucic acid – glucosinolates	69-9047	373	+0.074°	1.435
ITC – OZT	69-9047	406	+0.688***	19.073
ITC – OZT	69-3101	154	+0.597***	9.175
ITC (mother) – ITC (progeny)		154	+0.764***	14.601
OZT (mother) – OZT (progeny)		154	+0.589***	8.986

° = not significant   ITC = isothiocyanate
*** = significant at the 0.001 level   OZT = oxazolidinethione

TABLE 6:3

PER CENT OF SEED COAT AND FIBRE CONTENT: SEEDS WITH YELLOW COAT IN COMPARISON WITH
SEEDS WITH BROWN COAT

Species, variety Colour of seed	Seed coat % of whole seed	Fibre content % of dry matter
Brassica campestris		
68-620 yellow seeds	13.0	5.9
68-620 brown seeds	18.1	11.0
Bele brown seeds	18.4	11.5
Brassica juncea		
I.L. 2317 yellow seeds	16.3	7.2
I.L. 2317 brown seeds	21.0	10.5
Sinapis alba		
68-3603 yellow seeds	19.0	6.9
68-3603 brown seeds	24.7	9.1

TABLE 6:4

COMPARISON OF 1000-GRAIN WEIGHT, CONTENT OF OIL, SEED COAT AND CRUDE FIBRE IN PLANTS WITH YELLOW AND BROWN SEEDS FROM A POPULATION OF SUMMER TURNIP RAPE[62]

Characters	Population mean m_{x+y}	Yellow seeds (25 plants)			Brown seeds (25 plants)			$t_{diff.}$
		m_x	variation	s_x	m_y	variation	s_y	
1000-grain weight, g	2.94	2.95	2.26–3.65	0.38	2.92	2.37–3.91	0.40	0.31°
Oil content in dry matter, %	44.37	45.85	41.9–48.0	1.59	42.88	38.4–46.2	1.95	6.36***
Seed coat, %	14.62	12.76	11.3–14.4	0.99	16.48	13.8–18.0	1.23	12.50***
Crude fibre in meal, %	10.56	8.29	7.50–9.50	0.48	12.83	10.98–15.01	0.84	24.27***
Crude fibre in seed, %	5.90	4.49	4.03–5.04	0.25	7.31	6.46–8.63	0.51	26.61***

As is shown in Table 6:2, the content of glucosinolates is strongly determined genetically, and environmental conditions normally have little influence[59-61].

The pigmentation of the seed coat seems to be strongly related to the thickness of the seed coat. As shown in Table 6:3, yellow seeds have a lower percentage of seed coat than brown ones, and this also influences the content of fibre and oil (Table 6:4). Since the colour of the seed coat is determined by genetic factors, there seems to be quite a good basis for breeding varieties with high oil content and low content of pigments and fibre[62].

The third of these prerequisites is nowadays fulfilled for quality breeding, since rapid and sensitive methods have been developed for the analysis of fatty acid composition[49-51] and for glucosinolate content[52-54].

The new genetic combinations obtained after crossing and selection and which give progenies considerably superior to the previously grown varieties are quite rare, and so if the material is too small the very best combinations may escape the screening of the breeder. If the trial ground is uneven, lines which seem to be bad or good owing to environmental influence may get lost or be taken out for further testing. Even if the selected lines really are good under the conditions prevailing at the breeding station they may fail when tested in the growing area if the conditions are different there.

6.5 RESULTS OBTAINED AND EXPECTED

6.5.1 Seed yield

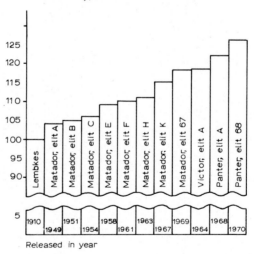

Fig.6:6. Increase in yield of crude fat of winter rape in Sweden as a result of plant breeding.

As shown in Fig.6:6, a considerable increase in the yield of *Brassica* oil crops can be achieved for a certain period of years, in this case by the interaction of improved cultivation techniques and improved varieties. Fig.6:6 shows the improvement of winter rape by plant breeding in Sweden. Similar progress has been achieved in other oil crops and by many plant breeders in different countries. In appendix I important varieties of *Brassica* oil crops are listed. Only varieties that hold an important share of the cultivated area of the crop in question are included. The improvement of seed yield, about 50 kg per annum in Sweden[6], will most probably continue at the same rate or faster in the future.

6.5.2 Oil content

In Sweden the oil seed has been priced according to oil content since 1956. The introduction of this system greatly increased interest in varieties with a high content of oil. It may be seen from Fig.6:7 that this rather soon resulted in

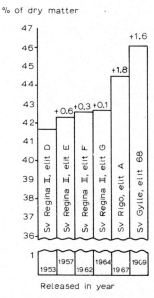

Fig.6:7. Increase in oil content of summer rape as a result of plant breeding.

varieties with considerably higher oil content than the old ones. As the newest varieties have an oil content which is 3–4 per cent units higher, this means a rather rapid increase in the yield of crude fat per unit area in recent years. The higher content of fat implies a greater technical output and consequently lower production costs per unit weight of oil.

It is possible that the protein in the near future will be of higher value than the oil, but as long as the seed is priced in relation to its oil content and not to its content of protein, plant breeders aim at increasing the oil content to higher and higher levels, especially since this seems to be possible without great difficulty and without any reduction in the seed yield.

6.5.3 Fatty acid composition

(a) Erucic acid content

In 1961 Canadian plant breeders reported the isolation of single plants of rapeseed with oil free from erucic acid[40]. The plants were isolated from the German fodder rape variety Liho and the result was accomplished by selection in two successive generations. The decrease in erucic acid was accompanied by a strong reduction in eicosenoic acid content, by an approximate doubling of the linoleic acid content and a 4–5 fold increase in the amount of oleic acid. The linolenic and palmitic acid contents were somewhat increased. These zero erucic acid plants have been used in breeding programs in Canada as well as in Europe to combine the absence or low levels of erucic acid with desirable agronomic characteristics of rapeseed[5, 24, 26, 29]. The fatty acid composition of one such line developed in Canada is presented in Table 6:5 and compared with two other Canadian summer

TABLE 6:5

FATTY ACID COMPOSITION OF IMPROVED OIL SELECTIONS COMPARED WITH STANDARD VARIETIES
GOLDEN AND ARLO[24]

Species and variety	Fatty acids, per cent						
	16:0	18:0	18:1	18:2	18:3	20:1	22:1
B. napus							
Golden	3.3	1.1	18.6	14.0	7.8	13.4	41.8
Nugget	3.3	1.5	22.8	12.2	*5.4*	14.2	40.6
Zero erucic	4.7	1.8	*63.3**	*20.0*	8.9	*1.3*	*0.0*
B. campestris							
Arlo	3.2	1.1	26.6	17.5	8.8	11.8	31.0
Yellow sarson	1.8	0.8	11.7	10.5	8.3	5.9	*61.0*
Zero erucic	4.3	0.1	*54.8*	*31.1*	9.7	*0.0*	*0.0*

* Figures in italics indicate major differences compared with standard varieties.

rape varieties. Later similar selection of zero erucic acid plants was made from *Brassica campestris*[20]. The fatty acid composition of a typical Canadian breeding line of turnip rape is presented in Table 6:5 along with data for a common summer turnip rape variety, Arlo, and the high erucic acid type, 'Yellow sarson', grown in India.

116

Crossing these zero erucic acid strains with conventional genotypes of *B. napus* and *B. campestris* demonstrated several interesting facts: The erucic acid content of the seed is determined by the genetic constitution of the embryo and not by that of the maternal plant[26, 29, 63, 64]. It appears that the level of erucic acid is mainly governed by two genes in the allotetraploid *B. napus*[64] and by one gene in the diploid *B. campestris*[63]. The absence of erucic acid does not influence the oil content, as shown in Table 6:6.

TABLE 6:6

OIL CONTENT AND FATTY ACID COMPOSITION OF SELF-POLLINATED SEEDS FROM RAPE PLANTS
GENETICALLY CAPABLE OF PRODUCING FIVE LEVELS OF ERUCIC ACID[23]

Plant geno-type*	Seed oil content, %	Fatty acid composition in percentage of total acids								
		16:0	16:1	18:0	18:1	18:2	18:3	20:0	20:1	22:1
++++	39.0	3.8	0.4	0.8	18.4	15.8	9.8	0.8	11.0	39.2
−+++	45.5	3.8	0.5	1.3	24.3	15.7	9.1	1.0	13.9	30.4
−−++	38.9	4.1	0.4	1.4	34.2	16.3	7.9	0.7	14.1	20.9
−−−+	39.4	4.8	0.7	0.8	41.8	18.7	7.3	1.0	12.1	12.8
−−−−	43.4	4.8	0.9	1.2	63.1	20.0	8.2	0.8	1.0	0.0

* (+) indicates a gene contributing 9–10% erucic acid to the seed oil.

Considerable interest has been focused on the zero erucic acid genotypes, and a commercial variety called Oro has been released in Canada. In Sweden, lines of winter rape and summer rape have been grown on an experimental scale (a few hundred acres). In this connection it is worth noting for the reader with no background in genetics and plant breeding, that the gene for zero erucic acid was isolated from a summer rape variety useful for green fodder production. As far as is known no other gene source for zero erucic has been used in the rapeseed zero-erucic breeding programs, although it has been reported that plants with no erucic acid have also been isolated from another summer rape variety[41]. In any crossing of zero erucic to high yielding varieties it is inevitable that several 'poor agronomic characters' are introduced in the offspring together with the desired property 'absence of erucic acid', and that it requires very great efforts (back-crosses) to combine absence of erucic acid with all or almost all of the other desired agronomic characters that are present in other, high yielding, parental strains.

The results from efforts to increase the erucic acid content of the seed oils from various *Cruciferae* were recently presented[65]. The conclusions drawn were as follows: By inbreeding and crossing followed by selection, it is possible to increase the erucic acid level in species such as *Brassica carinata* and *Sinapis alba* with intermediate levels of this acid (about 40–50 per cent) in natural populations. No further increase seems possible in species which already have high levels of erucic acid (about 60 per cent) in their seed oils, such as *Crambe abyssinica* and *Brassica campestris*, ssp. *trilocularis* (Yellow sarson). This is explained by the very strict positioning of oleic, linoleic and linolenic acid in the 2-position of the triacylglycerols. Thus the sum of saturated and C_{20} and C_{22} fatty acids cannot exceed 67 mole per cent (see further Chapter 7). Therefore, as the increase in erucic acid content of the seed oil of the first group of species approaches the 67 mole per cent level, the change is supposed to level off.

(b) Other fatty acid contents

Most plants and lines with no erucic acid obtained in the Canadian and Swedish breeding programs have slightly increased linolenic acid content and approximately doubled linoleic acid content compared to conventional lines (see Tables 6:5 and 6:7). Efforts to increase the linoleic and decrease the linolenic acid content

TABLE 6:7

THE FATTY ACID COMPOSITION OF LOW ERUCIC ACID LINES OF WINTER AND SUMMER RAPE IN COMPARISON WITH BRONOWSKI AND TYPICAL SWEDISH CULTIVARS DEVELOPED AT SVALÖF

Cultivar or line	Fatty acid composition, per cent						
	16:0	18:0	18:1	18:2	18:3	20:1	22:1
Winter rape							
Victor	3.3	1.0	10.3	13.4	9.1	7.5	51.2
Panter	3.5	0.9	10.4	12.5	9.7	7.5	51.0
Sv 66/609	4.5	0.7	56.7	21.9	10.1	2.6	2.3
Summer rape							
Rigo	3.6	1.1	12.2	14.3	9.3	9.4	45.8
Bronowski	3.9	1.8	37.6	16.4	10.5	16.5	11.8
Sv 64/108	5.2	1.7	53.0	22.0	12.2	1.6	0.5

have not met with the same success as efforts to decrease the erucic acid. Part of this is due to the effect of climate on the level of linoleic and linolenic acids, which is greater than that on erucic acid. Thus selection in a greenhouse is not highly effective for these acids compared to selection in the field[22]. It has, however, been reported that one selection from Liho had 71% oleic, 15% linoleic and 8% linolenic acid, whereas another had 43%, 36% and 11%, respectively. These levels seem to be genetically fixed since they remain similar in the offspring of the original seed.

TABLE 6:8

VARIATION IN THE PRESENT (1969) BREEDING MATERIAL OF RAPESEED IN CANADA AND AT SVALÖF, SWEDEN

Fatty acid	Shorthand designation	Variation in content of fatty acids in the breeding material		Typical values for winter rape
		Canada*	Sweden	
Palmitic	16:0	2.8–10.0	2–10	3
Oleic	18:1	7.8–78.0	7–70	10
Linoleic	18:2	9.3–35.5	10–40	13
Linolenic	18:3	< 4	4–25	9
Eicosenoic	20:1	1–14	0–20	7
Erucic	22:1	0–62	0–60	51

* According to refs. 5 and 24.

According to Canadian experience the range of variation that can be obtained by plant breeding is as follows: palmitic 2.8% to 10.0%; oleic 7.8 to 78.0%; and linoleic 9.3 to 35.5%[24]. These authors also report that it has been possible to reduce the linolenic acid level to approximately 4%. No such lines are, however, close to being released as new varieties. Similar variation has been found in the Swedish breeding material of rapeseed (Table 6:8).

TABLE 6:9

CONTENT OF GLUCOSINOLATES IN TYPICAL RAPE CULTIVARS IN COMPARISON WITH THAT IN BRONOWSKI AND IN LINES DERIVED FROM THE CULTIVARS

Sample	Content of		Total content
	Non-hydroxy-derivatives calculated as gluconapin	Hydroxy-derivatives calculated as progoitrin	
	per cent in dry, defatted seed		
Winter rape			
Victor	1.79	4.31	6.10
Matador	1.71	4.69	6.40
Sv 71-3828-6	0.41	0.85	1.26
Summer rape			
Rigo	1.31	3.72	5.03
Bronowski*	0.06	0.14	0.20
Line selected from Bronowski	0.07	0.04	0.11

* Mean of two samples from isolated multiplications of genetically 'pure' Bronowski. Other samples, outcrossed to other genotypes, had considerably higher contents of glucosinolates[59].

6.5.4 Glucosinolates

In 1967 it was found simultaneously in Canada and Sweden that the Polish summer rape variety, Bronowski, had very low levels of glucosinolates (see Tables 6:9 and 6:10). The agronomic characters of Bronowski are not so good as to suggest its direct use in the production of rapeseed meal with low levels of

TABLE 6:10

GLUCOSINOLATE CONTENT OF STANDARD RAPESEED VARIETIES AND SELECTIONS EXPRESSED AS MILLIGRAM PER GRAM OIL-FREE MEAL OF THE HYDROLYSIS PRODUCTS, 3-BUTENYL AND 4-PENTENYL ISOTHIOCYANATE AND OXAZOLIDINETHIONE (OZT)[24]

Species and variety	Isothiocyanate		Oxazolidinethione
	Butenyl	Pentenyl	
B. napus			
Tanka	2.5	0.6	11.5
Nugget	1.7	0.3	7.8
Bronowski	*0.3*	*t*	*0.3*
B. campestris			
Echo and Arlo	2.0	1.5	1.7
Sel. OZT	2.4	1.6	*0*
Sel. pentenyl	1.1	*0*	1.0
Sel. pentenyl + OZT	2.1	*0*	*0*

toxic substances, but it can be successfully utilized in crossing with other genotypes. It should, however, also be pointed out that the agronomic characters of Bronowski, from which the low glucosinolate gene can be obtained, are considerably better than those of Liho from which the 'zero erucic' gene was obtained. Besides crossing Bronowski with high yielding varieties of both winter and summer rape selection work has also been carried out on this variety. The levels of glucosinolates in selections from Bronowski and present varieties are given in Table 6:9. Selection work on *B. campestris* has yielded types without glucobrassicanapin or without progoitrin, as well as types without either of them[24]. Table 6:10 presents the levels of glucosinolates in typical Canadian varieties and in the breeding lines. It has been found that the level of glucosinolates is mainly controlled by the genotype of the maternal plant, and that the levels of erucic acid and glucosinolates are inherited independently of each other[60]. Thus there are good prospects for the development of genotypes of both rape and turnip rape without erucic acid and without—or with just a trace of—glucosinolates.

6.5.5 Other components

The breeding of varieties with lower contents of seed coat, fibre and pigments has recently started. As mentioned in Section 6.4.4, the possibilities for reaching this goal seem to be quite good, and the yield of yellow-seeded varieties most probably will be as high as in the brown-seeded ones[62]. Decreasing the content of fibre, pigments and seed coat may have disadvantages not yet revealed, but which must be overcome by the plant breeders.

As the composition of the protein of rapeseed seems to be quite good (see refs. 66 and 67 and Chapters 7 and 15), there is no urgent need of immediate improvement of the amino acid pattern. Increasing the protein content at the same time as the oil content is increased may be difficult, as the environmental influence causes the same decrease in oil content as increase in protein content[68]. The environmental influence upon the composition of amino acids is notable, but there are also significant differences between varieties[67].

REFERENCES

1 G. Andersson and G. Olsson, in H. Kappert, W. Rudorf (Eds.) *Handbuch der Pflanzenzüchtung*, Vol. 5, Paul Parey, Berlin/Hamburg, 1961, p. 1.
2 B. Lööf, *Sveriges Utsädesfören. Tidskr.*, 73:1–2 (1963) 25. (In Swedish with English summary)
3 B. Lööf and G. Andersson, in E. Åkerberg, A. Hagberg (Eds.) *Recent Plant Breeding Research, Svalöf 1946–1961*, (1963) 246. Almquist och Wiksell, Stockholm.
4 L.-Å. Appelqvist, in E. Åkerberg, A. Hagberg (Eds.) *Recent Plant Breeding Research, Svalöf 1946–1961*, (1963) 301, Almquist och Wiksell, Stockholm.
5 R. K. Downey, *Can. Food Industries*, 34 (1963) 34.
6 *Sveriges Oljeväxtodlares Centralförening 1943–1968*, 199 pp., Malmö 1968. (In Swedish)
7 W. Tingnell, *Svensk Frötidning*, 37:11 (1968) 155. (In Swedish)
8 Anon., *Bulletin C.E.T.I.O.M.*, 40 (1969) 1, Paris.
9 *Skånes Oljeväxtodlareförening 1943–1968*, 139 pp., Malmö 1968. (In Swedish)
10 R. Jönsson, *Sveriges Utsädesfören. Tidskr.*, 76:1 (1966) 54. (In Swedish with English summary)
11 R. Jönsson, *Sveriges Utsädesfören. Tidskr.*, 79:1–2 (1969) 29. (In Swedish with English summary)
12 R. Jönsson, *Svensk Frötidning*, 38:7–8 (1969) 106 (In Swedish).
12a B. Brunin, paper presented at *Journées Internationales sur la Colza, Paris, May, 1970*, C.E.T.I.-O.M., 174 Avenue Victor-Hugo, Paris.
13 E. Josefsson, *Zeitschrift für Pflanzenzüchtung*, 59 (1968) 384.
14 B. Lööf, *Zeitschrift für Pflanzenzüchtung*, 46:4 (1961) 405.
15 K.-H. Riemann, *Der Züchter*, 33:5 (1963) 217.
16 W. Schuster, *Fette, Seifen, Anstrichmittel*, 69 (1967) 831.
17 J. H. Bruun and J. R. Matchett, *J. Am. Oil Chem. Soc.*, 40 (1963) 1.
18 E. Åkerberg and A. Hagberg (Eds.), *Recent Plant Breeding Research, Svalöf 1946–1961*, (1963) Almquist och Wiksell, Stockholm.
19 C. F. Andrus, *Euphytica*, 12 (1963) 205.
20 R. K. Downey, *Can. J. Plant Sci.*, 44 (1964) 295.
21 R. K. Downey, *Qualitas Plantarum et Materiae Vegetabiles*, 13 (1966) 171.
22 R. K. Downey, *Agricultural Institute Review*, 21 (1966) 16.

23 R. K. Downey and B. M. Craig, *J. Amer. Oil Chem. Soc.*, 41:7 (1964) 475.
24 R. K. Downey, B. M. Craig and C. G. Youngs, *J. Amer. Oil Chem. Soc.*, 46:3 (1969) 121.
25 E. Knapp, *Vorträge für Pflanzenzüchter*, 1 (1958) 82.
26 J. Krzymanski and R. K. Downey, *Can. J. Plant Sci.*, 49 (1969) 313.
27 E. R. Leng, *Zeitschrift für Pflanzenzüchtung*, 47:1 (1962) 67.
28 B. Lööf, *Sveriges Utsädesfören. Tidskr.*, 73:1–2 (1963) 101 (In Swedish with English summary)
29 B. Lööf and L.-Å. Appelqvist, *Zeitschrift für Pflanzenzüchtung*, 52:2 (1964) 113.
30 R. Manner, *Meddelande från Gullåkers Växtförädlingsanstalt, Hammenhög*, 14 (1957) 45.
31 R. Manner, *Meddelande från Gullåkers Växtförädlingsanstalt, Hammenhög*, 14 (1957) 61.
32 R. Manner, *Zeitschrift für Pflanzenzüchtung*, 41 (1959) 395.
33 J. Morice, *Ann. de L'Amélioration des Plantes, I.N.R.A.*, 1 (1960) 85.
34 J. Morice, *Ann. de L'Amélioration des Plantes, I.N.R.A.*, 2 (1960) 155.
35 J. Morice, *Bulletin C.E.T.I.O.M.*, 35 (1967) 1.
36 G. Olsson, *Hereditas*, 46 (1960) 29.
37 G. Olsson and G. Andersson, in E. Åkerberg, A. Hagberg (Eds.) *Recent Plant Breeding Research. Svalöf 1946–1961*, (1963) 64, Almquist och Wiksell, Stockholm.
38 K.-H. Riemann, *Der Züchter*, 34 (1964) 156.
39 M. Rives, *Ann. de L'Amélioration des Plantes, I.N.R.A.*, 1 (1954) 21.
40 B. R. Stefansson, F. W. Hougen and R. K. Downey, *Can. J. Plant Sci.*, 41 (1961) 218.
41 B. R. Stefansson and F. W. Hougen, *Can. J. Plant Sci.*, 44 (1964) 359.
42 J. B. Free and P. M. Nuttall, *J. Agric. Sci. Camb.*, 71 (1968) 91.
43 G. Olsson, *Sveriges Utsädesfören. Tidskr.*, 59:5 (1949) 193. (In Swedish with English summary)
44 G. Olsson, *Sveriges Utsädesfören. Tidskr.*, 62:4 (1952) 311. (In Swedish with English summary)
45 G. Olsson, *Sveriges Utsädesfören. Tidskr.*, 65:6 (1955) 418. (In Swedish with English summary)
46 B. Lööf, *Sveriges Utsädesfören. Tidskr.*, 75:5 (1965) 270. (In Swedish with English summary)
47 G. Olsson, *Hereditas*, 40 (1954) 398.
48 G. Olsson, *Hereditas*, 46 (1960) 351.
49 L.-Å. Appelqvist, *Arkiv Kemi*, 28 (1968) 551.
50 L.-Å. Appelqvist and K.-A. Melin, *Lipids*, 2 (1967) 351.
51 R. K. Downey and B. L. Harvey, *Can. J. Plant Sci.*, 43 (1963) 271.
52 L.-Å. Appelqvist and E. Josefsson, *J. Sci. Food Agric.*, 18 (1967) 510.
53 E. Josefsson, *J. Sci. Food Agric.*, 19 (1968) 192.
54 C. G. Youngs and L. R. Wetter, *J. Amer. Oil Chem. Soc.*, 44 (1967) 551.
55 L.-Å. Appelqvist, *Acta Agric. Scand.*, 18 (1968) 3.
56 L.-Å. Appelqvist, *Physiol. Plant.*, 21 (1968) 455.
57 L.-Å. Appelqvist, *Physiol. Plant.*, 21 (1968) 615.
58 L.-Å. Appelqvist, *Hereditas*, 61 (1969) 9.
59 E. Josefsson and L.-Å. Appelqvist, *J. Sci. Food Agric.*, 19 (1968) 564.
60 E. Josefsson and R. Jönsson, *Zeitschrift für Pflanzenzüchtung*, 62 (1969) 272.
61 Z. P. Kondra, *Diss. Order No. 68-5910*, Univ. of Saskatoon (1967).
62 R. Jönsson and L. Bengtsson, *Sveriges Utsädesfören. Tidskr.*, 80:3 (1970) 149. (In Swedish with English summary)
63 D. G. Dorrell and R. K. Downey, *Can. J. Plant Sci.*, 44 (1964) 499.
64 B. L. Harvey and R. K. Downey, *Can. J. Plant Sci.*, 44 (1964) 104.
65 L.-Å. Appelqvist and R. Jönsson, *Zeitschrift für Pflanzenzüchtung*, 63 (1970) 000.
66 R. W. Miller, C. H. Van Etten, C. McGrew, I. A. Wolff and Q. Jones, *Agric. Food Chem.*, 10:5 (1962) 426.
67 D. R. Clandinin and L. Bayly, *Can. J. Animal Sci.*, 43 (1963) 65.
68 G. Andersson, O. Hall and B. Lööf, *Sveriges Utsädesfören. Tidskr.*, 79:3–4 (1969) 248. (In Swedish with English summary)

CHAPTER 7

Chemical Constituents of Rapeseed

L.-Å. APPELQVIST

Division of Physiological Chemistry, Chemical Center, University of Lund (Sweden)

CONTENTS

7.1 INTRODUCTION

In this chapter the chemical components of seeds from *Brassica napus*, rape, and *Brassica campestris*, turnip rape (both often called rapeseed; see Chapter 3), and some related species are to be discussed. As shown in Chapters 3 and 6, there are differences in the chemical composition among individual seeds in a sample (a population), as well as among the different morphological sub-units of the seed: the seed coat, the hypocotyl (including other embryonic structures) and the two cotyledons. Furthermore, each organ is composed of more than one cell type and each cell contains various organelles with differences in chemical composition, of key interest to the biochemist. However, the predominant amount of data available on the composition of rapeseed has been obtained with bulk samples of whole seeds, and this provides relevant information for the food technologist.

Since there are often marked differences in composition between different cultivars or lines of a species, it is noted with considerable regret that chemists often report extensive analyses of seed samples of unknown biological status. In some instances the composition of the seed coat (or hull) and that of the major part of the seed (the embryo) are reported separately. Such data will be discussed, since they are of interest both from the industrial and biological points of view. Compositional data for seeds are also presented in Chapters 5 and 6 whereas data for the three products of industrial processing, rapeseed oil, lecithin and meal are given mainly in Chapters 10, 13 and 15.

7.2 GROSS COMPOSITION

The moisture content of 'naturally dry' rapeseed, *i.e.* when the water in the seed is in equilibrium with the atmospheric moisture, generally ranges from 6 to 8 per cent (*cf*. Fig.5:11). This rather low natural moisture content is a consequence of the high oil content of rapeseed compared with, for example, soyabeans. Since there is a rather rapid exchange of water between the surrounding atmosphere and the seed when stored in open small bags in the laboratory it may be assumed that compositional data reported on an 'as is' basis refer to seeds of *ca*. 7 per cent moisture.

Major components in dry rapeseed are lipids (oil), proteins, carbohydrates and glucosinolates. The oil content of seeds of *Brassica napus* and *Brassica campestris*, as determined by extraction with a hydrocarbon solvent, varies over a wide range. Typical figures for winter rape are 42–50 per cent (see Fig.5:16), for summer rape 37–47 per cent (ref. 1 and Fig.5:16), for winter turnip rape 40–48 per cent[2], and for summer turnip rape 36–46 per cent in dry matter[1, 2]. The variability in oil content among samples of Canadian and Swedish crops is discussed in Chapter 5. The influence of genotype and environment on the oil content of rapeseed was

TABLE 7:1

SPECIES AND CULTIVAR DIFFERENCES IN PROXIMATE COMPOSITION OF SEED MEALS OF RAPE AND TURNIP RAPE[8]

Species and cultivar	Proximate composition of the meal (oil-free, dry basis) in per cent			
	Crude protein	Crude fiber	Ash	Nitrogen-free extract
Brassica campestris				
Polish	40.8	16.7	8.4	34.1
Echo	39.8	15.6	8.1	36.5
Zero erucic	44.4	12.6	7.0	36.0
Brassica napus				
Argentine	45.5	14.0	8.0	32.5
Target	43.5	14.6	7.8	34.1
Oro	42.4	13.6	7.3	36.7

TABLE 7:2

AMOUNT OF HULL AND COMPOSITION OF HULL AND EMBRYOS ('MEATS') FROM SEEDS OF RAPE AND TURNIP RAPE[9]

Constituent	*Brassica napus* cv. Tanka	*Brassica campestris* cv. Arlo
Hull (per cent)*	16.5	18.7
Oil*		
Seed	41.5	40.0
Hull	16.0	16.2
Meats	47.1	45.0
Protein**		
Seed	44.7	44.2
Hull	18.7	20.6
Meats	53.6	53.4
Crude Fibre**		
Seed	11.8	11.7
Hull	34.3	31.6
Meats	3.0	3.6

* Moisture-free basis.
** Moisture-free, oil-free basis.

recently treated in detail[3], and thus no such discussion will be repeated here.

The residue from hydrocarbon extraction of rapeseed, called rapeseed meal, contains about 40 per cent protein. There are wide range variations in protein content. Among many samples of a certain cultivar a negative correlation of oil and protein content is recorded[4]. Since the oil has always been considered the most valuable part of the seed, much more data are available on variation in oil content than in protein content. Canadian and Swedish studies have demonstrated considerable variation in protein content between species, cultivars and samples of a certain cultivar[5-7]. Figures for protein content of seeds and seed parts are shown in Tables 7:1 and 7:2. It should be remarked that 'nitrogen' $\times$ 6.25 does not yield a true figure for protein content. Recent investigations suggest that a factor of 5.53 is a more correct figure for rapeseed[10]. By application of the latter factor the true protein level would be *ca.* 12 per cent lower than figures generally reported. This agrees with other reports in the literature (see ref. 11).

From surveys of the Canadian rapeseed crops a total span in protein content of defatted meal from 33.0 to 47.9 per cent involving specific, varietal and environmental influences has been recorded for the year 1969 (see Table 5:7). Carbohydrates and miscellaneous organic compounds are in proximate analyses named 'nitrogen-free extract', an expression that is typical in food tables (see Chapter 15). The figure is arrived at by difference between 100 per cent and the percentage of other analysed compounds; moisture, fat, protein, fibre and ash. As seen from Table 7:1, this class of compounds amounts to almost as large proportions as the proteins. No marked differences occurred among the species or cultivars investigated[8].

The content of 'crude fibre', mainly cellulose and hemicellulose, amounts to *ca.* 15 per cent in the defatted seed meals (Tables 7:1 and 7:2). This figure is high compared with that for many oil crops, *e.g.* soyabeans. The large amount of crude fibre in rapeseed meal is a consequence of the small seed size with large proportions of seed coat, rich in fibre (Table 7:2). No marked difference in content of crude fibre is noted for the eight Canadian samples shown in Tables 7:1 and 7:2. Variations in fibre content in seed populations are discussed in Chapter 6 (see Tables 6:9 and 6:10). The content of minerals, called 'ash', is typically 7–8 per cent of the meal and the variation among samples recorded in Table 7:1 appears to be rather small.

As seen in Table 7:2, there are large differences in content of oil, protein and crude fibre between the seed coats and the 'meat' fraction (cotyledons + hypocotyl). No marked differences in hull composition between a typical cultivar of summer rape and another of summer turnip rape are observed[9]. A considerable variation in hull content (as percentage of total seed), has been recorded which is of interest since the lowest possible hull content would be advantageous in the production of higher quality rapeseed meal and especially rapeseed protein concentrates.

Widely different definitions have been used for the term 'lipids'. One is to name all compounds as lipids, which are soluble in certain organic solvents such as diethyl ether and which are insoluble in water. Another is to define lipids as derivatives of fatty acids. This discrepancy is discussed in many texts (see, for example, ref. 12). The classification used in this chapter is a partial merger of the two viewpoints by using the terms 'acyl lipids', 'other lipids' and 'lipid-soluble substances'. By the name 'other lipids' compounds entirely made of long-chain derivatives of poly-acetate or poly-isoprenoid types are distinguished from compounds in which a major part of the molecule has a hydrophilic structure, and which are denoted 'lipid-soluble substances' *e.g.* chlorophyll with a porphyrin nucleus and a polyisoprenoid side chain. Like any other classification of lipids, this one suffers from some shortcomings, but, hopefully, also several advantages.

At this point it should also be emphasized that most of the lipid components, except the surface waxes and the storage lipids (the triacylglycerols) occur *in vivo* more or less firmly bound to proteins[12]. Therefore, in most cases, the isolation of lipids from a tissue starts with the disruption of lipoprotein bonds.

7.3.1 Acyl lipids

Compositional information on any seed oil is generally given as content and composition of total fatty acids and content of so-called nonsaponifiables (see also 8.4). However, since both the fatty acids and the nonsaponifiables are the result of preliminary treatment of the natural constituents occurring in the seed tissues (saponification of the total lipids), such information should be successively replaced by data on the true constituents. In this chapter an attempt will be made to treat the rapeseed lipids in a more natural context. This undoubtedly leads to some confusion since, for example, the fatty acid or sterol composition of each sterol derivative might not have been studied, but only the proportion of sterol derivatives—such as sterol esters, sterol glucosides (SG), esterified sterol gluco-sides (ESG)—and the qualitative composition of the total sterol fraction.

(a) Triacylglycerols
These components, which until recently were named triglycerides, generally account for 95–98 per cent of the total lipids of mature *Brassica* seeds. Therefore the large amount of data presented in the literature on the total fatty acid pattern of *Brassica* seeds, discussed under (g) actually gives a good picture of the triacyl-glycerol fatty acid composition. Under this heading some of the papers that actually report on the structure of the isolated triacylglycerols of such seeds will be discussed.

Although the fatty acids are not randomly arranged but strongly associated

TABLE 7.3

CONTENT OF FATTY ACIDS IN TRIACYLGLYCEROLS AND IN MONOACYLGLYCEROLS FROM SOME CRUCIFEROUS SEEDS, AS WELL AS PROPORTION OF EACH FATTY ACID IN THE 2-POSITION[20]

Species and cultivar		Fatty acid composition (mole %)									22:0+	
		16:0	16:1	18:0	18:1	18:2	18:3	20:0	20:1	20:2	22:1	22:2
Brassica campestris, Yellow Sarson	Triacylglycerol	2.1	0.5	1.1	12.6	14.8	8.9	1.2	4.1	0.8	53.6	0.4
	2-position	2.2	0.8	1.0	31.6	38.5	22.1		1.5		2.3	
	proportion[a]	35	53	30	84	87	83		12		1	
Brassica napus, Victor	Triacylglycerol	3.3	0.3	0.7	11.4	15.4	10.7	0.7	8.1	0.5	48.1	0.9
	2-position	2.0	0.6	0.9	28.0	40.6	25.1		0.6		2.1	
	proportion[a]	20	67	43	82	88	78		3		2	
Brassica napus, Regina	Triacylglycerol	4.0	0.4	1.1	13.9	17.5	10.2	1.5	10.7	0.7	39.6	0.4
	2-position	0.6	0.5	0.1	27.9	43.0	25.8		0.9		1.2	
	proportion[a]	5	42	3	67	82	84		3		1	
Brassica napus, zero erucic acid line	Triacylglycerol	5.0	0.3	0.9	52.7	24.9	12.8	1.0	2.1		0.3	
	2-position	1.4	0.4	0.2	40.6	37.6	19.7		0.2			
	proportion[a]	9	44	7	26	49	51		32			
Sinapis alba, Seco	Triacylglycerol	2.4	0.3	0.8	26.3	10.4	12.1	1.3	9.1	0.6	35.6	1.1
	2-position	1.2	0.6	0.1	45.7	21.2	26.7		1.3		3.1	
	proportion[a]	17	67	4	58	68	74		5		5	

[a] (2-position/triacylglycerol × 3) × 100 = Proportion, i.e. percentage of fatty acid type esterified at the 2-position.

with the 1- or 2- or 3-carbon of the triacylglycerol[13], because of the large number of fatty acids that occur in *Brassica* oils (15 fatty acids in amounts of 0.5 per cent or more in typical cultivars[14]) it is easily understood that no sample of such oils has been fully characterized with regard to its triacylglycerol composition.

The methods commonly used for the analysis of triacylglycerols are discussed in some detail in Chapter 8.

Triacylglycerols can be separated by gas chromatography into classes with equal numbers of carbon atoms, by argentation thin layer or column chromatography into classes with equal numbers of double bonds and by reversed phase thin layer or column chromatography into other classes based on a combination of the number of carbon atoms and double bonds. Utilizing a combination of these techniques, it is theoretically possible to fractionate the complex mixture of triacylglycerols of *Brassica* seeds into groups of so-called 'molecular species', one group containing, for example, only palmitic, oleic and erucic acid. The positioning of these fatty acids could then be disclosed by application of specific enzymes, followed by further reactions of the lipolysis products[15, 16]. It has, however, been reported that great losses occur during the many analytical steps involved and since the losses are, to some extent, selective (see Chapter 8) no such experiments have, to our knowledge, been successfully carried out. Very useful information on the triacylglycerol behaviour is also obtained by X-ray crystallographic and DSC (differential scanning calorimetry) measurements of the natural triacylglycerol mixtures, as discussed in Chapter 11.

Rather much interesting information has, however, been accumulated on the triacylglycerol structure by application of the methods for positional analysis (pancreatic lipase techniques) to unfractionated triacylglycerols of various brassicas and *Sinapis alba*[17–20] and by GLC analyses of such triacylglycerols. An example of the results of such lipase studies is presented in Table 7:3. From these and other studies that have reported data of well-defined seed materials, the following patterns are disclosed.

(1) All saturated and very long-chain (20–24 carbon atoms) mono-unsaturated fatty acids are exclusively or almost exclusively esterified at the 1- and 3-positions, whereas the unsaturated C18-acids are preferentially located at the 2-position[13, 17–20].

(2) Also the di-unsaturated C20- and C22-acids appear at the 1- and 3-positions[20].

(3) There appears to be no relationship between the erucic acid content of the triacylglycerols and that recorded for the 2-position[17, 19, 20, 22]. This is further discussed below.

(4) There is a slight difference in strictness in positioning of the various fatty acids of rape and white mustard seeds, the latter having a slightly more random pattern[20, 22]. Although small, this difference is of technological importance, since the solidification range of these oils is close to 0 °C as discussed in Chapter 11.

(5) In 'zero-erucic acid' oil, there is a preference for linoleic and linolenic acids at

STEREOSPECIFIC ANALYSIS OF A SAMPLE OF RAPESEED OIL[13]

| Position | Fatty acid composition (mole %) | | | | | | | | |
	16:0	16:1	18:0	18:1	18:2	18:3	20:1	22:0	22:1
1	4.1	0.3	2.2	23.1	11.1	6.4	16.4	1.4	34.9
2	0.6	0.2		37.3	36.1	20.3	2.0		3.6
3	4.3	0.3	3.0	16.6	4.0	2.6	17.3	1.2	51.0

the 2-position[20], which might be of importance for the technological value of such oils in comparison with soybean oil, which has a random distribution of linolenic acid (see, for example, ref. 13). The relation between triacylglycerol structure and oxidative stability is further treated in Chapter 11.

Whereas these results were arrived at by application of the classical pancreatic lipase method, yielding information about the fatty acids in position 2 and those in positions 1 and 3 as a group, there is information available on one sample of rapeseed oil treated with a newer method that differentiates between positions 1 and 3[13]. As shown in Table 7:4, the fatty acids of the 3-position are indeed different from those in the 1-position. The large amounts of erucic acid in the 3-position should be noted. It has recently been suggested that more unusual fatty acids, e.g. erucic acid in Cruciferae, preferentially occupy the 3-position[23].

The inequality of the 1- and 3-positions is also mirrored in the GLC data on the intact triacylglycerols of known Swedish cultivars shown in Table 7:5[22]. Thus the low-erucic sample of Bele summer turnip rape (ca. 30 per cent C22:1) had only

TABLE 7:5

TRIACYLGLYCEROL PATTERNS BY CARBON NUMBER OF SEED OILS FROM RAPE, TURNIP RAPE AND WHITE MUSTARD[24]

| Species and cultivar | Triacylglycerol pattern (wt. %) Carbon number | | | | | | | | |
	50	52	54	56	58	60	62	64	66
Brassica napus cv. Victor	tr	0.6	1.7	6.0	7.8	17.6	60.8	5.3	tr
B. napus cv. Regina	0.2	1.2	2.8	8.4	15.4	20.9	47.1	4.0	tr
B. campestris cv. Duro	tr	1.2	2.1	3.3	17.5	22.8	50.0	3.0	tr
B. campestris cv. Bele	tr	2.3	7.8	13.5	38.4	20.1	16.2	1.7	
Sinapis alba cv. Seco	0.3	1.8	3.3	8.0	23.3	18.5	39.4	5.4	tr

16 per cent of C62 (predominantly mono-C18, di-C22), whereas the high erucic acid sample of Victor winter rape (53 per cent C22:1) had 61 per cent of C62. The presence of only a trace of C66 (tri-C22 or mono-C20, mono C22, mono-C24 or mono-C18, di-C24) is highly significant and probably means that the small and variable amounts of erucic acid reported to occur in the 2-position[17–20] are the result of imperfections in the pancreatic lipase technique, as recently discussed[20], although it can not be excluded that 22:1 in the 2-position is associated with C18 in the 1- and 3-positions. Other GLC analyses of rapeseed oil have given similar results to those discussed above[21]. The presence of no more than a trace of C66 demonstrates the limited value of the calculations of triacylglycerol composition that are rather frequently published, calculations according to which 2–3 per cent of trierucylglycerol should be present in rapeseed oil[18, 19]. It has been pointed out that the triacylglycerols of most seed oils comprise a mixture of oils from different morphological subunits[25], and such heterogeneity is a further objection to the application of distribution rules. Since it is known that the fatty acid patterns of the cotyledons, the hypocotyl and the seed coat of rapeseed are different (see Table 7:11), this objection certainly applies to the triacylglycerols extracted from entire seeds of rape and turnip rape.

The various 'distribution rules' that have been put forward and their utilization in approximate technological calculations are discussed in detail in Chapter 8, which also reviews some older literature in the triacylglycerol field. The absence of erucic acid from the 2-position is obviously under strict genetic control, since no single plants of *Brassica napus, B. campestris, Crambe abyssinica* and *Sinapis alba* with more than 67 mole per cent of 'outer-position' acids were found in recent Swedish studies, although considerable efforts were made to increase the erucic acid content of such seeds by plant breeding[26].

(b) Mono- and diacylglycerols

The amount of these compounds in high-quality seeds is generally very low. About 2 per cent of the total lipids of summer rape, cv. Golden and a 'zero-erucic' acid genotype, have been reported as 'partial glycerides'[27]. The fatty acid pattern of the partial glycerides of 'zero-erucic' acid rape was rather similar to that of the triacylglycerols of the same seeds, whereas the former from cv. Golden had large amounts of saturated fatty acids. It may be remarked that partial glycerides can be natural metabolites of the seeds or breakdown products from phospholipids and triacylglycerols generated during the seed extraction.

(c) Phospho-, galacto- and sulpholipids

The total amount of phospholipids and other 'polar' lipids is rather low in mature seeds, when related to the total lipid content, since their absolute quantities calculated per seed are constant or slightly decrease, while the content of triacylglycerols increases manifold during the seed maturation[27, 28].

About 0.5–1 per cent of the total lipid extract of mature seeds is a mixture of polar lipids. In industrial processing, a major part of the phospholipids is extracted from the 'cooked' seeds by hexane (since the 'cooking' disrupts the lipoprotein bonds) and precipitated by the addition of water to the oil solution (see Chapter 9).

The dried product is called 'rapeseed lecithin', the composition of which is treated in Chapter 13. Compared with the rather detailed knowledge available on the chemical composition of rapeseed lecithin, there are rather few and incomplete data reported on the phospho-, sulpho- and galactolipids of well-defined rapeseed samples.

The following compounds have been identified from extracts of mature rapeseed[25]: phosphatidyl choline (PC), phosphatidyl ethanolamine (PE) phosphatidyl inositol (PI), monogalactosyl diglyceride (MGDG), digalactosyl diglyceride (DGDG), sterol glucoside (SG) and esterified sterol glucoside (ESG).

The fatty acid spectra of these complex acyl lipids are, as a group, characterized by large amounts of palmitic and linoleic acids and low amounts of eicosenoic and erucic acids even in seeds where the triacylglycerols contain large amounts of these acids[14, 27, 29]. Whereas the fatty acid spectra of the individual phospholipids of rapeseed lecithin have been determined (see Table 13:3), similar data for mature seeds are rare. Some unpublished Swedish analyses[28] of different lipid classes of

TABLE 7:6

THE FATTY ACID COMPOSITION OF VARIOUS LIPID CLASSES ISOLATED FROM SLIGHTLY IMMATURE SEEDS OF *Brassica napus*[28]

Genotype	Lipid class	Content of major fatty acids (per cent of total)						
		16:0	18:0	18:1	18:2	18:3	20:1	22:1
Gylle	P-lipids and SQDG	21.7	1.1	23.1	38.0	9.4	1.1	2.6
	MGDG	12.7	8.9	14.5	13.7	14.0	6.0	18.9
	DGDG	12.8	0.7	27.1	40.1	11.7	1.8	2.0
	Neutral lipids, mainly triacylglycerols	4.3	0.9	18.1	16.6	9.8	11.9	36.9
Zero-erucic, line 802	P-lipids and SQDG	18.3	0.6	22.3	47.9	7.4		
	MGDG	25.4	8.7	42.2	11.2	3.0	2.1	
	DGDG	19.6	6.2	21.9	28.5	13.1		
	Neutral lipids, mainly triacylglycerols	6.3	0.9	40.2	37.4	12.1	1.6	

SQDG = sulphoquinovosyl diglyceride, MGDG = monogalactosyl diglyceride, DGDG = digalactosyl diglyceride.

slightly immature rapeseed, cv. Gylle and a zero-erucic acid line, are shown in Table 7:6. The phospholipids, which constitute the major portion of the polar lipids, have essentially similar fatty acid spectra in the two genotypes regardless of the great differences in triacylglycerol fatty acid spectra. It is noteworthy that the fatty acid spectra of phospholipids are, in general, rather similar regardless of species and plant family[30]. The fatty acids of the monogalactosyl diglyceride appear to be significantly different from those of the digalactosyl derivative, the latter being rather similar to the phospholipids. (This topic is treated in detail elsewhere[29].)

(d) Sterol esters

The relative proportions of free sterol and various sterol derivatives (sterol ester, sterol glucoside and esterified sterol glucoside) have apparently not been exactly determined as yet in *Brassica* seeds. A sample of Canadian summer rape cv. Golden was reported to contain free sterols and sterol esters in the proportions 1 to 7. Nothing was mentioned about the quantities of SG and ESG[27], although their presence was established by thin layer chromatography. On the other hand, rapeseed of Canadian and Japanese origin was reported to contain predominantly free sterol (70 and 67 per cent) and the remainder essentially as sterol glucoside (24 and 28 per cent)[31]. The reason for this discrepancy is not known. The figure given for the content of sterol esters as per cent of total lipids (3.5 per cent) appears very high, when compared with other data in the same paper, and about 1 per cent seems to be a more realistic figure[27].

The sterol esters of summer rape cv. Golden were characterized by rather high levels of palmitic, stearic and linoleic acids (about 20 per cent each) and rather low contents of oleic, eicosenoic and erucic acids, compared with the triacylglycerols of the same seeds. As expected, the sterol esters of 'zero-erucic' rapeseed contained no eicosenoic and no erucic acid. Instead they were rich in oleic and linoleic acids[27]. The qualitative and quantitative compositions of the total sterol fraction are presented under 7.3.2(c).

(e) Waxes

Nothing seems to be reported on the true waxes of rapeseed, *i.e.* the esters of long-chain fatty acids with long-chain aliphatic alcohols. The waxes of the leaves of a related species, *B. oleracea*, have, however, been studied in detail[32].

(f) Free fatty acids

The content of free fatty acids is low, *ca.* 0.3 per cent in the oil of fully mature rapeseed, when adequately handled after harvest[33]. Improper handling causes elevated levels of free fatty acids, as discussed in detail in Chapter 5.

The quantitative composition of the free fatty acids is not exactly the same as that of the triacylglycerols and depends on the type of damage that caused the rise in free fatty acid level[34]. This also means that the residual triacylglycerols are not representative of the total triacylglycerols, and thus only seeds with very low levels of free fatty acids should be used in studies on the triacylglycerol structure (*cf.* ref. 20). Generally, the free fatty acids are characterized by more palmitic and less erucic acid than the triacylglycerols of the same seeds[34].

(g) Total fatty acid composition
Qualitative composition. As mentioned above, the data often used to characterize seed oils are their fatty acid compositions. The qualitative compositions of seeds of rape, *(Brassica napus)* turnip rape *(B. campestris)*, brown (yellow) mustard *(B. juncea)* and white (yellow) mustard *(Sinapis alba)* are similar and complex compared with those of major vegetable oils, such as soyabean oil, cottonseed oil, groundnut oil and sunflowerseed oil. Fifteen fatty acids are present in levels from 0.5 per cent and higher. They range in chain length from 16 to 24 carbons, and in number of double bonds from zero to three. The long-chain monoenes, eicosenoic and erucic acids, are considered typical fatty acids for the family Cruciferae, since they are present in considerable amounts in about three-fourths of all crucifers studied[30, 35–38], and outside this family they have to date been found in large amounts only in one plant species, *Tropaeolum majus*[30].

By plant breeding it has been possible to develop cultivars of *Brassica napus* and *B. campestris* with very low levels of eicosenoic acid, and zero or small amounts of erucic acid, as discussed in Chapter 6 (see Tables 6:2 and 6:4). Therefore, it is likely that within a few years, commercial crops and oils of rape and turnip rape from several countries will no longer have the characteristic fatty acid patterns of classical crucifers. In this connection it may be of interest that the sterol patterns of the new cultivars appear to be very similar to those of the classical cultivars. Thus the patterns of sterols, and possibly also triterpenoid alcohols, could be used as 'specific tags' for *Brassica* oils, regardless of their erucic acid content.

At least three of the monoenes of *Brassica* oils (18:1, 20:1 and 22:1) occur in two isomers with the only difference that the double bond is located 7 or 9 carbon atoms from the methyl end[14, 41, 42]. These isomers should, according to recent nomenclature rules, be named (n–7)- and (n–9)-fatty acids instead of the earlier designation ω–7 and ω–9 fatty acids. The (n–9)-isomers predominate but the isomer proportions vary with species and chain length[14] as further discussed below. Typical levels for the uncommon isomers are 0.5–2 per cent of the total fatty acids. If fatty acids that are normally present in levels below 0.5 per cent are also included, the total number of fatty acids of *Brassica* oils probably amounts to about 50, since a total of 49 fatty acids has been tentatively identified in extracts of *Sinapis alba* (white mustard)[43]. The latter report did not recognize the presence of isomers of the 18:1, 20:1 and 22:1 fatty acids, and thus the number could be 52, but on the other hand, some of the trace components reported might have been extracted from 'contaminants' rather than from the seed material itself[44]. Among

134

TABLE 7:7

TYPICAL RANGES OF VARIATION IN CONTENT OF COMMON FATTY ACIDS IN THE OILS OF SOME CULTIVARS OR BREEDING LINES OF SOME CRUCIFEROUS SEEDS

Species and type	Ranges in percentage content of						Ref.
	Palmitic	Oleic	Linoleic	Linolenic	Eicosenoic	Erucic*	
Brassica campestris							
Winter turnip rape	2–3	14–16	13–17	8–12	8–10	42–46	36, 49, 52, 53, 55
Summer turnip rape, classical cultivars	2–3	17–34	14–18	9–11	10–12	24–40	49, 54, 56
Summer turnip rape, low-erucic acid lines**	4–7	48–55	27–31	10–14	0–1	0	14, 56
Sarson and Toria	2–3	9–16	11–16	6–9	3–8	46–61	53, 56
Brassica juncea	2–4	7–22	12–24	10–15	6–14	18–49	36, 53
Brassica napus							
Winter rape, classical cultivars	3–4	8–14	11–15	6–11	6–10	45–54	36, 49, 53, 55, 57, 60
Winter rape, low-erucic acid cultivars or lines**	4–5	40–48	15–25	10–15	3–19	3–11	34, 60
Summer rape, classical cultivars†	3–4	12–23	12–16	5–10	9–14	41–47	51, 54, 56
Summer rape, low-erucic acid cultivars or lines**	5	52–55	24–31	10–13	0–2	0–1	34, 56
Brassica Tournefortii	2–4	6–12	11–16	10–16	6–8	46–52	37, 38
Sinapis alba	2–3	16–28	7–10	9–12	6–11	33–51	17, 19, 20, 37, 43, 49, 52, 53, 59

* Refs. 35 and 56 contain data for variation in erucic acid content among many breeding samples. These data are not included in this table.
** Results from only a few samples available. The range of variation in fatty acid compositions is expected to increase as new cultivars are being released.
† Except for the Polish cultivar Bronowski, which is not grown to any significant extent. This cultivar has *ca.* 10% erucic acid (see Table 6:4).

minor components safely established, the following generally occur in levels from 0.1 to 0.5 per cent: 14:0, 16:1, 16:3, 20:3, 22:3 and 24:0.

Whereas the structures of some of the major fatty acids of rapeseed were elucidated many years ago[45], those of the uncommon monoene isomers were established recently[41]. The only di- and triunsaturated acids that occur in considerable amounts in rapeseed are linoleic and linolenic acid whose structures are well-known. It has long been assumed that the positions of the double bonds in the di- and triunsaturated C20- and C22-fatty acids found in minor or trace amounts in rapeseed were the same as in linoleic and linolenic acid, calculated from the methyl end. Quite recently it was conclusively demonstrated that this is the case for the 20:2, 20:3, 22:2 and 22:3 acids[46]. All double bonds in unsaturated fatty acids of *Brassica* oils appear to have the *cis* configuration.

The structure of 16:3, which was first isolated from rapeseed leaves[47], was later found to be 7, 11, 13-hexadecatrienoic acid[48]. This acid is present in large proportions in the monogalactosyl diglycerides of rapeseed leaves and occurs as a trace component in the seed triacylglycerols[29]. It is reasonable to assume that the trace of 16:3 which is found in the triacylglycerols of *Brassica napus* seeds has the same structure.

Variations in quantitative composition. A large number of papers on the fatty acid composition of *Brassicas* and related species can be found in the literature. Since a considerable portion of the older, and also some more recent fatty acid data were recently critically reviewed[49], another critical review of such data will not be repeated here. Furthermore, a tabulation of rapeseed fatty acid data reported before 1964 has been published[50]. The criteria used in this chapter to evaluate the literature data will be briefly mentioned followed by a discussion of typical fatty acid patterns for the species involved and established ranges of variation. Since the various methods used for the determinations of fatty acid composition and their limitations with respect to accuracy and precision are discussed in Chapter 8, only a few key points will be mentioned here.

(1) The classical, fractional distillation *in vacuo* of fatty acid methyl esters seems to have yielded quite reliable data[45]. The figures arrived at by using such techniques have been reviewed earlier[30, 49]. Since the figures are not outside the ranges reported for more recent GLC analyses, they are not referred to in Table 7:7.

(2) Thin layer or paper chromatography is considerably less precise than gas chromatography and, furthermore, only the latter can adequately resolve the complex fatty acid mixture of *Brassica* oils into the major and several of the minor components. For this reason compositional data reported from paper or thin layer chromatographic analyses will not be discussed.

(3) When adequately packed columns are used, gas chromatography can resolve all of the fatty acids that occur in *Brassica* seed oils at levels from about 0.5 per cent and higher, except for the positional isomer pairs 9- and 11-octadecenoic acids, 11- and 13-eicosenoic acids, and 13- and 15-docosenoic acids. It is regrettable that

Brassica seed oils are sometimes analyzed on too polar polyester columns, on which linolenic and eicosenoic acids overlap (see Fig.8:1). As a consequence, too high figures for linolenic acid (15–20 per cent) and no eicosenoic acid are sometimes reported. An uncritical compilation of even all modern gas chromatographic data would thus be misleading.

A final criterion for inclusion in Table 7:7 has been that the original paper should refer to laboratory-extracted seeds and not to commercial oils, since the latter are often mixtures of *Brassica napus* and *B. campestris* oils. Although several papers have not reported any 24:0 or 24:1, which undoubtedly are present to a total of *ca.* 1.5 per cent in most *Brassica* oils[49], the errors involved by taking the rest as 100 per cent is so small that such data are considered representative and are included in the table.

As seen from Table 7:7, the total range of variation in fatty acid composition among a large number of samples of *Brassica napus*, winter type, is rather small. This crop type is characterized by large amounts of erucic acid, typically 48–53 per cent. The range of variation in erucic acid content of *B. napus*, summer type, is much larger, from 10 per cent in the Bronowski cultivar to *ca.* 45 per cent in common Canadian and European cultivars. *B. campestris*, winter type, is, on the average, slightly lower in erucic acid content than winter rape, but otherwise, the two crops have similar fatty acid patterns. Rather few figures are available on winter turnip rape, and the range of variation observed is small. The summer annuals of *B. campestris*, including the types common in India and Pakistan, namely var. *trilocularis* and *dichotoma* (see Table 3:1), display a very large range of variation in content of erucic (30–55 per cent) and oleic acids (10–27 per cent). The summer turnip rape, common in Canada but a very minor crop in Europe, is characterized by comparatively low erucic acid content (*ca.* 30 per cent). The Indian 'Sarson' (var. *trilocularis*) and 'Toria' (var. *dichotoma*) have, on the other hand, typically 50–55 per cent erucic acid. Such genotypes of *Brassica juncea*, called brown or yellow mustard, which are grown for oil production in India (see Chapter 2), have rather high erucic acid levels, *ca.* 50 per cent. There are, however, other genotypes of this species which have only *ca.* 25 per cent erucic acid in the seed oil. This observation was recently discussed in a paper giving detailed data for many cultivars obtained from India, Japan and Europe[53]. *Brassica Tournefortii*, which according to the rules of *Codex Alimentarius* may be included in the term 'rapeseed', is characterized by high levels of erucic acid, *ca.* 50 per cent.

The fatty acid pattern of white mustard, *Sinapis alba*, is included in Table 7:7 because plants of this species are also grown for oil production. The erucic acid content of common cultivars is about 40 per cent with a reported range of 33–44 per cent.

Whereas the total range of variation in oleic and erucic acid contents among all the species discussed is of considerable magnitude, that in linoleic and linolenic acid contents is generally rather small. Since both comparatively recent original papers

TABLE 7:8

THE INFLUENCE OF ENVIRONMENTAL CONDITIONS ON THE FATTY ACID COMPOSITION OF SEEDS OF *Brassica napus* AND *B. campestris*

Sample	Locality	Fatty acid percentages						Ref.
		16:0	18:1	18:2	18:3	20:1	22:1	
Brassica napus Winter type, cv. Victor	Bologna, Italy	3.3	13.1	13.2	7.2	10.1	47.9	52a
	Svalöv, Sweden	3.1	10.1	13.6	9.7	7.4	51.1	
Winter type, cv. Valois	Versailles, France	3.2	11.0	12.6	9.3	7.4	52.0	53
	Svalöv, Sweden	3.0	8.7	12.2	10.7	6.6	53.7	
B. napus Summer type, cv. Czyzowskich	Czechoslovakia	3.0	14.4	13.6	8.7	9.6	46.0	53
	Svalöv, Sweden	3.1	14.2	13.6	8.7	10.3	45.5	
B. campestris Winter type, cv. Lembkes	Izmir, Turkey	2.7	17.1	13.8	8.8	10.4	42.0	52
	Svalöv, Sweden	2.3	13.1	13.1	10.5	9.4	46.1	
B. campestris Summer type, cv. Bele	Izmir, autumn-sown	2.6	28.7	15.2	7.8	13.1	27.8	52
	Izmir, spring-sown	3.0	29.9	17.0	7.3	11.6	26.7	
	Svalöv, Sweden	2.4	24.5	16.8	10.1	11.4	30.0	

TABLE 7:9

FATTY ACID COMPOSITION OF SEEDS FROM SOME CULTIVARS AND BREEDING LINES OF *Brassica napus*, *B. campestris*, *B. juncea* AND *Sinapis alba*

Sample	Fatty acid composition (%)														Isomer proportions (n-9)/(n-7)			Ref.
	16:0	16:1	18:0	18:1	18:2	18:3	20:0	20:1	20:2	22:0	22:1	22:2	24:0	24:1	18:1	20:1	22:1	
Brassica napus, cv. Victor	3.0	0.3	0.8	9.9	13.5	9.8	0.6	6.3	0.5	0.7	52.3	1.1	0.2	1.0	24/1	4/1	50/1	14
B. napus, cv. Regina II	3.9	0.4	1.2	12.6	15.6	7.6	0.7	10.1	1.3	0.7	43.6	1.0	0.2	1.1	13/1	5/1	50/1	14
B. napus, Zero-erucic acid line	4.9	0.5	2.0	47.9	25.2	15.2	1.2	1.9		1.1					7/1			14
B. campestris, cv. Duro	2.0	0.2	1.0	12.9	13.4	9.1	0.7	8.9	0.7	0.2	49.0	0.8	tr	1.1	12/1	6/1	35/1	14
B. campestris, cv. Bele	2.8	0.4	1.3	26.9	16.6	9.4	1.0	11.8	0.8	0.4	26.3	0.4	0.3	1.6	18/1	14/1	45/1	14
B. campestris, Zero-erucic acid line	7.2	1.2	1.5	48.2	26.6	13.5	0.6	1.2		tr	tr				8/1			14
B. juncea, mean of 5 Indian samples	2.5	0.3	1.2	8.0	16.4	11.4	1.2	6.4	0.9	1.2	46.2	1.6	0.7	1.9	Not determined			53
Sinapis alba, cv. Seco, mean of 6 samples	2.7	0.3	0.9	22.8	8.6	10.5	0.6	8.4	0.3	0.5	41.1	0.4	0.2	2.7	Not determined			49

and textbooks (see ref. 49) indicate much greater variation in linolenic acid content, it is most likely that their results were based on inadequate analytical methods. It has often been assumed that climatic effects on the fatty acid composition, especially the linolenic acid content of rapeseed oil, are considerably greater than shown in Table 7:7. Hence some data from fatty acid studies on seeds of known cultivars produced in places with great differences in climate are shown in Table 7:8. From the data shown in this table it is obvious that seeds produced at more southern latitudes are generally lower in linolenic and erucic and higher in oleic acids. It seems important, however, to emphasize that the effect of any extreme field conditions is rather small. On the other hand, plantings in growth chambers have yielded seeds with dramatically different fatty acid spectra compared with outdoor plantings, *e.g.* only 2–4 per cent linolenic acid[61, 62].

The effect of environmental factors, such as supply of nutrients, day temperature, night temperature and photoperiod on the fatty acid composition of rape and turnip rape were recently reviewed[63].

Table 7:9 presents the contents of a total of 17 fatty acids in seed samples of Swedish cultivars and breeding lines. In addition, mean values for the content of 14 fatty acids in typical samples of *Brassica juncea* and *Sinapis alba* are given. The only noteworthy difference in levels of minor acids among the species studied is the relatively large content of 20:0, 20:2, 22:0, 22:2 and 24:1 in *B. juncea* and the relatively high content of 24:1 in *Sinapis alba*. The proportions of (n–9)- to (n–7)-isomers of 18:1, 20:1 and 22:1 fatty acids differ with fatty acid chain length and apparently also with genotype (see Table 7:9). A relationship between extent of elongation in the two isomer series has previously been suggested[64]. Thus the high erucic acid sample 'Victor' had relatively much of (n–7)-20:1, a fatty acid which is probably derived from palmitoleic acid, (n–7)-16:1, by two-fold acetate addition in the same fashion as erucic acid is most likely derived from oleic acid, (n–9)-18:1, by two-fold elongation. On the other hand, the low erucic acid sample 'Bele' had comparatively more of vaccenic acid, (n–7)-18:1, and (n–9)-20:1, both acids supposedly elongated once. The zero-erucic acid lines obviously have rather high levels of vaccenic acid, but no 13-eicosenoic acid. Thus in these samples, fatty acids elongated once do occur.

As demonstrated in Table 7:10, there are considerable differences in fatty acid spectra of the three morphologically easily distinguishable parts, the seed coat (to which a one- to two-cell-wide layer of endosperm often adheres), the two cotyledons and the remainder of the embryo, which is mainly hypocotyl and is therefore denoted 'hypocotyl'. Although the differences in linoleic and erucic acid contents between cotyledons and hypocotyl are quite sizable, the most striking difference is seen between these two and the seed coat in the content of palmitoleic acid and (as a consequence?) in the proportions of the (n–9)- to (n–7)-isomers of the octadecenoic, eicosenoic and docosenoic acids (*cf.* ref. 62). It should be emphasized that the predominant lipid class in these tissues is that of the triacylglycerols,

TABLE 7:10

FATTY ACID COMPOSITION OF COTYLEDONS, HYPOCOTYL AND SEED COAT OF *Brassica napus*, CV. REGINA II

Percentage composition and total amounts of fatty acids determined by GLC analysis on a packed column; monoene isomer proportions estimated from analysis on a capillary column[14].

Seed part	Total amounts of fatty acids obtained from 100 seed parts (mg)	Fatty acid composition (%)														Isomer proportions (n–9)/(n–7)		
		16:0	16:1	18:0	18:1	18:2	18:3	20:0	20:1	20:2	22:0	22:1	22:2	24:0	24:1	18:1	20:1	22:1
Cotyledons	61.5	3.3	0.3	1.2	12.4	15.2	9.0	0.8	9.5	0.7	0.5	44.6	1.0	0.4	1.1	12/1	8/1	25/1
Hypocotyl	11.0	6.6	0.6	1.6	15.3	21.3	9.8	0.8	10.3	0.8	0.5	29.9	0.6	0.4	1.5	8/1	8/1	16/1
Seed coat	4.8	5.2	2.0*	1.2	18.7	17.0	6.1	0.6	14.4	0.6	0.7	30.0	0.7	0.6	1.1	0.9/1	0.3/1	3/1

* Also 1.1 per cent of 16:3.

since large variations occur in the fatty acid spectra among various lipid classes (see Table 7:6). The large amounts of the uncommon (n–7)-isomers in the seed coats are of interest from a biosynthetic point of view (see ref. 62 for references). It might also be of interest from an industrial point of view, to provide a source of uncommon fatty acids for the chemical industry (see Chapter 12).

7.3.2 Other lipids

As previously mentioned, seed oils are often characterized by their fatty acid composition and their content of so-called non-saponifiable lipids (see also Chapter 8). The non-saponifiable lipids comprise a heterogeneous group of substances, some of which are natural constituents of the seed oil (e.g. hydrocarbons, free sterols and triterpenoid alcohols), whereas others are modified by the saponification (e.g. sterol esters converted to sterols and fatty acids). Most of the data reported to date on sterols, aliphatic and triterpenoid alcohols, tocopherols etc. of rapeseed are from studies of the non-saponifiables rather than the natural oils. Methods have, however, been developed for the non-destructive isolation of minor constituents from the bulk of triacylglycerols[65] and therefore the amount of available information on the natural constituents is expected to increase.

The total content of non-saponifiables in some German cultivars of rapeseed was estimated to be 0.5 per cent of the oil for winter rape, 0.6 per cent for summer rape and winter turnip rape and 0.8 per cent for summer turnip rape[58]. The somewhat higher values reported for Swedish cultivars—0.9 per cent for winter rape, 1.0 per cent for summer rape, 0.8 per cent for winter turnip rape and 1.1 per cent for summer turnip rape[49]—might indicate differences in analytical methods used rather than varietal or environmental effects. The relatively higher values for the seeds lowest in oil content, summer turnip rape and white mustard, are, however, probably significant and may mirror the preferential location of the non-saponifiables on the seed coat (cf. ref. 66). The recent data on Swedish and German seed agree with handbook data on rapeseed and mustardseed oil[67].

The quantitative composition of rapeseed oil non-saponifiables has been reported as follows: hydrocarbons (non-polyenes) 8.7 per cent, squalene 4.3 per cent, aliphatic alcohols 7.2 per cent, triterpenoid alcohols 9.2 per cent and sterols 63.6 per cent[65]. Generally, the analyses reported for non-saponifiables refer to commercial oils rather than seeds of known origin.

(a) Hydrocarbons

Oil from mature summer rapeseed, cv. Golden, is reported to contain 0.3 per cent hydrocarbon[27]. The hydrocarbon fraction of a sample of rapeseed oil (origin not reported) was separated into 36 components: n-hydrocarbons from C11 to C31, iso- and/or anteiso-hydrocarbons from C11 to C17, C19 to C21 (see Fig.8:8). The major component, ca. 8–10 per cent of the total hydrocarbons, was the $C_{29}H_{60}$ n-alkane[68].

(b) Carotenoids

No data on the carotenoids of seeds of known cultivars of rape or turnip rape seem to be available. Some Polish samples of rapeseed oil were reported to contain 4.3–7.0 p.p.m. of total carotenoids, calculated as β-carotene[69]. According to other authors, a sample of rapeseed oil of unreported origin contained *ca.* 35 p.p.m. of carotenoids, of which less than 2 per cent was β-carotene and about 52, 14 and 19 per cent lutein, neo-lutein A and neo-lutein B respectively[70].

(c) Aliphatic alcohols

The total amount of aliphatic alcohols, isolated from the non-saponifiables of rapeseed oil, appears to be very small[71]. No data seem to be available on the chain length distribution of these alcohols of seed oils of brassicas, whereas detailed studies have been undertaken on these compounds in leaves of a related species *viz*. *B. oleracea*[32].

(d) Triterpenoid alcohols

The qualitative composition of the triterpenoid alcohols of rapeseed is very complex (see Fig.8:7). No oil investigated by a pioneering laboratory in this field of lipid chemistry had exactly the same triterpenoid alcohol pattern[65]. Therefore, this fraction is well suited to serve in the identification of an unknown vegetable oil. Actually, the triterpenoid fraction of rapeseed oil is found to be the most complex of all oils studied[72]. At least 13 components have been detected, and the major ones were identified as β-amyrin, cycloartenol and 24-methyl-cyclo-artenol[72].

(e) Sterols

The qualitative and quantitative compositions of the total sterols, as obtained after saponification, will be treated in this section. The content of sterols is generally about 0.5–1 per cent of the oil, but there is considerable uncertainty about the proportions of various sterol derivatives; see 7.3.1. (d). The qualitative and quantitative compositions of the total sterols of seeds of well-known origin were recently reported from two laboratories[73, 74]. From the data, which are shown in Table 7:11, it is obvious that β-sitosterol is the major component and that no marked differences exist in sterol patterns among the cultivars hitherto studied. Earlier Canadian data also fit this pattern[27]. The identification of chole-sterol was performed in both laboratories by analysis in a gas chromatograph joined with a mass spectrometer. Further evidence for the identity of cholesterol of plant and human origin has been presented, together with a survey of older literature on 'cholesterol-like' compounds in extracts of higher plants[75]. The demonstration of the occurrence in a vegetable oil of a sterol, which has been considered so typical of animal fats, is noteworthy and may be parallelled by the demonstration of arachidonic acid in mosses and ferns. Thus two of the com-

TABLE 7:11

THE QUANTITATIVE COMPOSITION OF TOTAL STEROLS (FREE AND ESTERIFIED) IN SEEDS OF *Brassica napus*, *B. campestris* AND *Sinapis alba*

| Sample | Brassicasterol | Campesterol | Sterol composition (%) | | | | | Ref. |
			Cholesterol	β-Sitosterol	Stigmasterol	Others		
Brassica napus cv. Matador	13.0	31.9	1.7	53.4				73
B. napus cv. Regina II	10.1	25.6	3.8	60.6				73
B. napus summer rape	16.1	27.8	Trace	52.3	1.1	2.7		74
B. campestris cv. Golden Ball*	13.4	22.4	2.7	61.5	Trace	Trace		74
B. campestris cv. Wallace*	19.6	25.2	0.3	53.4		Trace		74
Sinapis alba	5.2	34.6	3.2	43.8		13.2		74

* In the original paper named *Brassica rapa*, a designation now considered less appropriate for the subspecies *Brassica campestris*, var. *rapifera*.

144

ponents previously considered 'specific tags' in animal fats actually occur in plant lipids as well.

Recent German studies have indicated that the percentage of cholesterol in rapeseed might be considerably higher[76] than that reported in Table 7:11. Obviously further samples have to be studied, using various methods, to obtain a better understanding of the sterols of rapeseed and the variation with species, cultivar and growing conditions.

7.3.3 Miscellaneous lipid-soluble compounds

(a) Chlorophylls and related pigments

In textbooks of biochemistry, the chlorophylls are generally not treated together with lipids, but with other metalloporphyrins. Since they are lipid-soluble and are of technological signifiance for the oil quality, they will be discussed in this section.

The term 'related pigments' refers to any breakdown products of chlorophylls, which still possess chromophoric groups (see a recent monograph[77]). Very little seems to be known on the qualitative composition of the chlorophyllous pigments of well-defined seeds of rape and turnip rape.

From the analyses of many samples of pressed or extracted Polish rapeseed oil by a direct spectrophotometric method, it was concluded that almost all of the pigment was present in the magnesium-free form, i.e. as pheophytins rather than as chlorophylls[69]. No chlorophyll a was found, but 0.1–1.8 p.p.m. chlorophyll b. The content of pheophytin a ranged from 18.0 to 25.6 p.p.m. and that of pheophytin b from 0.5 to 6.2 p.p.m. It should be noted, however, that the oils analysed had been stored at the laboratory for about 6 months before analysis, and that the decomposition of chlorophylls to pheophytins could have occurred at the laboratory. With the same analytical technique, other authors have also arrived at similar results for a sample of commercial rapeseed oil[70]. It may be remarked that the pheophytins easily form complexes with other metals, e.g. copper and iron[77].

From recent studies, it appears very likely that immediate analysis of the lipid extract from viable, high-quality rapeseed will show it to contain mainly chlorophylls and very little, if any, pheophytins. Older seeds and seeds damaged in various ways contained variable amounts of chlorophylls and pheophytins[78].

The total content of chlorophyll (and related pigments) should be as low as possible from an industrial point of view (see Chapter 10). Therefore, extensive studies have been undertaken on the effect of agricultural techniques during harvest and drying of rapeseed on the chlorophyll content, as discussed in Chapter 5.

(b) Tocopherols

The tocopherols are generally isolated after careful saponification of the oil. The errors involved in the determination of these oxidation-sensitive compounds are discussed in Chapter 8. Older literature data reporting from *ca.* 300 to *ca.* 800 p.p.m. tocopherol in rapeseed oil should be evaluated with these analytical difficulties in mind; see a compilation[79] from 1965. A recent study of rapeseed oil gave a level of 680 p.p.m. of total tocopherol with the following pattern: 35 per cent α-tocopherol, 63 per cent γ-tocopherol and 2 per cent δ-tocopherol[80]. The same report contains information about the structures and names of the various tocopherols and tocotrienols found in nature[80]. From unpublished Swedish studies[81] presented in Table 7:12, it appears that there are greater differences in

TABLE 7:12

TOCOPHEROL CONTENTS IN INDUSTRIALLY PRODUCED CRUDE OILS FROM SEEDS OF RAPE, TURNIP RAPE AND WHITE MUSTARD[81]

Species and cultivar	Tocopherol content, (p.p.m.)	Tocopherol pattern		
		α-Tocopherol (%)	γ-Tocopherol (%)	δ-Tocopherol (%)
Brassica napus cv. Victor	910	43	57	
B. napus cv. Regina	911	27	73	
B. campestris cv. Duro	795	36	64	
B. campestris cv. Bele	836	36	64	
Sinapis alba cv. Seco	876	10	86	4

tocopherol pattern between species and genotypes than expected from the analytical errors involved. This study is apparently the only one at present available that presents tocopherol data on *Brassica* seeds of known origin.

(c) Lipid-soluble metals
The heavy-metal ions found in the vegetable oils are mainly associated with the phospholipids, but may be bound to the pheophytins[77]. Since the amounts of iron and copper are of great significance for the oxidative stability of rapeseed oil and soya-bean oil[81a], some data on the content of these metals in seeds of well-defined origin will be discussed in this section. Unpublished Swedish observations have indicated that the seed pretreatments and extraction conditions exert a major influence on the contents of iron and copper in the oils extracted[82]. Grinding seeds of *B. napus*, cv. Victor in a porcelain mortar followed by Soxhlet extraction with hexane gave higher figures for iron and copper, 0.3–1.4 and 0.2–2.8 p.p.m., respectively, compared to extraction in the so-called Swedish steel tubes[83], 0.05–0.4 and 0.01–0.19 p.p.m., respectively. With the more polar solvent, hexane–ethanol (3:1) the extracts from the steel tubes were markedly higher in iron and copper, 0.5–1.0 and 0.01–0.60 p.p.m., which is expected since such a solvent dissolves phospholipids to a considerably greater extent. Four of the samples of the

TABLE 7:13

SUGARS DETECTED AFTER HYDROLYSIS OF POLYSACCHARIDE FRACTIONS FROM WHITE MUSTARD SEED[86]

For each fraction, sugars are listed in approximately descending order of concentration; those in parentheses are relatively minor components

Solvent used for extraction	Seed coats		Embryos	
	Yield (%)	Sugars detected after hydrolysis	Yield (%)	Sugars detected after hydrolysis
Cold water	6.4	Galactose, arabinose, galacturonic acid (xylose, glucose, mannose)	2.4	Arabinose, glucose (xylose, galactose, galacturonic acid)
Hot water	8.6	Arabinose, galactose, galacturonic acid (xylose, rhamnose, glucose, mannose)	3.4	Arabinose (xylose, galactose, galacturonic acid)
Ammonium oxalate	16.1	Galacturonic acid, xylose (arabinose, mannose, galactose)	1.7	Arabinose (xylose, galactose, galacturonic acid, glucose)
Hot water (after delignification)	14.0	Arabinose, galacturonic acid (xylose, galactose)	8.5	Arabinose, galacturonic acid (xylose, galactose, rhamnose)
5% KOH	6.4	Xylose, galacturonic acid (glucose, galactose, mannose, arabinose)	9.8	Xylose, arabinose (glucose, galacturonic acid, mannose)
20% KOH	5.4	Xylose, mannose, glucose, galacturonic acid (galactose, arabinose)	1.7	Xylose, glucose (mannose, arabinose)
Residue	16.0	Glucose (galacturonic acid)	2.4	Glucose (arabinose, galacturonic acid)

References pp. 169–173

genotype studied had rather similar levels, but one was exceptionally rich in copper, which necessitates further studies in this field[82]. Typical values for degummed rapeseed oil are 0.05 p.p.m. copper and 1.2–3.1 p.p.m. iron (Table 8:4). This probably indicates that a considerable portion of the lipid-soluble iron of commercial rapeseed oil stems from the equipment.

7.4 CARBOHYDRATES

7.4.1 Mono-, oligo- and polysaccharides

Japanese studies on the qualitative and quantitative composition of the mono- and oligosaccharides of *Brassica napus* seeds gave the following pattern: Fructose 0.51 per cent, glucose 0.21 per cent, sucrose 1.11 per cent, raffinose 0.15 per cent and stachyose 0.19 per cent, calculated on defatted meals[84]. In more recent European studies, the following ranges of variation were reported: Fructose 0.8–1.0 per cent, glucose 1.2–1.3 per cent, sucrose 1.6–1.7 per cent, raffinose 0.2–1.7 per cent and stachyose 0.3–0.4 per cent[85].

Studies of the polysaccharides of *Brassica campestris* were recently initiated at the Food Research Institute in Ottawa, Canada. So far, however, none of this work has been published. In view of the lack of detailed information on rapeseed polysaccharides, the information available on a related crucifer, white mustard *(Sinapis alba)* will be discussed.

Sucrose and stachyose together with traces of raffinose were detected in aqueous extracts of 'grocery-quality' white mustard[86]. As seen from Table 7:13 there are large differences in both quantity and composition of various polysaccharides of the seed coat compared to the embryo (mainly cotyledons). It is easily observed that those sugars that are typical constituents of pectic materials, *i.e.* galactose, arabinose and galacturonic acid, predominate in the more easily soluble materials. From such extracts a pure homopolysaccaride, an araban, was isolated and characterized. More severe extraction conditions dissolved 'hemi-cellulosic' and cellulosic fractions, with xylose and glucose as dominating sugars (Table 7:13). No evidence was found for the presence of starch or fructans by the same authors. It should be remarked that the fractions isolated contained substantial amounts of protein, and therefore the weight data in the table can only give an estimate of the proportions. The mustard seed araban appeared very similar to pectic arabans, except that it is more highly branched than usual. These more recent data are supported by older studies, reporting the isolation by cold-water extraction of a mucilage in *ca.* 2 per cent yield (calculated on whole seed). The aqueous extract was fractionated into cellulose and two acidic polysaccharides, both containing galactose, arabinose and galacturonic acid[87, 88].

From recent studies it appears that the acidic polysaccharides and the cellulose

148

are not randomly associated in the solution but exist as a 'particle' with specific properties. The cellulose was in the crystalline form (IB) probably encapsulated with the acidic polysaccharides, since upon denaturation, the 'globuline-type' particles gave insoluble cellulose I in a yield of approximately 50 per cent by weight[89]. It is suggested that this might represent cell-wall material in a true biological form[89]. Most extraction methods used to purify bio-polymers (proteins and nucleic acids) involve the risk of breaking natural associations, and this is especially the case with methods used in the isolation of polysaccharides.

The inclusion of cellulose in crystalline form in the water-soluble polysaccharides, if it occurs in rapeseed, might be of significance in the spinning of protein fibres from rapeseed extracts. Such experiments have been reported to be considerably more complicated than the spinning of soyabean protein[90].

7.4.2 Glucosinolates

(a) Nomenclature and introduction
Since the first paper to report the isolation of sinigrin[91] appeared in 1840, these compounds have been given various names, mustard oil glucosides, thioglucosides and, recently, glucosinolates[92]. Since the latter name was suggested by a pioneering laboratory of glucosinolate chemistry[93], it is used throughout this text. The enzyme which at neutral pH splits the parent compounds into isothiocyanates, glucose and bisulphate has also been named differently in the past. An early name was myrosin, and then myrosinase was adopted, followed by thioglucosidase[92]. According to the rules of the International Union of Biochemistry, this enzyme should be named thioglucoside glucohydrolase (E.C.3.2.3.1). Accordingly, in this chapter, there will be a slight inconsistency, since the parent compounds will be named glucosinolates and the enzyme thioglucoside glucohydrolase (occasionally myrosinase). This question of nomenclature obviously has to be settled by authorities in the field.

The glucosinolates of rapeseed and related crucifers are far better known than

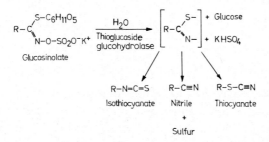

Fig.7:1. The general structures of glucosinolates and products formed by enzymatic hydrolysis, followed by chemical rearrangement reaction (see text and refs. 94 and 96 for details). The thiocyanates are generally not formed in the autolysis of *Brassica* seed meals.

any other non-lipid material of such seeds; likewise, the enzyme thioglucoside glucohydrolase has been better characterized than any other enzyme of *Brassica* seeds. Since several excellent reviews on the glucosinolates have been published during the past 10 years[93-96], these compounds will not be treated in detail in this chapter, although they are of utmost significance for the feed value of rapeseed meal. The glucosinolates are also discussed in Chapters 6 and 15.

(b) Structure and qualitative composition

The general structure of the glucosinolates, which was definitely established for one compound (sinigrin) in 1956[97], thereby revising the structure proposed in 1897, is shown in Fig.7:1. Whereas some cruciferous seeds have very simple glucosinolate patterns, others contain at least 9 different glucosinolates. Table 7:14 summarizes the knowledge hitherto obtained on the glucosinolate patterns of rape, turnip rape and some related crucifers[98-109].

TABLE 7:14

GLUCOSINOLATES PRESENT IN SEEDS OF SOME *Brassica* SPECIES

The figures refer to references in which the presence of the glucosinolate in the species indicated is reported.

Glucosinolate	B. campestris	B. juncea	B. napus
Allylglucosinolate (sinigrin)		99, 105	
3-Butenylglucosinolate (gluconapin)	107, 108, 109	99, 105	99, 107 108, 109
4-Pentenylglucosinolate	108, 109		102, 108
4-Methylthiobutylglucosinolate	108		
5-Methylthiopentylglucosinolate	108		
4-Methylsufinylbutyl-glucosinolate	103, 109		103
5-Methylsulfinylpentyl-glucosinolate	103, 109		103
2-Phenylethylglucosinolate	108, 109		108, 109
2-Hydroxy-3-butenyl-glucosinolate (progoitrin)	109		98, 109
2-Hydroxy-4-pentenyl-glucosinolate	106, 109		106, 109

No qualitive differences in glucosinolate patterns among Swedish cultivars were found from paper chromatographic studies of *B. napus* and *B. campestris*[109]. In view of the allotetraploid nature of *B. napus* (see Fig.3:1), and the finding that some strains of the allotetraploid *B. juncea* contained only allylglucosinolate (typical of *B. nigra*), others contained only 3-butenyl glucosinolate (typical of *B. campestris*), and still others contained a mixture of both[105], the glucosinolate

150

patterns of *B. napus* could be expected to vary from that of *B. oleracea* to that of *B. campestris*. It is, therefore, amazing that all genotypes of *B. napus* so far reported in the literature[60, 108–110] contain hydroxylated glucosinolates as the predominating compounds (including the low glucosinolate type Bronowski[110]) which are very low in both *B. campestris*[107–110] and *B. oleracea*[109] hitherto studied. For information on the pathways of biosynthesis of glucosinolates, see recent reviews[94, 111].

(c) Quantitative composition and methods of determination

Generally, the exact quantities of different glucosinolates are not reported. The information given is often expressed as major, minor and trace spots from paper chromatograms. Instead of a quantitative determination of intact glucosinolates, one or more of the split products have been determined quantitatively. If only the total amount of glucosinolates is to be determined, the glucose or the sulphate released can be assayed. Recently, a micromethod was described for the determination of enzymatically released glucose, which allows the estimation of the glucosinolate content of single cotyledons, and is thereby useful in plant-breeding work[112]. The assay of the sulphate moiety has been undertaken at a number of laboratories, mainly to demonstrate equimolarities of the glucosinolate constituents[108, 110, 113–115]. For the separate determination of isothiocyanates and oxazolidinethiones in seed digests, two somewhat different methods were

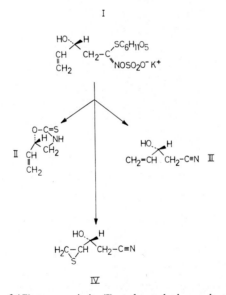

Fig.7:2. The structures of (*R*) – progoitrin (I) and autolysis products in the presence of untreated rapeseed meals, *viz.* (*S*) – goitrin (II), also called oxazolidinethione, (*R*) – 1-cyano-2-hydroxy-3-butene (III) and two diastereomeric (2 *R*) – 1-cyano-2-hydroxy-3,4-epithiobutanes IV[117].

TABLE 7:15

CONTENT OF GLUCOSINOLATES IN SEED MEALS OF SOME EUROPEAN CULTIVARS OF RAPE AND TURNIP RAPE, GROWN AT VARIOUS LOCALITIES[110, 115]

Species and cultivar	Origin	Glucosinolate content, (per cent in dry matter)		
		Gluconapin	Progoitrin	Gluconapin + progoitrin
Brassica napus, winter type				
Matador	Mean of 3 samples, grown in Sweden	1.71	4.69	6.40
Heimer	Mean of 7 samples, grown in Sweden	1.71	4.51	6.24
Victor	Mean of 11 samples, grown in Sweden	1.79	4.31	6.11
Rapol	Germany 1965	1.66	4.92	6.58
Diamant	Germany 1965	1.45	4.98	6.43
Tonus	Nord, France 1965	1.73	4.16	5.89
Nain de Hamburg	Nord, France 1965	1.21	2.64	3.85
Sarepta	Nord, France 1965	1.10	2.81	3.91
Sarepta	Versailles, France unknown	2.30	3.28	5.58
Sarepta	Svalöf, Sweden 1961	1.38	2.94	4.32
Valois	Versailles, France unknown	2.22	5.55	7.77
Valois	Svalöf, Sweden 1961	1.43	5.45	6.88
Dublański	Svalöf, Sweden 1965	1.81	4.12	5.93
Niemierczański	Svalöf, Sweden 1965	2.00	4.73	6.73
Skrzeszowicki	Svalöf, Sweden 1965	1.77	4.04	5.81
Warszawski	Svalöf, Sweden 1965	1.87	4.46	6.33
Poświcki	Svalöf, Sweden 1965	2.06	4.68	6.74
Górczański	Svalöf, Sweden 1965	2.03	4.30	6.33
Brassica napus, summer type				
Regina II	Mean of 8 samples, grown in Sweden	1.27	3.04	4.31
Rigo	Mean of 3 samples, grown in Sweden	1.31	3.72	5.03
Zollerngold	France 1965	1.76	3.18	4.94

Czyżowskich	Czechoslovakia	1965	1.50	4.05	5.55
Czyżowskich	Svalöf, Sweden	1965	1.27	3.78	5.05
Mtochowski	Czechoslovakia	1965	1.80	3.59	5.39
Bronowski I.L. 1997	Unknown (received from Poland)		0.15	0.79	0.94
Bronowski I.L. 2243	Unknown (received from Czechoslovakia)		0.87	2.78	3.65
Bronowski Sv 64–1055	Grown in field at Svalöf in 1964		1.29	2.35	3.64
Bronowski IHAR	Grown in Poland 1967		0.11	0.53	0.64
Bronowski IHAR	Grown in France		0.06	0.12	0.18
Bronowski 2178 L 5	Grown in France		0.05	0.15	0.20
Brassica campestris, winter type					
Duro	Mean of 3 samples, grown in Sweden		3.10	0.22	3.32
Rapido II	Mean of 3 samples, grown in Sweden		2.95	0.50	3.44
Gruber	Tikkurila, Finland	1965	3.84	0.46	4.30
Rapido I	Hahbiala, Finland	1965	3.07	0.57	3.64
Brassica campestris, summer type					
Bele	Mean of 2 samples, grown in Sweden		1.78	1.26	3.30
Sv 58/4	Mean of 5 samples, grown in Sweden		2.15	1.15	3.31

recently described[108, 115]. One is characterized by yielding information on the amounts of individual isothiocyanates[108], whereas the other yields data for a group of isothiocyanates[115]. Both determine the oxazolidinethiones as a group and are well suited to microquantities. Other methods have also been described. Advantages and disadvantages of various methods were recently discussed[115], and thus no such discussion will be repeated here.

Studies in two laboratories revealed that the yield of aglucone was irregular and far from quantitative at the pH of optimal activity in a pure glucosinolate–thioglucoside glucohydrolase system, when the glucosinolates were autolyzed in the presence of the seed meal[115, 116]. Among the autolysis products of rapeseed meal, (R)-1-cyano-2-hydroxy-3-butene and (2R)-1-cyano-2-hydroxy-3,4-epithio-butane have been identified[117]. The complicated reaction patterns displayed in seed meals of rapeseed and *Crambe abyssinica* have been elegantly elucidated at the Northern Regional Laboratory of the USDA, and thoroughly discussed in several papers[96, 116, 117]. Figures 7:1 and 7:2 summarize some of this work. It has been found that the toxicity of the cyanohydroxy butene is higher than that of iso-thiocyanates[96].

In view of the considerable losses of isothiocyanates that occur under auto-lytic conditions in rapeseed meals, heat treatments were found to be necessary prerequisites for a quantitative assay of rapeseed glucosinolates yielding iso-thiocyanates[108, 115, 116]. The considerable losses which occurred with an earlier, well-established method[119, 120] makes the discussion of some older quantitative data less meaningful.

The finding of a cultivar of summer rape, 'Bronowski', with markedly reduced levels of glucosinolates (about 10 per cent of those of typical rapeseed cultivars), and the implications of this finding for the production and utilization of rapeseed meal, are obvious from Chapters 6 and 15. Table 7:15 presents some data on the quantitative composition of glucosinolates in typical European cultivars. It is easily seen that a rather small variation exists between common cultivars within any crop type, but typical differences do exist between crop types. Thus *Brassica napus* is high in progoitrin and *B. campestris* high in gluconapin, but the total amount is lower in *B. campestris*. The spring-sown crops are generally lower in glucosinolates than the fall-sown ones. The marked difference in gluco-sinolates between winter rape, the predominant crop type in Europe, and summer turnip rape, the predominant one in Canada (see Table 6:7), sheds light on the different results and recommendations among animal nutritionists from the two continents (see Chapter 15). It should also be pointed out that the variation in glucosinolate content with cultivation conditions, such as availability of sulphur in the soil[115, 120] is much greater than any environmentally effected variation in fatty acid composition (*cf.* Tables 7:7 and 7:8).

The ratio of 3-butenyl- to 4-pentenylglucosinolate displays some variation with species and genotype, *Brassica napus* being relatively richer in the former and

154

B. campestris having approximately equal quantities of both glucosinolates. (see Table 6:7 and refs. 60, 121, 122, 123).

The concentration of free isothiocyanates in cruciferous seeds seems to be negligible[94]. It appears that the glucosinolates and the degradative enzyme, thioglucoside glucohydrolase, are present in different cells in mature seeds[94]. Grinding of the seeds brings substrate and enzyme in contact. A discussion of the properties of the enzyme is presented under 7.5.4.

7.5 PROTEINS AND AMINO ACIDS

The rapeseed proteins can, in the same way as any other seed protein, be considered to have a structural function (membrane components), a catalytic function (membrane-bound or free) and/or a 'storage' function. In view of the small amounts of enzyme protein generally involved in enzymatic reactions, it is reasonable to assume that the major portion of the proteins reported under 7.5.2 and 7.5.5 represent storage proteins with no enzymatic activity. In rapeseed the storage protein is located in specific organelles, the so-called protein bodies or aleurone grains (see Fig.3:9 and 3:10).

Under this heading the gross amino acid composition will also be discussed, since the amounts of free amino acids and peptides are very small in mature rapeseed[124, 125]. Whereas some enzymatic activity is membrane-bound, other enzymes seem to be located in the cytoplasm. In the case of rapeseed enzymes nothing seems to be known with certainty about their subcellular location, nor has any attempt apparently been made to isolate membrane protein from dormant rapeseed. Therefore, this discussion will be limited to enzymes and 'bulk proteins', supposedly mainly storage protein.

It may be remarked that compared with the relatively large number of papers published on rapeseed lipids, those on rapeseed protein are indeed very few.

7.5.1 Free amino acids

Mature rapeseed of Canadian origin is reported to contain only traces of free amino acids, *ca.* 0.08 μmoles of alanine and 0.15 μmoles of glutamic acid per g defatted seeds[124]. Highly immature seed, on the other hand, contained considerable amounts of free amino acids and peptides in aqueous extracts. Three weeks before maturity, 11 per cent of the nitrogen in an aqueous extract was free amino acids, with glutamic acid, proline and alanine as major components[124].

Extensive Danish studies have recently been reported on the non-protein amino acids of many Cruciferae[125], but neither *Brassica campestris* nor *B. napus* was included. A related crucifer, *Brassica nigra*, is reported to contain aspartic and glutamic acids, arginine and lysine, alanine, asparagine, γ-aminobutyric acid, pipecolic acid, glycine, leucine and/or isoleucine, proline, serine, threonine, valine, and 2-aminoethanol[125]. The quantitatively dominating amino acids were alanine, aspartic acid and glutamic acid. Other studies on *B. napus* reported the presence of all common amino acids, as well as asparagine and glutamine[85]. In *Sinapis alba* seeds, a rather similar pattern of free amino acids was reported. The *S. alba* seeds also contained *p*-hydroxybenzylamine[125], which might be meta-bolically related to the glucosinolate sinalbin present in such seeds (see Fig. 15:3). From extensive studies of the total amino acid patterns of cruciferous seeds[126] the presence of 3,4-dihydroxyphenylalanine in seeds of *Brassica rapa*, (= *Brassica campestris*, ssp. *rapifera*) has been reported (see ref. 125).

7.5.2 Bulk ('storage') proteins

(a) Solubility characteristics

The influence of a large number of parameters on the extraction and subsequent precipitation of proteins from rapeseed has been reported in a series of publications from a Czechoslovakian laboratory[128–132].

The first two reports concerned the optimal conditions for protein extraction as judged from the N content of the extract and filtration velocity, properties of technological rather than chemical interest[128, 129]. The following conditions were deemed optimal: Use of a 10-fold excess of 0.15–0.30 per cent NaOH with moderate shaking for 20–40 min. It is emphasized that very high concentrations of alkali are undesirable because larger amounts of non-protein compounds dissolve, and furthermore, some protein degradation may occur. It should, however, be pointed out that the starting material in this case was commercially produced rapeseed meal and that it has been found that the extractability and precipitability of protein from rapeseed is strongly influenced by the previous heat-treatment of the seed or meal[90].

In a third paper, the precipitiation of protein was reported to be only 40–50 per cent (depending on the protein concentration) at the isoelectric point, which was as low as pH 3.5. The yield of protein was independent of the temperature between 18 and 40 °C and the time of storage of the protein solution, 1–80 h[130]. Later, the efficiencies of water and various salt solutions in protein extraction were compared. Extensive data on the pH and gross composition (dry matter nitrogen, reducing sugars, iron and phosphorus) of successive extractions with eight different solvents were recorded[131]. Of the total nitrogen of the meal, about 19 per cent was soluble in water, about 64 per cent in 10 per cent NaCl and about 81 per cent in 0.15 per cent NaOH. The nitrogen content of the extract from the NaOH treatment was the highest of all tested.

Recently, the solubility characteristics of the proteins of laboratory-defatted

PEPTIZATION OF PROTEINS IN SEED MEALS OF RAPE AND TURNIP RAPE BY THE OSBORNE SERIES OF FOUR SOLVENTS[8]

Species and cultivar	H_2O (%)	Total meal nitrogen soluble in			Nitrogen in residue (%)
		5 per cent NaCl (%)	70 per cent EtOH (%)	0.2 per cent NaOH (%)	
B. campestris					
Polish	44.6	24.0	4.1	6.3	21.0
Echo	44.5	25.0	4.4	6.6	19.5
Zero erucic	44.7	24.7	4.3	5.5	20.8
B. napus					
Argentine	51.3	20.5	3.9	8.1	16.2
Target	50.6	20.5	4.0	9.1	15.8
Oro	48.4	22.4	3.3	8.5	17.4

seed meals of three cultivars each of rape and turnip rape were reported[8]. As seen from Table 7:16, about 50 per cent of the rapeseed proteins were water-soluble compared to *ca.* 45 per cent of the turnip rape proteins. These figures are substantially lower than those for soyabeans, but higher than those for the industrially processed rapeseed meal discussed above[131]. The amount of insoluble nitrogen appeared to be of similar magnitude (about 20 per cent) in both laboratory and industrially processed meals[8, 131]. The lower proportion of water-soluble proteins in turnip rape (by classical terminology called albumin) was balanced by a higher proportion of salt-soluble proteins (globulins), there also appears to be a species difference in percentage of insoluble nitrogen, with turnip rape having the higher proportion. It is, however, emphasized that each species and cultivar shows some differences in solubility characteristics[8]. In this connection, it should also be pointed out that commercial samples of oil seeds can have been exposed to rather high temperatures during the artificial drying, resulting in loss of viability (see pp.72–73) which could affect the solubility characteristics. Therefore, only seed samples with a very high percentage of viable seeds should be used in protein studies on oil seeds if reproducible results are to be expected. Even then, seed lots might vary in solubility characteristics, since mineral nutrition may produce large differences in the proportions of various single proteins [133, 134]. The alkali-soluble proteins of the six seed samples had a minimum solubility at pH 4.4–4.6, substantially higher than that recorded for similar extracts of commercial rapeseed meal[132]. The yield was, however, only *ca.* 25 per cent of the meal on a dry-matter basis, or *ca.* 50 per cent of the meal nitrogen (Table 7:17). The remainder of the nitrogen was made up of slightly less than 20 per cent

SPECIES AND CULTIVAR DIFFERENCES IN NITROGEN SOLUBILITY INDEX OF THE MEAL, YIELD AND
PROPERTIES OF THE ISOLATED PROTEIN AND WHEY NITROGEN CONTENT[8]

Species and cultivar	Nitrogen content of dry meal (%)	Nitrogen solubility index (%)	Isolated protein from meal		Nitrogen content of whey (% of total)
			Yield of product, (g/100g meal)	Nitrogen content (%)	
B. campestris					
Polish	6.2	81.3	24.5	13.4	27.4
Echo	6.1	83.2	26.3	12.9	25.8
Zero erucic	6.9	79.5	24.5	13.6	29.2
B. napus					
Argentine	7.1	82.3	25.2	13.7	32.3
Target	6.7	83.4	25.3	13.7	30.3
Oro	6.4	85.7	26.6	12.8	31.3

insoluble and slightly more than 30 per cent in the 'whey', the soluble, non-precipitable nitrogenous substances, which besides proteins contains nucleic acids and glucosinolates. The yield of precipitated protein was exceptionally low in rape and turnip rape compared with other major oilseed crops[8]. The nitrogen content of the 'whey' is lower for turnip rape than for rape, which might be due to the less complex protein pattern of the diploid turnip rape compared to that of the allotetraploid rape (cf. Fig.3:1), and/or to the lower percentage of glucosinolates in the former species (cf. Table 7:15).

By a short treatment of seeds of rape and turnip rape in boiling water, other authors found a substantial reduction in the amount of water-soluble protein[135]. The differences between heat-treated rape and turnip rape seem to be more accentuated, however, in comparison with untreated seeds (Table 7:17), since 26 per cent of the former and only 14 per cent of the latter were removed by three-fold extraction by water at 80°C. This investigation also reported the effect of various operating parameters on the extraction of protein and glucosinolates. A thorough discussion of various processes recently reported for the preparation of edible protein products from rapeseed is outside the scope of this chapter. The detoxification of rapeseed meal to produce feedstuffs of superior quality is discussed in Chapter 15, and the novel and expanding field of the production of edible proteins from cruciferous seeds has been reviewed recently[136]. It is noteworthy that the inter-specific differences in losses of solids and protein by aqueous extraction of laboratory-processed meals were reflected in similar differences between commercial meals, although the total losses were markedly higher[135].

This might reflect a less-strong heat coagulating effect on the seed proteins *in situ* by the dry heat treatment used industrially (*ca.* 80 °C for 20–30 min at *ca.* 8 per cent seed moisture, *cf.* Chapter 9) compared with that caused by a short immersion in boiling water. It should be remarked that several important differences in solubility characteristics have been reported for the two species *Brassica campestris* and *B. napus* which are traded as a mixture under the name rapeseed.

(b) Isolation of individual proteins or groups of proteins
It seems to be rather well established that the proteins of a hybrid, such as the allotetraploid *Brassica napus*, are the sum of those of the two parents, in this case *B. campestris* and *B. oleracea*. Only in rare cases do protein 'hybrids' appear (see discussion in ref. 137). From studies on β-galactosidase, β-glucosidase and esterases of the aforementioned species, British scientists reported many electrophoretically separated bands with these enzymatic activities[137]. In some cases the hybrid, *B. napus*, had observable bands which were not visually detected in the parent species. This, however, could indicate that the amount was increased in the hybrid, so that it was more easily detectable visually, or it could also be really absent in the parents (*cf.* the situation with progoitrin on p.151). The great complexity of the protein solutions of rape, turnip rape and related species (see Fig.3:1) was noted in studies on the 'albumins' and 'globulins' of these species, using acrylamide electrophoresis in conjunction with serological techniques, general protein dyeing, or tests for enzymatic activities called esterase, β-glucosidase, β-galactosidase and myrosinase[137–140]. The myrosinase studies are discussed on p.165, together with other reports on myrosinase. The serological studies revealed several proteins of each of the three diploid species that were not present in the other two. In immunoelectrophoresis, up to 24 precipitin arcs were observed. Many interesting observations of taxonomic significance were made in these studies, a discussion of which is outside of the scope of the present chapter. Suffice it to mention that these studies point to a greater complexity of the seed proteins of *Brassica napus* than those of *B. campestris*.

The determination of the solubility characteristics of defatted meals from maturing rapeseed demonstrated a gradually increasing proportion of proteins that are soluble in dilute ammonia solution after the removal of water-soluble proteins[124].

These Canadian studies also showed that the N-terminal amino acids of a protein fraction separated by Sephadex chromatography were glutamic acid, glycine and alanine. Comparison of total amino acid composition with the amount of N-terminal acids allowed the authors to postulate that a major portion of the storage protein is synthesized by successive addition of amino acids at the C-terminal ends of smaller proteins as maturation proceeds, concomitant with the synthesis of new terminal residues. It did not appear from their data that larger amounts of polypeptides were being condensed as maturation proceeds. Some variation in amino acid pattern with stage of maturation has been recorded for a chromatographically isolated protein fraction[124].

Further attempts to isolate pure proteins from rapeseed cv. Nugget using combinations of gel filtration and ion exchange chromatography, resulted in the isolation of one basic protein (named A-IV-S) with a molecular weight of about 14,000 and at least 8 other chromatographic peaks which might represent pure protein or mixtures of proteins, and which account for 35 per cent of the sodium pyrophosphate soluble nitrogen[143]. A major component which had a rather large molecular weight, sedimentation coefficient 12 S (see ref. 144 for explanation), appeared homogeneous in alkaline solution, but dissociated into subunits with the sedimentation coefficients 3 S and 7.2 S in acidic conditions. This component was denoted B-I. It was salt-soluble, but precipitated upon dialysis against distilled water, and is thus a globulin according to classical nomenclature. It may be remarked that a globulin with similar characteristics accounts for *ca.* 65 per cent of the soyabean proteins, whereas the B-I of Canadian summer rapeseed accounted for only *ca.* 35 per cent of the rapeseed protein. A water-soluble protein with similar characteristics (denoted A-I-bs) also appeared in the pyrophosphate extract.

These two fractions, B-I and A-IV-S according to the Canadian nomenclature, were subsequently isolated from well-defined seed samples representing two species, rape and turnip rape, and different cultivars as well as one cultivar with different supplies of nutrients[133]. Dramatic differences in content of these two

TABLE 7:18

AMOUNTS OF PROTEINS B-I AND A-IV-S ISOLATED FROM SEEDS OF DIFFERENT CULTIVARS OF RAPE AND TURNIP RAPE

The weights are based on 10.0g whole seed[133]

Sample	Proteins				
	B-I			A-IV-S	
	Weight (mg)	N (%)		Weight (mg)	N (%)
B. napus					
Target 1	322.0	16.8		108.0	16.1
Target 2	275.0	17.0		121.0	16.2
Bronowski	480.0	16.5		60.5	16.2
Nugget 1	120.0	17.4		100.0	16.8
Nugget 2	144.0	17.0		340.0	15.0
B. campestris					
Echo	221.0	17.5		172.0	17.4
S-7165	350.0	16.5		43.6	14.0
Yellow Sarson	46.0	16.4		2.0–3.0	

160

proteins were recorded, as shown in Table 7:18. The amount of B-I protein varied from 46 mg in 'Yellow Sarson' *(Brassica campestris*, ssp, *trilocularis)* to 480 mg per 10 g whole seed of 'Bronowski' *(Brassica napus)*, and that of A-IV-S from 2–3 mg in 'Yellow Sarson' to 340 mg in the well-nourished 'Nugget-2' sample. It has recently been reported that sulphur nutrition greatly affects the amounts of B-I and A-IV-S, their proportions being reversed between sulphur-deficient and sulphur-adequate cultures[134].

The effect of mineral nutrition on the elution profiles of the 'albumins' chromatographed on CM-cellulose and Sephadex G-100 is shown in Fig.7:3. The

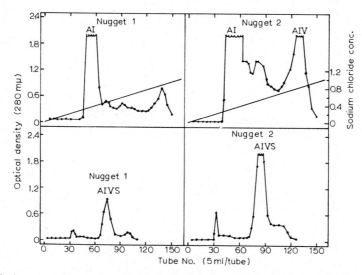

Fig.7:3. Elution curves for the chromatography of the water-soluble fraction of the molar sodium chloride extract of the two Nugget rapeseed samples on CM-cellulose (top figures). The elution curves for the chromatography of fraction A-IV on Sephadex G-100 using 0.01 M phosphate (pH 6.9) as eluant are given in the bottom figures[133].

change in proportion of various proteins was not accompanied by any major differences in chromatographic and electrophoretic properties, except in a few cases[134]. Relatively large differences in amino acid composition of the A-IV-S proteins from different species and cultivars were recorded (Table 7:19), especially for *B. napus* compared with *B. campestris*. Within the latter species, Yellow Sarson was different from the other two. The very small difference in amino acid composition between 'Nugget 1' and 'Nugget 2', although the amount of A-IV-S protein in the former was 3.4 times that of the latter, would be expected from what is known about the strong genetic control of any 'template-governed' synthesis of a macromolecule.

Greater differences were observed in the B-I protein composition[133]. Since this protein is composed of several subunits, which are held together by non-covalent forces, one could expect some flexibility in the proportions of different

TABLE 7:19

AMINO ACID ANALYSES OF A-IV-S PROTEINS FROM DIFFERENT SAMPLES OF RAPE AND TURNIP RAPE

mmoles/g protein (17.0% N)[133].

Amino acid	Target 1	Target 2	Brassica napus Bronowski	Nugget 1	Nugget 2	Yellow Sarson	Brassica campestris Echo	S-7165
Aspartic acid	0.16	0.18	0.15	0.15	0.15	0.18	0.14	0.17
Threonine	0.32	0.33	0.21	0.26	0.26	0.14	0.22	0.39
Serine	0.52	0.55	0.28	0.34	0.37	0.17	0.34	0.37
Glutamic acid	2.56	2.66	1.64	2.06	2.14	0.22	1.84	1.75
Proline	0.91	0.92	0.53	0.76	0.77	0.38	0.66	0.57
Glycine	0.56	0.56	0.34	0.46	0.48	0.28	0.38	0.39
Alanine	0.49	0.49	0.32	0.44	0.42	0.24	0.36	0.37
Valine	0.48	0.49	0.34	0.40	0.39	0.16	0.42	0.44
Cystine	0.31	0.32	None detected	0.39	0.41	None detected	None detected	None detected
Methionine	0.07	0.08	0.02	0.12	0.13	0.01	0.10	0.11
Isoleucine	0.30	0.32	0.20	0.29	0.28	0.12	0.23	0.22
Leucine	0.63	0.68	0.39	0.52	0.54	0.16	0.49	0.49
Tyrosine	0.07	0.07	0.04	0.06	0.07	0.13	0.08	0.04
Phenylalanine	0.16	0.17	0.11	0.14	0.15	0.18	0.12	0.11
Ammonia	1.60	1.76	1.36	1.55	1.64	0.18	1.44	1.60
Lysine	0.63	0.65	0.42	0.54	0.54	0.14	0.35	0.54
Histidine	0.30	0.31	0.20	0.26	0.25	0.13	0.23	0.18
Arginine	0.39	0.40	0.24	0.36	0.34	0.12	0.25	0.15
Total N (mmoles)	11.15	10.91	8.43	10.64	10.76	*	9.19	9.21
% N recovery	92	90	70	88	89		76	77

* Sample too small for accurate determination of protein N.

subunits, each with a 'template-fixed' amino acid pattern. There were differences both in the elution profiles and the amino patterns of the B-I protein complex of seeds of various species, cultivars and levels of mineral nutrition, although the sedimentation constants were similar. It was suggested that the limiting factor for the synthesis of A-IV-S of Nugget-1 was the sulpho-amino acids, causing a reduction in the amount of A-IV-S and a modification of subunit proportions in the B-I complex[133].

Further studies of the 12 S 'globulin' of Canadian rape and turnip rapeseed revealed great similarities in composition of this protein aggregate from the two species. Compositions were given for one of the subunits, a glycoprotein, that was isolated from both species[141].

In unpublished Swedish work, the presence of a great number of proteins in aqueous and alkaline extracts of *B. napus*, cv. Panter, has been demonstrated. Thus, about 20 weakly acidic proteins, about 20 neutral proteins and 5 basic proteins were detected. The basic proteins accounted for *ca.* 20 per cent of the total and had molecular weights in the range 15–20,000. Only *ca.* 5 per cent of the total soluble proteins had molecular weights of 50–75,000, and a major portion had molecular weights from *ca.* 120,000 to *ca.* 150,000[142].

Since there are considerably less dramatic effects of nutrient supply on the fatty acid patterns of rapeseed[51], one might suggest that a nutrient deficiency affects the amount of enzyme protein to a smaller extent that it does the 'storage' protein.

7.5.3 Enzymes

Since a discussion of the metabolism of rapeseed constituents in general is outside the scope of this book, only those catabolic enzymes known to exert an effect on oil or meal quality will be discussed: lipase, lipoxygenase (lipoxidase) and thio-glucoside glucohydrolase (myrosinase). Other enzymes reported to occur in dormant rapeseed are choline kinase[147], glyceric acid kinase[148] and glutamic acid decarboxylase[149]. Furthermore, protein fractions associated with enzymatic activities of the somewhat unspecific designations β-galactosidase, β-glucosidase and esterase have been studied[137, 139]. Enzymatic activity in fully mature, dry seeds is generally very low, and often not at all observable. At the onset of germination, there is, however, a rapid activation of the enzymes that are necessary to catabolize the storage materials to provide energy and precursors for anabolic reactions. Even dormant seeds have active catabolic enzymes, since a disintegration of the cells will give rise to split products of the storage material.

Rapeseed lipase activity is quite low in dormant seed, but increases rapidly at germination to a maximum activity that is 100 times that of dormant seeds[150]. An interesting species difference in lipase activation during germination has been reported, namely that the activity of *Brassica campestris* extracts in a certain assay system was five times greater after 5 and 7 days compared with *B. napus* extracts[150]. This difference may be of technological significance (*cf.* Table 5:5). The pH optimum for the lipase from 3-days-germinated rapeseed was 8.5, and the Michaelis–Menton constant 4.1×10^{-4} M with tributyrin as substrate. No studies were made on the specificity, but such studies have been undertaken with another crucifer, *Crambe abyssinica*, in which it was shown that the attack was essentially random with some preference for palmitic and non-preference for erucic acid at low extents of lipolysis[151]. This slight specificity is in accord with recently reported analyses of the composition of the free fatty acids of germinated rapeseed, which showed higher contents of palmitic and linoleic acids and a lower content of erucic acid in the free fatty acids of germinated seeds, compared with the pattern in the triacylglycerols of the dormant seeds[34].

The lipase activity of maturing rapeseed has been shown to decrease during the seed maturation, provided the activity was calculated on a dry-matter basis. If calculated in proportion to the amount of seed water, it was constant during the period studied[152].

Little is reported on rapeseed lipoxygenase, but there seems to exist lipoxygenase activity in dormant rapeseed, even though typical assay systems show no activity[153]. This probably depends on the presence of inhibitor(s) in dormant rapeseed. It has been reported that the activity rapidly increases upon germination[153].

The thioglucoside glucohydrolase (E.C. 3.2.3.1), often called myrosinase, of rape and mustard seeds has been studied in many laboratories. In fact it seems to be the most extensively studied of all enzymes in dormant rape or mustard seed. For early papers in this field, one as old as from 1840, the reader is referred to a comprehensive review[94].

As discussed under 7.4.2 in connection with the glucosinolates, they are separated from the degradative enzyme in the intact tissue. Actually, the thioglucoside glucohydrolase is located in special cells, so-called idioblasts, whereas the glucosinolates appear to be present in almost all cells and tissues of a *Brassica* plant (see ref. 94). Some years ago, the question of whether myrosinase is one or two enzymes was under debate[154–157]. Now there seems to be general agreement that there is only one enzymatic step in the splitting of the glucosinolate, namely a hydrolytic cleavage of the β-thioglucosidic bond, followed by spontaneous (chemical) rearrangements of the 'glucose-less' glucosinolate ion to produce isothiocyanate and bisulfate at pH 7 in dilute solutions, as originally suggested some time ago[158, 159]. Under other conditions various split products develop, as discussed in detail on p.154. In view of the non-stereospecificity of several typical autolysis products, it has been suggested that the factors governing the production of aglucone derivatives other than isothiocyanate are non-enzymic[118]. Whereas there is thus no evidence for more than one enzymatic step in the breakdown of glucosinolates, there appears to be more than one form of thioglucoside glucohydrolase present in seeds. Thus, for example, certain protein fractions with myrosinase activity are stimulated by sodium ascorbate addition whereas others are not[157, 163]. The ascorbate stimulation of the enzyme is reported from several other studies[115, 157, 159]. On the other hand, an enzyme with very similar properties, isolated from *Aspergillus sudowii*, displayed no activity changes upon ascorbate addition[161]. Since *Aspergillii* have been found on rapeseed (see p.78), the so-called sinigrinase of *Aspergillus* might deserve some attention from the oilseed chemists' point of view. It should be noted that recent serological studies have also revealed the presence of two proteins of *B. juncea* (brown mustard) with thioglucoside glucohydrolase activity[162]. In unpublished Swedish studies, two protein fractions from *B. napus* and four from *Sinapis alba* with thioglucosides glucohydrolase activity have been isolated[163]. Furthermore, other recent studies have provided evidence for the presence of isoenzymes of thioglucoside gluco-hydrolase in seeds of *B. napus*[164]. There were also differences in isoenzyme patterns of various organs[164].

For a long time, it has been assumed that 'myrosinase' was specific for the splitting of β-thioglucosidic bonds (see ref. 94). Recently, it has been shown that purified preparations of *B. juncea* thioglucoside glucohydrolase also attack other glucosides, provided they possess a β-glucosidic bond[165]. The rate of breakdown of glucosinolates is, however, much greater. Studies on *B. nigra* thioglucoside glucohydrolase have revealed that this enzyme can also transfer the D-glucopy-ranosyl residue to an accepter other than water[166].

The pH and temperature optima, the temperature stability both *in situ* and in solution, the Michaelis–Menton constants and the inhibition of enzymatic activity have been studied on preparations of different purities and from various seed materials[115, 157, 160]. In view of the far-from-clear situation with regard to enzyme identities, such data will not be presented here in detail. It should, how-ever, be mentioned that the enzyme preparations generally display a broad pH optimum around 7[115, 157].

7.5.4 Total amino acid composition

Since the content of free amino acids in mature seeds is negligible, an analysis of a hydrolysate of the seed meal from *Brassica campestris* or *Brassica napus* provides a measure of the average amino acid composition of the proteins, soluble and insoluble (*cf.* Table 15:4). Although this information is primarily of interest in the estimation of the nutritional quality of rapeseed meal (see Chapter 15), any specific, varietal or environmental differences in gross amino acid composition could indicate differences in proportions of proteins with specific amino acid patterns or differences in amino acid composition of individual proteins. Such information would undoubtedly be of chemical and biochemical interest.

The inter-specific variation in content of total nitrogen (in proximate analysis synonymous with protein) among the species of interest is considerable on a

PROTEIN CONTENT OF SOME *Brassica* SEED MEALS AND NITROGEN DISTRIBUTION AFTER ACID HYDROLYSIS[126]

Species	Protein per cent		Nitrogen distribution as per cent of total nitrogen			
	Whole seed	Extracted meal	Amino acids	Ammonia	Insoluble	Unknown
Brassica campestris*	23.8	41.5	74.6	12.8	2.9	9.7
B. juncea**	27.5	43.8	73.1	12.5	3.5	10.9
B. napus*	23.8	43.1	71.9	12.4	2.4	13.3
Sinapis alba†	31.2	43.4	71.5	14.3	2.3	11.9

* Mean of four seed samples.
** Mean of two seed samples.
† In the original paper named *Brassica hirta*[113], a name now considered less appropriate for the species studied.

'whole seed' basis, but rather small on an 'extracted meal' basis (Table 7:20). Only 72–75 per cent of the total N can be accounted for as amino acids, with slightly more than 10 per cent each of 'ammonia-N' and 'unknown-N'. A considerable portion of the 'unknown-N' could stem from the glucosinolates.

Two different investigations have indicated that the turnip rape proteins are richer in lysine than those in rapeseed[5, 126]. Generally, the content of essential amino acids seemed somewhat higher in turnip rape, as reflected in higher quality indices for nutritional value[126]. Varietal differences have been observed, but no statistically designed studies seem to have been reported so far[5]. A highly significant effect of growth locality on the lysine content and a significant effect on the content of histidine, arginine, phenylalanine and leucine were noted in Canadian studies[5]. From Swedish studies, utilizing soil-free cultures, very moderate effects of nitrogen fertilization on the amino acid pattern could be observed, although seed yield was markedly increased by nitrogen addition and the nitrogen content of the meal was significantly increased[121]. On the other hand, very substantial differences were observed on addition of sulphate to the nutrient solution, notably for arginine, cystine and aspartic acid[121]. In this experiment, the differences in seed yield and protein content were, however, more drastically changed by sulphate addition[110]. It should be noted that the amount of cystine increased relatively more on sulphate addition than did methionine.

Of considerable interest is the observation that the amount of methionine used for the biosynthesis of glucosinolates, in proportion to that present in protein (or free?), is drastically reduced in the 'minus'-S seeds compared to 'highest'-S seeds, 0.05 *versus* 1.29[121]. In several studies on other plants, it has been found that the addition of mineral fertilizers generally affects the content of protein and the amount of non-protein amino acids, but more seldom the protein amino acid pattern (see references cited in ref. 121).

7.6 NUCLEIC ACIDS

Apparently no detailed studies on rapeseed nucleic acids have as yet been reported. It has, however, been mentioned that aqueous ammonia solutions of defatted rapeseed contain considerable amounts of nucleic acids[124]. Nucleic acid in such an extract of slightly immature seeds accounted for approximately 10 per cent of the nitrogen. It appears that proportionately more of the total nitrogen present in aqueous solutions of immature seeds refers to nucleic acid, resulting in increasing recovery of nitrogen as amino acids, as the maturation proceeds[124].

7.7 OTHER ORGANIC COMPOUNDS

7.7.1 Sinapine

By far the most studied of all non-lipid constituents of rape and turnip rape are the glucosinolates. Whereas the major glucosinolates of rape and turnip rape were isolated as their potassium salts, the p-hydroxyphenyl glucosinolate of white mustard was isolated as the sinapine salt[92]. Probably because of the early association of the glucosinolates with sinapine, this compound seems to be the longest known and best characterized of the minor constituents of defatted meals of crucifers. The structure of sinapine, the choline ester of 3,5-dimethoxy-4-hydroxy-cinnamic acid, or sinapic acid, is shown in Fig.7:4, together with the natural hydrolysis products. It has been suggested that both hydrolysis products are im-

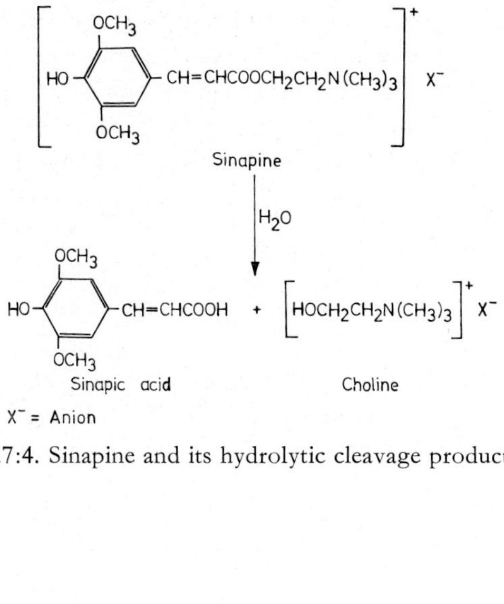

Fig.7:4. Sinapine and its hydrolytic cleavage products[169].

portant metabolites in higher plants, sinapic acid for the biosynthesis of lignins and flavonoids, and choline in methylation reactions. Sinapine is rapidly hydrolyzed in germinating white mustard seed[167] but no degradation takes place when ground seeds are autolyzed. It has a bitter taste and one might therefore suppose it to influence the palatability of rapeseed meal (see Chapter 15), but no toxic effects were found in feeding experiments with chickens[168] and rats[169]. The content of sinapine was reported to vary between 0.78 per cent and 1.33 per cent for 31 genotypes of rapeseed, and was markedly influenced by soil and climatic conditions[170]. A single sample of summer turnip rape studied in another investigation had 0.92 per cent sinapine in the seed meal[169]. It was found that some of the sinapine of *Crambe abyssinica* was present as the glucoside, 1-sinapoyl-β-D-glucose [169]. Most probably, also some of the sinapine of rape and turnip rape is present as glucoside. There does not appear to be any proportionality between the amount of glucosinolates and the amount of sinapine, since the proportion of the two compounds is about 15 to 1 for crambe but 5 to 1 for rapeseed and white mustard.

7.7.2 Flavonoids

Recent Canadian studies have revealed the presence of oligomeric phenols in aqueous extracts of seed coats of *Brassica campestris*[171]. The major part of the phenolic polymers was, however, strongly bound to other structures of the seed coat. Hot acidic butanol liberated alcohol-soluble pigments, three of which were identified as cyanidin, pelargonidin and malvidin. No free pigments of the anthocyanidin class seemed to be present. Several unknown phenolic acids were also revealed in alcoholic extracts, both after acid and after alkaline hydrolysis of seed coats. It was therefore concluded that a condensed tannin occurs in the seed coats of turnip rape[171]. Apparently little is known about the polyphenols of *B. napus*. Unpublished studies have indicated a difference in polyphenol content of rapeseed *versus* white mustard seed[172]. Indications of polyphenols have also been reported from studies on *Brassica napus* and *B. nigra* (ref. 173 and *loc. cit.*). The presence of flavonol glucosides in seeds of *Brassica campestris* has been recorded[174]. The only flavones present were rutin and its hydrolysis products.

So far nothing seems to be reported on rapeseed lignins, although indirect evidence for the presence of appreciable amounts of lignins has been recorded.

7.8 INORGANIC CONSTITUENTS

Data on individual elements have generally been obtained by analysis of commercial rapeseed meals (see, for example, Table 15:1). Very recently, detailed studies on the mineral composition of seed meals of the two cultivars Regina II and

CONTENT OF SOME ELEMENTS IN DEFATTED SEED MEALS OF *Brassica napus*[175]

Element	Cultivar	
	Regina II	Bronowski
S (%)	1.72	0.77
P (%)	1.10	1.22
Ca (%)	0.70	1.11
K (%)	1.61	1.25
Mg (%)	0.56	0.57
Na (mg/kg)	7	6
Cu (mg/kg)	4.7	5.2
Zn (mg/kg)	68	62
B (mg/kg)	18	24
Fe (mg/kg)	7	8
Mn (mg/kg)	37	37

Bronowski of *Brassica napus* have been published[175]. The latter cultivar is characterized by its low content of glucosinolates (see 7.4.2), whereas the former has a
'normal' level of these compounds (Table 7:21). Statistical analysis demonstrated
that the Regina cultivar was very significantly higher in S and K, but lower in
P, Ca, and B. Other differences were not significant. Since the glucosinolate
anion is generally balanced by potassium as a cation[94] the concomitant reduction
of S and K in the low glucosinolate genotype Bronowski is not surprising.

REFERENCES

1   H. R. SALLANS, *J. Am. Oil Chem. Soc.*, 41 (1964) 215.
2   B. LÖÖF, *Svensk Frötidning* 38 (1969) 123. (In Swedish).
3   W. SCHUSTER *Fette, Seifen, Anstrichmittel*, 69 (1967) 831.
4   G. STOLLE, *Züchter*, 24 (1954) 202.
5   D. R. CLANDININ AND L. BAYLY, *Can. J. Animal Sci.*, 43 (1963) 65.
6   A. T. H. GROSS AND B. R. STEFANSSON, *Can. J. Plant Sci.*, 46 (1966) 389.
7   G. ANDERSSON, O. HALL AND B. LÖÖF, *Sveriges Utsädesforen. Tidskr.*, 79: 3–4 (1969) 248.
8   F. W. SOSULSKI AND A. BAKAL, *Can. Inst. Food. Technol. J.*, 2 (1969) 28.
9   C. G. YOUNGS, *Fats and Oils in Canada, Semi-Annual Rev.*, 2:2 (1967) 39, Department of
Industry, Ottawa, Canada.
10   R. TKACHUK, *Cereal Chem.*, 46 (1969) 419.
11   L. R. WETTER, in *Rapeseed Meal for Livestock and Poultry—A Review*, The Canada Department of Agriculture, Publication 1257, Ottawa, 1965, p. 32.
12   D. CHAPMAN, *Introduction to Lipids*, McGraw-Hill, London, 1968.
13   H. BROCKERHOFF AND M. YURKOWSKI, *J. Lipid Res.*, 7 (1966) 62.
14   L.-Å. APPELQVIST, *Hereditas*, 61 (1969) 9.
15   H. BROCKERHOFF, *J. Lipid Res.*, 6 (1965) 10.

16 W. E. M. LANDS, R. A. PIERINGER, P. M. SLAKEY AND A. ZSCHOCKE, *Lipids*, 1 (1966) 444.
17 F. H. MATTSON AND R. A. VOLPENHEIN, *J. Biol. Chem.*, 236 (1961) 1891.
18 G. JURRIENS AND L. SCHOUTEN, *Rev. Franc. Corps Gras*, 12 (1965) 505.
19 H. GRYNBERG AND H. SZCZEPANSKA, *J. Am. Oil Chem. Soc.*, 43 (1966) 151.
20 L.-Å. APPELQVIST AND R. J. DOWDELL, *Arkiv Kemi*, 28 (1968) 539.
21 R. D. HARLOW, C. LITCHFIELD AND R. REISER, *Lipids*, 1 (1966) 216.
22 U. PERSMARK, *Paper presented at World Fat Congress Chicago, 1970.*
23 C. LITCHFIELD, *Chem. Ind. (London)*, (1970) 341.
24 U. PERSMARK, unpublished data.
25 C. LITCHFIELD AND R. REISER, *J. Am. Oil Chem. Soc.*, 42 (1965) 757.
26 L.-Å. APPELQVIST AND R. JÖNSSON, *Z. Pflanzenzuecht.*, 63 (1970) 340.
27 M. E. McKILLICAN, *J. Am. Oil Chem. Soc.*, 43 (1966) 461.
28 L.-Å. APPELQVIST, unpublished observations.
29 L.-Å. APPELQVIST, *Paper presented at World Fat Congress Chicago, 1970.*
30 T. P. HILDITCH AND P. N. WILLIAMS, *The Chemical Constitution of Natural Fats*, Chapman and Hall, London, 4th ed., 1964.
31 T. KIRIBUCHI, C. S. CHEN AND S. FUNAHASHI, *Agr. Biol. Chem. (Tokyo)*, 29 (1965) 265.
32 S. J. PURDY AND E. V. TRUTER, *Proc. Royal Soc. (London)*, Ser. B, 158 (1963) 544.
33 L.-Å. APPELQVIST, *Sveriges Utsädesfören. Tidskr.*, 71:1 (1961) 74 (In S With E.s)
34 L.-Å. APPELQVIST in A. RUTKOWSKI (Ed.), *International Symposium for the Chemistry and Technology of Rapeseed Oil and other Cruciferae Oils*, Warszawa, 1970, p. 45.
35 K. J. GOERING, R. ESLICK AND D. L. BRELSFORD, *Econ. Botany*, 19 (1965) 251.
36 K. L. MIKOLAJCZAK, T. K. MIWA, F. R. EARLE, I. A. WOLFF AND Q. JONES, *J. Am. Oil Chem. Soc.*, 38 (1961) 678.
37 R. W. MILLER, F. R. EARLE, I. A. WOLFF AND Q. JONES, *J. Am. Oil Chem. Soc.*, 42 (1965) 817.
38 L.-Å. APPELQVIST, in manuscript.
39 P. CAPELLA AND E. FEDELI, unpublished observations.
40 D. FIRESTONE, unpublished observations.
41 D. F. KUEMMEL *J. Am. Oil Chem. Soc.* 41 (1964) 667.
42 R. G. ACKMAN, *J. Am. Oil Chem. Soc.*, 43 (1966) 483.
43 J. L. IVERSON, *J. Assoc. Offic. Anal. Chemists*, 49 (1966) 332.
44 L.-Å. APPELQVIST, *Arkiv Kemi*, 28 (1968) 551.
45 T. P. HILDITCH, T. RILEY AND N. L. VIDYARTI, *J. Soc. Chem. Ind.*, 46 (1927) 457.
46 E. W. HAEFFNER, *Lipids*, 5 (1970) 430.
47 F. B. SHORLAND, *Nature*, 156 (1945) 269.
48 J. K. HEYES AND F. B. SHORLAND, *Biochem. J.*, 49 (1951) 503.
49 L.-Å. APPELQVIST, *Acta Agr. Scand.*, 18 (1968) 3.
50 A. JAKUBOWSKI AND P. KRASNODEBSKI, *Inst. Przemyslu Tluszczowego*, 16 (1966) 105, (In Polish).
51 L.-Å. APPELQVIST, *Physiol. Plantarum*, 21 (1968) 455.
52 L.-Å. APPELQVIST, *Physiol. Plantarum*, 21 (1968) 615.
52a L.-Å. APPELQVIST, *Riv. Ital. Sostanze Grasse*, 46 (1969) 478.
53 L.-Å. APPELQVIST, *Fette, Seifen, Anstrichmittel*, 72 (1970) 783.
54 B. M. CRAIG, *Can. J. Plant Sci.*, 41 (1961) 204.
55 F. DEMBIŃSKI, P. KRASNODEBSKI AND T. ORLOWSKA, *Pamietnik pulawski-Prace Iung*, 25 (1967) 5.
56 R. K. DOWNEY, *Can. Food Ind.*, 34 (1963) 34.
57 R. DELHAYE AND A. GUYOT, *Bull Rech. Agron. Gembloux*, 4 (1969) 44.
58 F. G. SIETZ, *Fette, Seifen, Anstrichmittel*, 69 (1967) 325.
59 D. GRIECO AND G. PIEPOLI, *Riv. Ital. Sostanze Grasse*, 41 (1964) 283.
60 J. KRZYMANSKI, *Hodlowa Rósl. Aklim. Nasien.*, 14 (1970) 95.
61 D. CANVIN, *Can. J. Botany*, 43 (1965) 63.
62 M. BECHYNE AND Z. P. KONDRA, *Can. J. Plant Sci.*, 50 (1970) 151.
63 L.-Å. APPELQVIST, *Paper presented at DGF-Tagung, Düsseldorf, 1970.*
64 L.-Å. APPELQVIST, *Acta Univ. Lund. Sectio II*, No 7 (1968).

65   G. Jacini, E. Fedeli and A. Lanzani, *J. Assoc. Offic. Anal. Chemists*, 50 (1967) 84.
66   F. R. Earle, J. E. Peters, J. A. Wolff and G. A. White, *J. Am. Oil Chem. Soc.*, 43 (1966) 330.
67   E. W. Eckey, *Vegetable Fats and Oils*, A.C.S. Monograph No. 123, Reinhold, New York, 1954.
68   P. Capella, E. Fedeli, M. Cirimele and G. Jacini, *Riv. Ital. Sostanze Grasse*, 40 (1963) 603.
69   H. Niewiadomski, I. Bratkowska and E. Mossakowska, *J. Am. Oil Chem. Soc.*, 42 (1965) 731.
70   J. A. G. Box and H. A. Boekenoogen, *Fette, Seifen, Anstrichmittel*, 69 (1967) 724.
71   P. Capella, G. de Zotti, G. S. Ricci, A. F. Valentini and G. Jacini, *J. Am. Oil Chem. Soc.*, 37 (1960) 564.
72   E. Fedeli, A. Lanzani, P. Capella and G. Jacini, *J. Am. Oil Chem. Soc.*, 43 (1966) 254.
73   P. Capella and G. Losi, *Ind. Agr.*, 6 (1968) 277.
74   D. S. Ingram, B. A. Knights, I. J. MacEvoy and P. McKay, *Phytochemistry*, 7 (1968) 124.
75   A. Seher and E. Homberg, *Fette, Seifen, Anstrichmittel*, 70 (1968) 481.
76   A. Seher and E. Homberg, *Paper presented at DGF-Tagung, Düsseldorf, 1970.*
77   L. P. Vernon and G. R. Seely, *The Chlorophylls*, Academic Press, New York, 1966, 679 pp.
78   L.-Å. Appelqvist and S.-Å. Johansson, in manuscript.
79   M. W. Dicks, *Vitamin E Content of Foods and Feeds for Human and Animal Consumption, Bulletin 435*, Agr. Exptl. Station, Univ. of Wyoming, Laramie, USA, 194 pp.
80   K. J. Whittle and J. F. Pennock, *Analyst*, 92 (1967) 423.
81   U. Persmark, unpublished results.
81a J. C. Cowan, *J. Am. Oil Chem. Soc.*, 43 (1966) 300A.
82   L.-Å. Appelqvist, unpublished results.
83   L.-Å. Appelqvist, *J. Am. Oil Chem. Soc.*, 44 (1967) 209.
84   T. Mizuno, *Nippon Nogeikagaku Kaishi*, 32 (1958) 340. (In Japanese)
85   J. Hrdliča, H. Koztowska, J. Pokorny and A. Rutkowski, *Nahrung*, 9 (1965) 71.
86   E. L. Hirst, D. A. Rees and N. G. Richardson, *Biochem. J.*, 95 (1965) 453.
87   K. Bailey and F. W. Norris, *Biochem. J.*, 26 (1932) 1609.
88   K. Bailey, *Biochem. J.*, 29 (1935) 2477.
89   G. T. Grant, Carol McNab, D. A. Rees and J. Skerrett, *Chem. Commun.*, (1969) 805.
90   B. Törnell, personal communication.
91   O. Robiquet and Z. Z. Boutron, *J. Pharm. Chim.*, 17 (1831) 279.
92   M. G. Ettlinger and G. P. Dateo, Jr., *Studies of Mustard Oil Glucosides, Final Rept. Contract DA19-129-QM-1059*, U.S. Army Natick Laboratories, Natick, Massachusetts, 1961.
93   M. G. Ettlinger and A. Kjaer in T. J. Mabry (Ed.), *Recent Advances in Phytochemistry*, Vol 1. Appleton-Century-Croft, New York, 1968, 100 pp.
94   A. Kjaer in L. Zechmeister (Ed.), *Progress in the Organic Chemistry of Natural Products*, Vol. 18, Springer-Verlag, Vienna, 1960, pp.122–176.
95   A. Kjaer, in T. Swain (Ed.), *Comparative Phytochemistry*, Chapter 11, Acaedmic Press London, 1966.
96   C. H. Van Etten, M. E. Daxenbichler and I. A. Wolff, *J. Agr. Food. Chem.*, 17 (1969) 483.
97   M. G. Ettlinger and A. J. Lundeen, *J. Am. Chem. Soc.*, 78 (1956) 1952.
98   E. B. Astwood, M. A. Greer and M. G. Ettlinger, *J. Biol. Chem.*, 181 (1949) 121.
99   A. Kjaer, J. Conti and K. A. Jensen, *Acta Chem. Scand.*, 12 (1953) 1271.
100 K. A. Jensen, J. Conti and A. Kjaer, *Acta Chem. Scand.*, 12 (1953) 1267.
101 A. Kjaer, J. Conti and I. Larsen, *Acta Chem. Scand.*, 7 (1953) 1276.
102 A. Kjaer and R. B. Jensen, *Acta Chem. Scand.*, 10 (1956) 1365.
103 M. G. Ettlinger and C. P. Thompson, *Final Report to Quartermaster Research and Engineering Command, Contract DA 19-129-QM-1689 (1962), AD-290 747*, Office of Technical Services, U.S. Dept. of Commerce.
104 A. Kjaer and K. Rubinstein, *Acta Chem. Scand.*, 8 (1954) 598.

171

105 J. G. Vaughan, J. S. Hemingway and H. J. Schofield, *J. Linn. Soc. (Botany)*, 58 (1963) 435.
106 B. A. Tapper and D. B. MacGibbon, *Phytochemistry*, 6 (1967) 749.
107 M. E. Daxenbichler, C. H. Van Etten, F. S. Brown and Q. Jones, *J. Agr. Food Chem.*, 12 (1964) 127.
108 C. G. Youngs and L. R. Wetter, *J. Am. Oil Chem. Soc.*, 44 (1967) 551.
109 E. Josefsson and C. Mühlenberg, *Acta Agr. Scand.*, 18 (1968) 97.
110 E. Josefsson and L.-Å. Appelqvist, *J. Sci. Food Agr.*, 19 (1968) 564.
111 E. Josefsson, Pattern, content and biosynthesis of glucosinolates in some cultivated cruciferae, *Dissertation*, University of Lund, 1970.
112 K.-A. Lein and W. J. Schöön, *Angew. Botanik* 43 (1969) 87.
113 C. H. Van Etten, M. E. Daxenbichler, J. E. Peters, I. A. Wolff and A. N. Booth, *J. Agr. Food Chem.*, 13 (1965) 24.
114 J. E. McGhee, L. D. Kirk and G. C. Mustakas, *J. Am. Oil Chem. Soc.*, 42 (1965) 889.
115 L.-Å. Appelqvist and E. Josefsson, *J. Sci. Food Agr.*, 18 (1967) 510.
116 C. H. Van Etten, M. E. Daxenbichler, J. E. Peters and H. L. Tookey, *J. Agr. Food Chem.*, 14 (1966) 426.
117 M. E. Daxenbichler, C. H. Van Etten, W. H. Tallent and I. A. Wolff, *Can. J. Chem.*, 45 (1967) 1971.
118 H. L. Tookey and I. A. Wolff, *Can. J. Biochem.*, 48 (1970) 1024.
119 L. R. Wetter, *Can. J. Biochem. Physiol.*, 33 (1955) 980.
120 L. R. Wetter, *Can. J. Biochem. Physiol.*, 35 (1957) 293.
121 E. Josefsson, *J. Sci. Food. Agr.*, 21 (1970) 98.
122 Z. P. Kondra and R. K. Downey, *Can. J. Plant Sci*, 49 (1969) 623.
123 I. Zeman, D. Zemanová and J. Nováková, in A. Rutkowski (Ed.), *International Symposium for the Chemistry and Technology of Rapeseed Oil and other Cruciferae Oils*, Warszawa, 1970, p.499.
124 A. J. Finlayson, *Can. J. Biochem.*, 45 (1967) 1225.
125 P. O. Larsen, *Free Amino Acids in Cruciferae and Resedaceae, Risö Report No. 189, Dissertation*, Copenhagen, 1969.
126 R. W. Miller, C. H. Van Etten, C. McGrew, I. A. Wolff and Q. Jones, *J. Agr. Food Chem.*, 10 (1962) 426.
127 J. Hrdlička, J. Pokorny, A. Rutkowski and M. Wojciak, *Nahrung*, 9 (1965) 77.
128 J. Pokorny, Scient. Papers from Inst. Chem. Techn. Prague, *Food Technol.*, 7:2 (1963) 149.
129 J. Pokorny, Scient. Papers from Inst. Chem. Techn. Prague, *Food Technol.*, 7:2 (1963) 157.
130 J. Pokorny, M. Vodicka and J. Zalud, Scient. Papers from Inst. Chem. Techn. Prague, *Food Technol.*, 7:2 (1963) 167.
131 J. Pokorny and Z. Sefr, Scient. Papers from Inst. Chem. Techn. Prague, *Food Technol.*, 8:2 (1964) 234.
132 J. Pokorny, Z. Sefr and J. Zalud, Scient. Papers from Inst. Chem. Techn. Prague, *Food Technol.*, E 22 (1968) 113.
133 A. J. Finlayson, R. S. Bhatty and C. M. Christ, *Can. J. Botany*, 47 (1969) 679.
134 L. R. Wetter, *Paper Read at the International Conf. for the Science, Techn. and Marketing of Rapeseed and Rapeseed Products, Ste. Adéle, Canada, Sept., 1970*.
135 K. E. Eapen, N. W. Tape and R. P. A. Sims, *J. Am. Oil Chem. Soc.*, 46 (1969) 52.
136 R. P. A. Sims, *J. Am. Oil Chem. Soc.*, 48 (1971) 733.
137 J. G. Vaughan and A. Waite, *J. Exptl. Botany*, 18 (1967) 269.
138 J. G. Vaughan, A. Waite, D. Boulter and S. Waiters, *J. Exptl. Botany*, 17 (1965) 332.
139 J. G. Vaughan and A. Waite, *J. Exptl. Botany*, 18 (1967) 100.
140 J. G. Vaughan, E. Gordon and D. Robinson, *Phytochemistry*, 7 (1968) 1345.
141 L. A. Goding, R. S. Bhatty and A. J. Finlayson, *Can. J. Biochem.*, 48 (1970) 1096.
142 J. C. Jansson, S. Å. Liedén, B. Lönnerdal and J. Porath, unpublished results.
143 R. S. Bhatty, S. L. McKenzie and A. J. Finlayson, *Can. J. Biochem.*, 46 (1968) 1191.
144 W. J. Wolf, *Cereal Sci. Today*, 14:3 (1969) 75.
145 A. J. Finlayson, *Can. J. Plant Sci.*, 45 (1965) 184.
146 R. Tkachuk and G. N. Irvine, *Cereal Chem.*, 46 (1969) 206.

147   T. Ramasarma and L. R. Wetter, *Can. J. Biochem. Physiol.*, 35 (1957) 853.

148   K. Ozaki and L. R. Wetter, *Can. J. Biochem. Physiol.*, 38 (1960) 125.

149   L.-Å. Appelqvist, unpublished studies.

150   L. R. Wetter, *J. Am. Oil Chem. Soc.*, 34 (1957) 66.

151   H. L. Tookey and I. A. Wolff, *J. Am. Oil Chem. Soc.*, 41 (1964) 602.

152   A. Rutkowski and Z. Makus, *Oleagineux*, 13 (1958) 203.

153   W. Franke and H. Frehse, *Z. Physiol. Chem.*, 298 (1954) 1.

154   R. D. Gaines and K. J. Goering, *Biochem. Biophys. Res. Commun.*, 2 (1960) 207.

155   R. D. Gaines and K. J. Goering, *Arch. Biochem. Biophys.*, 96 (1962) 13.

156   P. Calderon, C. S. Pedersen and L. R. Mattick, *J. Agr. Food Chem.*, 14 (1966) 665.

157   I. Tsuruo, M. Yoshida and T. Hata, *Agr. Biol. Chem. (Tokyo)*, 31 (1967) 18.

158   M. G. Ettlinger and A. J. Lundeen, *J. Am. Oil Chem. Soc.*, 78 (1956) 4172; 79 (1957) 1764.

159   Z. Nagashima and M. Uchiyama, *Bull. Agr. Chem. Soc. Japan*, 23 (1959) 555.

160   M. G. Ettlinger, G. P. Dateo, Jr., B. W. Harrison, T. J. Mabry and C. P. Thompson, *Proc. Natl. Acad. Sci. U.S.*, 47 (1961) 1875.

161   M. Ohtsuru, I. Tsuruo and T. Hata, *Agr. Biol. Chem. (Tokyo)*, 33 (1969) 1315.

162   J. G. Vaughan and E. I. Gordon, *Phytochemistry*, 8 (1969) 883.

163   B. Lönnerdal, J. C. Jansson and J. Porath, unpublished results.

164   D. B. MacGibbon and R. M. Allison, *Phytochemistry*, 9 (1970) 541.

165   I. Tsuruo and T. Hata, *Agr. Biol. Chem. (Tokyo)*, 32 (1968) 1425.

166   G. A. Howard and R. D. Gaines, *Phytochemistry*, 7 (1968) 585.

167   A. Tzagoloff, *Plant Physiol.*, 38 (1963) 202.

168   D. R. Clandinin, *Poultry Sci.*, 40 (1961) 484.

169   F. L. Austin and I. A. Wolff, *Agr. Food Chem.*, 16 (1968) 132.

170   P. Schwartze, *Naturwissenschaften*, 36 (1949) 88.

171   A. B. Durkee, *Phytochemistry*, 11 (1971) 1583.

172   P. Capella, personal communication.

173   E. C. Bate-Smith and P. Ribereau-Gayon, *Qualitas Plant. Mater. Vegetabiles*, 5 (1958) 189.

174   M. T. Francois and L. Chaic, *Bull. Soc. Pharm. Nancy*, 46 (1960) 21.

175   E. Josefsson, *Physiol. Plant*, 24 (1971) 150.

Analysis of Rapeseed Oil

U. PERSMARK

AB Karlshamns Oljefabriker, Karlshamn (Sweden)

CONTENTS

8.1 INTRODUCTION

The analytical methods, techniques, and instruments used for the evaluation and characterization of the properties of rapeseed oil are basically the same as those commonly used for other oils and fats. Therefore, the intention of this chapter is not to give any detailed descriptions of such methods*, but instead information on the most appropriate methods for accurate analysis of specific constituents of rapeseed oil will be presented.

8.2 FATTY ACIDS

Although the triacylglycerols are the natural components that normally constitute

---

* See Appendix II and refs. 1–3.

about 97–98 per cent of crude rapeseed oil, it is by tradition and for practical analytical reasons the fatty acids which represent the constitutional units of oils and fats.

The most characteristic fatty acid constituents in the plants of the Cruciferae family are, as discussed in detail in Chapter 7, the long chain monoenoic acids, *cis*-11-eicosenoic and *cis*-13-docosaenoic (erucic) acid. Knowledge about erucic acid in rapeseed oil dates from the past century[4]. As an example of classical and fundamental research on fatty acid analysis and the composition of fats and oils including rapeseed oil before the advent of the technique of gas chromatography, work by Hilditch *et al.*[5, 6] may be mentioned. Their analytical data were usually based on separation methods, *e.g.* low temperature crystallization, vacuum fractionation of methyl esters, and fractionation of fatty acid salts, and subsequent identification by various chemical and physical means, *e.g.* oxidation/fragmentation, derivatisation, refractometry, and spectroscopy. Compared to methods nowadays available for fatty acid separation and identification, the earlier methods were of course more crude and imprecise, and some components, for example eicosenoic acid, were usually not idntified. Furthermore, from a practical point of view the earlier methods were very time-consuming and manipulative, and the performance of fatty acid analyses was therefore limited to large, well-equipped laboratories. For this reason, data on rapeseed oil are not very frequent in the early literature. In general, however, several of the data reported from earlier investigations of rapeseed fatty acids[4] agree rather well with more recent data on the contents of the major components.

The introduction of chromatographic methods, especially gas chromatography, radically changed the possibilities for obtaining compositional data from biological materials. The continuous increase in the refinement and exactness of the chromatographic separations has made them a powerful analytical tool[7]. Their simplicity, together with their rapidness, has resulted in extensive amounts of compositional data on rapeseed fatty acids in only a few years, as discussed in detail in Chapter 7.

Fatty acid analyses of oils by gas chromatography are usually carried out after transesterification of the glycerides to their methyl esters by one of the numerous esterification methods which have been described. The choice of a suitable method is however more a question of individual preference than any advantage in analytical accuracy[8]. The instrumental parameter that primarily determines the degree of separation of the fatty acids is the column with its stationary phase. For the fatty acids of rapeseed as well as other fatty acids, polar polyester phases, *e.g.* DEGS, EGS, BDS, EGA, DEGA and EGSS-X(Y)*, and non-polar type phases, *e.g.* silicon rubber gums, and Apiezons*, are most frequently used[9]. In

---

* The development of new or modified stationary phases with increased temperature stability and intermediate separation properties is continuously in progress.

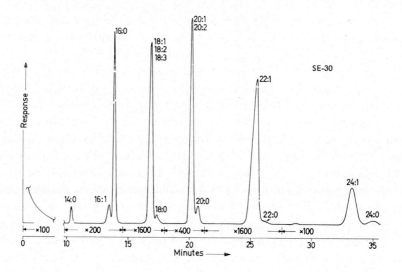

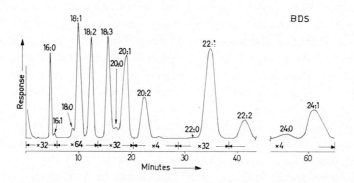

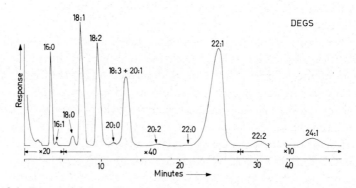

Fig.8:1. Rapeseed fatty acid methyl esters separated by gas chromatography[11].
A. SE-30 stationary phase (*ca.* 3 per cent). Temperature program: 120–220 °C (5 °C/min).
B. BDS stationary phase (*ca.* 10 per cent). Temperature: 190 °C.
C. DEGS stationary phase (*ca.* 10 per cent). Temperature: 180 °C.

practice the purpose of the analysis determines the choice of stationary phase. So far there exists no ideal stationary phase for rapeseed fatty acids. Principally the non-polar phases separate the components roughly according to chain-length, while polar phases are also able to separate according to unsaturation. It can not how-ever be taken for granted that the most highly polar phase is always preferable in an overall analysis. It is well known that with the most highly polar phases, overlap of peaks of fatty acids with chain lengths differing by two carbons is possible[10]. For rapeseed fatty acids such effects are especially noticeable between the eighteen and twenty carbon length fatty acids. Fig.8:1 illustrates the separation of rapeseed fatty acids on three common columns of varying polarity. The DEGS column is the most polar one, and it gives an almost complete separation between C18:0 and C18:1. Furthermore linolenic acid is retained so strongly that it appears together with the methyl ester with chain length two carbons longer. The less polar BDS column gives an incomplete separation between C18:0 and C18:1, but C18:3 is eluted ahead of C20:0. The non-polar SE-30 column separates the saturated acids from the unsaturated of the same chain length, while practically no separation is obtained between the unsaturated. These relations are of course valid for other chain lengths, *e.g.* C16/C18 and C20/C22, but the acids involved are present in smaller amounts, and are therefore mostly of minor importance (*cf*. Table 8:2). A common error in the identification of the main components of Cruciferae fatty acids (*cf*. Chapter 7) as measured from its frequency in the literature is confusion between C18:3, C20:0 and C20:1 (*cf*. p. 137).

The identification of components from GLC analysis is usually based on retention data, and consequently must be regarded as tentative. To circumvent the variables originating from the instrumental parameters, relative retention data are preferred. Various systems have been worked out to facilitate identifications based on such data. Farquhar *et al*.[12] used a linear graph which was constructed from the logarithms of the relative retention times from two different columns (so-called log–log diagram). Fatty acid composition data based on this system are compiled in ref. 13. Another system nowadays primarily used for fatty acids is the equivalent chain length (ECL) value[14] or the carbon number[15]. Both methods are basically the same: A graph is constructed by plotting the retention time (on a logarithmic scale) *versus* the number of carbon atoms from at least two normal saturated fatty acid methyl esters. From this graph equivalent chain length values or carbon numbers can be obtained for other methyl esters in the sample from their retention data. These values are primarily dependent upon the stationary phase. By comparison with known values from the literature or with model substances from one or several different stationary phases a fairly accurate identifi-cation may be achieved. This system has been further developed with regard to the relationship between ECL values and fatty acid structures[16-22]. Evidence has been provided for linearity between esters of similar 'end carbon chain' (ω-number)[17], and so-called separation factors have also been used for the identifi-

TABLE 8:1

ECL-VALUES OF FATTY ACID METHYL ESTERS OCCURRING IN CRUCIFERAE OILS

Stationary phase		DEGS	EGS	-CDX-Ac	Apiezon-L	BDS	Reseflex 446	BDS
Temperature °C		226	180/200	234	240	200	190/210	178/180
Ref.		23	23	23	23	25	14	11
Fatty acid								
16:1	ω7	16.55	16.56	16.55	15.70		16.4	16.38*
16:2	ω6	17.50	17.32	17.25	15.47			
16:3	ω3	18.52	18.33	18.10	15.47			
17:1	ω8	17.60	17.55	17.56	16.73			
18:0						18.00		
18:1	ω9	18.51	18.50	18.55	17.71		18.4	18.35
18:2	ω6	19.30	19.22	19.23	17.53	18.91	19.0	18.90
18:3	ω3	20.40	20.13	20.10	17.51	19.64	19.8	19.63
20:0						20.00		
20:1	ω9	20.44	20.38	20.32	19.78	20.32		20.33
20:2	ω6	21.36	21.13	21.13	19.48			20.86
22:1	ω9	22.28	22.30	22.27	21.57		22.4	22.36
24:1	ω9	24.27	24.40	24.20	23.67			24.26

* Mean value of 6 determinations (variation less than 1 unit).

178

cation of isomeric polyunsaturated fatty acids[17, 18]. Apart from very extensive fatty acid retention data reported by Ackman et al.[19-22], such data have been provided by Hofstetter et al.[23], by Haken[24] and by Jamieson and Reid[25] (cf. Table 8:1).

As mentioned above, GLC is not capable of resolving or detecting all of the fatty acids present in rapeseed oils under the more standardized instrumental conditions of the analysis. Minor components or isomers are lost or shaded by the major ones. Supplemental procedures and/or more sophisticated GLC runs are necessary to determine minor fatty acid components. Thus Bhatty and Craig[26] prefractionated rapeseed fatty acids by silicic acid/silver nitrate column chromatography and the fractions were subsequently analyzed by GLC.

A similar analysis based on another complexation technique has been described by Kuemmel[27]. The rapeseed methyl esters as their mercury derivatives were first fractionated according to unsaturation and then submitted to oxidative cleavage and analysis by GLC. The amounts of the minor unsaturated isomers reported were however in some cases too small to permit a definite identification.

The results reported by Ackman[28] on the same oil are in good agreement with those reported by Keummel[27]. Ackman[28] used very high resolution open tubular gas chromatography columns (Fig.8:2) in order to achieve resolution and

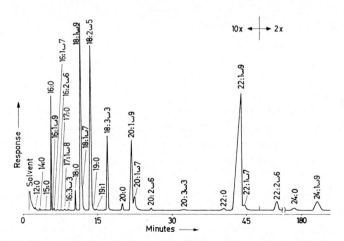

Fig.8:2. GLC of rapeseed oil methyl esters on BDS capillary column, 150 ft. × 0.01 in.[28].

detection of the minor fatty acids. With the aid of linear/log plot graphs (p.177) the author was able to tentatively identify trace amounts of structural isomers of several hitherto unknown fatty acids of rapeseed. Iverson[29] used urea fractionation on mustard seed fatty acids, in order to enrich the minor components. The GLC identifications were established with the aid of linear/log plot graphs based on GLC retention data. Long chain fatty acid homologues with 26 carbon atoms were identified in trace amounts (cf. Chapter 7). Haeffner[29a] recently described a method for the separation of the unsaturated C20 fatty acids, *viz*. C20:1, C20:2 and C20:3, using countercurrent distribution. The same author[29b] also reported the further isolation and structural determination of the C20 and C22 fatty acids using ozonolysis.

In terms of relative errors it is well known that the accuracy in the determinations of the minor components is normally much less than that of the major components. This tendency will be more accentuated if the minor components also have long retention times. Consequently rapeseed fatty acids such as 20:0, 20:2, 22:0, 22:2, 24:0 and 24:1 are especially likely to be inaccurately determined, which has often been demonstrated in the literature (*cf*. Chapter 7). When more elaborate methods are used together with gas chromatography, *e.g.* prefractionations for determining trace components, it must also be emphasized that it is the accuracy of the 'overall analysis' which is essential.

Many instrumental parameters are potential sources of poor accuracy and low precision. Empirical correction factors may be useful, but it is important to note that they can be very doubtful if used as 'instrumental or column constants' without strict control. No definite and general formulas can be given regarding the accuracy of fatty acid analysis by gas chromatography, but the discussion given by Horning *et al*.[30] seems realistic. Even if not completely perfect, the superiority of GLC to other existing techniques, such as paper and thin layer chromatography[31-34] is easily realized.

8.3 TRIACYLGLYCEROLS

The structures of triacylglycerols are characterized by the three esterified fatty acids and their positions in the glycerol molecule. Consequently, the elucidation of glycerol structures is primarily a question of analyzing the fatty acids and their positions. Theoretically this seems rather simple, but applied to natural fats* this is not the case.

Most reported data concerning triacylglycerol compositions are based on statistical distribution theories based on fatty acid data. From rather simple theories there has been a gradual development of more elaborate rules, along with the experimental evidence obtained (reviews: refs. 35–37). The most fruitful theory was set up when it was shown from pancreatic lipase hydrolysis that the distribution within the glycerol is non-random as far as the central positions are concerned, and furthermore that a general system for positional preferences exists according to the type of fatty acid involved. The 1,3-random-2-random

---

* Number of possible triacylglycerols $= \dfrac{n^3 + 3n^2 + 2n}{6}$ (n = number of fatty acids), without consideration of positional isomers.

distribution rule formulated by Van der Waal[38] and Coleman and Fulton[39] has in general shown close agreement with experimental data for most common vegetable oils even if exceptions have been found.

The basic requirements for the separation of triacylglycerols are the same as those for the fatty acids. That is, separation will depend on the degree of unsaturation and/or on the sum of fatty acid chain lengths given by the carbon number. The combined effect of both unsaturation and carbon number refers to the partition number (polarity number) which, according to Litchfield[40], is carbon number −2 (number of double bonds).

Classical methods for separation, *e.g.* fractional crystallization with or without solvent, have been replaced to a large extent by modern chromatographic techniques. The separation of acylglycerols according to unsaturation by silver-ion complexation chromatography[41] on columns was first described by De Vries[42], and the use of thin-layer chromatography has been described by Morris[43] and Basett *et al.*[44]. Various reversed-phase systems for acylglycerols for column, thin-layer and paper chromatography have been reported[45−48]. Separation according to carbon number is nowadays best accomplished by using gas chromatography, as reported by Kuksis[49], Litchfield *et al.*[50] and Harlow *et al.*[51]. The separational conditions with respect to the triacylglycerols will always be less advantageous in comparison to a mixture of the included fatty acids. In many cases the individual differences between the fatty acids in the triacylglycerols are cancelled by opposing effects of the other fatty acids in the same triacylglycerol

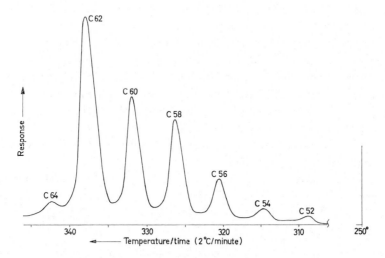

Fig.8:3. Carbon number separation of rapeseed triacylglycerols (*B. campestris*, Sv. Duro) by gas chromatography[52].
Column: 1/8″ × 50 cm. Stationary phase: OV-1 (*ca.* 2 per cent).
Temperature program: 250–350 °C (2 °C/min).
Injection temperature: 295 °C.

moiety, *e.g.* Ei Lo O* is equivalent to E Le P* with regard to unsaturation (4 C=C) carbon number ('56'), and partition number ('48'). All of the fatty acids can however easily be separated from each other on a polar column by gas chromatography. In this respect rapeseed oils are more complicated to elucidate than many other vegetable oils, which mainly contain C16 and C18 fatty acids. The example above clearly demonstrates the very complicated separation problem involved, and the necessity of using consecutive separation systems, which implies very tedious experimental work.

One of the main problems in the use of the multiple separation systems which in most cases are required for triacylglycerol analyses is the avoidance of losses due to irreversible adsorption on chromatographic supports, destruction due to oxidation or heat, etc. This may seriously lower the overall recovery and consequently quantification is made difficult since comparatively small amounts of sample are normally handled. Further complications arise when intermediate preparation recoveries must be calculated in absolute amounts either by direct measurements or indirectly using internal standards. The latter method can not so easily be applied due to the complex composition of the sample. Besides the quantitative aspects there are also qualitative aspects connected with the separation. An important feature of silver-ion adsorption chromatography is that the 'constant of adsorption' is not simply one double bond, *i.e.* the method distinguishes between different types of double bonds.

One of the many factors which may affect adsorption is the well-known difference between unsaturated geometrical isomers, *i.e. cis-* and *trans-*isomers. Other factors are the positions of the double bonds in the fatty acid chain[53-55], the relative positions of the double bonds in polyunsaturated acids[56], non-linearity between single and multiple double bonds[57, 58] and to a certain extent even the fatty acid position in the glycerol molecule[59, 60]. It is also evident that the adsorptive forces are affected by the chain length. A shorter monoenoic acid is more strongly adsorbed than the longer ones, *i.e.* in rapeseed oils decreasingly in the order palmitoleic—oleic—eicosenoic—erucic—nervonic acid. In reversed-phase systems corresponding effects appear which have not been so well studied in literature, and the partition number as described on page 181 is in fact only approximately valid.

Hilditch *et al.*[61] separated rapeseed triacylglycerols by fractional crystallization from acetone. The main classes of triacylglycerols reported were S-U(C18)-U (C22:1) (18%), U(C18)-di-U(C22:1) (54%) and di-U(C18)-U(C22:1)** (28%).

---

* Abbreviations for triacylglycerols containing the following fatty acids:

P	= palmitic acid	Le	= linolenic acid
O	= oleic acid	Ei	= eicosenoic acid
Lo	= linoleic acid	E	= erucic acid

** Qualitative differentiation of the fatty acids is here simplified to saturated, unsaturated C:18 and erucic acid, which are designated S, U(C18) and C22:1, respectively.

Rapeseed triacylglycerols have been separated by Kaufmann *et al.*[62] with the aid of argentation chromatography and reversed-phase thin-layer chromatography. Argentation separated the glycerols into ten fractions, each subsequently separated on reversed phase, giving a total of at least 50 different acylglycerols. The separated fractions were not fully identified and no quantification was reported (Fig.8:4).

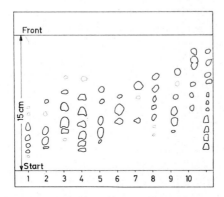

Fig.8:4. Reversed phase thin layer chromatography of rapeseed triacylglycerols[62].
1–10 are fractions of a rapeseed oil sample, separated by AgNO₃-chromatography.

Grynberg *et al.*[63] separated rapeseed triacylglycerols using essentially the same systems as Kaufmann[62]. Argentation chromatography gave in this case only four fractions, which after separation on reversed phase resulted in a total of 11 glyceride fractions. The small number of fractions obtained by Grynberg *et al.*[63] was probably due to larger sample sizes. In parallel work the oil was analysed by pancreatic lipolysis and the triacylglycerol composition subsequently calculated according to the 1,3-random-2-random distribution theory of Coleman[39]. 108 different triacylglycerols (> 0.1%) were calculated, which according to the authors were approximately confirmed by the results from the thin-layer chromatography.

Essentially the same results based on the same methods have been reported by Jurriens and Schouten[64]. Subbaram and Youngs[65] analysed rapeseed triacylglycerols by gas chromatography after oxidative fission of the double bonds, but the results reported were not quite in agreement with lipolysis data. The main triacylglycerols were tri-C18-unsaturated and mono-C22:1-di-C18-unsaturated (20%), mono-C20:1-di-C18-unsaturated (15%) and mono-C22:1-di-C18-unsaturated (35%). Corresponding data based on lipolysis gave a higher ratio between tri-C18-unsaturated and mono-C22:1-di-C18-unsaturated triacylglycerols*.

---

* Erucic acid content of the oil: 22 per cent.

Grynberg *et al.*[63] and Jurriens *et al.*[64] reported fairly good agreement with the distribution rule as mentioned above. Their statements are however based on comparatively approximate figures and considerable deviations can therefore not be excluded. One apparent disagreement can be explicitly pointed out: According to pancreatic lipolysis data, triacylglycerols (carbon number 66) account for 2–3 per cent, whereas gas chromatographic analysis of C66[66, 67] has shown such triacylglycerols to be absent or present only in trace amounts.

Mattson and Volpenheim[68] postulate from pancreatic analysis of rapeseed oil that saturated fatty acids and those longer than C18 regardless of unsaturation are preferentially positioned into outer positions[69–72]. Consequently unsaturated acids of chain length C18 and shorter are preferentially* in the central position.

Appelqvist and Dowdell[71] investigated especially the positions of the minor long chain fatty acids. Compared to those found for erucic and eicosenoic acids, higher ratios were found in the 2-position probably due to larger experimental errors in analyzing small amounts. The major part of the latter acids was, however, still found to be in the outer positions. On this basis the authors[71] suggested that the number of carbon atoms between the carboxylic end and the first double bond is a more determining factor than the total number of double bonds. This proposal does of course not rule out the total chain length as the determining factor which directs the positioning in the biosynthetic acylation reactions.

Smaller fractions of the long chain fatty acids found in the 2-position seem to indicate no restriction to the 2-position. Although the absolute selectivity of the pancreatic lipase analysis is not finally settled (*cf.* discussion in ref. 71) this seems to be the most appropriate hypothesis.

The recently reported relationships[72a] between the fatty acids and the corresponding distribution patterns in rapeseed oils with highly variable fatty acid compositions (erucic acid content ranging from 1 to 50%) indicate a very complex qualitative and quantitative fatty acid dependence which can not be formulated in terms of a general distribution rule.

The possible existence of optical antipodes and analytical techniques for distinguishing between the outer positions (1- and 3-positions) developed by Brockerhoff[73] suggest a third dimension of the triacylglycerol structure, *i.e.* steric configurations. Considering the biosynthetic formation of triacylglycerols from L-α-glycerophosphoric acid[74], a selective acylation would result in the formation of antipodes but restricted to those cases in which the 1- and 3-positions are occupied by different fatty acids (asymmetric central carbon atom). By the method described above Brockerhoff *et al.*[70] showed that distribution in the outer positions of vegetable oil acylglycerols more or less deviates from the random

---

* The concept of preference must of course generally be considered in relative terms and with a dualistic aspect, *i.e.* preference for one position is simultaneously also non-preference for another position.

rule. In rapeseed triacylglycerols substantial deviations between the 1- and 3-positions were found[70]. Other vegetable oils showed less irregularities, while more pronounced deviations have been found in acylglycerols from animal sources, containing more heterogenous fatty acid patterns of long chain fatty acids [75, 76]. The 'sterospecific' analyses have however been performed on unresolved

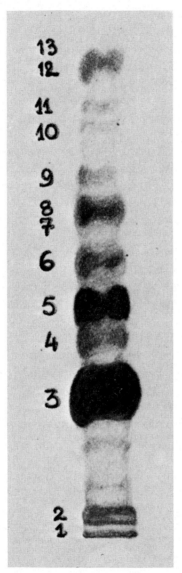

Fig.8:5. Class separation of rapeseed unsaponifiables by thin-layer chromatography.[84]
1. Unidentified, 2. Pigment (hydroxylic), 3. Sterols, 4. Alcohols (aliphatic), 5. Alcohols (terpenic), 6 and 7. Unknown, 8. Tocopherols, 9. Ketones (terpenic), 10. Carotenes, 11. Unknown, 12. Hydrocarbons (polyunsaturated), 13. Hydrocarbons (paraffinic).

acylglycerols and therefore conclusive evidence about configurations must await definite proof from individual acylglycerols[77].

8.4 UNSAPONIFIABLES

The triacylglycerols, the major lipid class of the oil, constitute 97–98 per cent of crude rapeseed oils. The remaining 2–3 per cent of other lipid materials consists of free fatty acids (about 1%), mono- and diacylglycerols, phosphatides and so-called unsaponifiables. The qualitative composition of free fatty acids and mono- and diacylglycerols can be related to those of the overall fatty acids and the triacylglycerols. Rapeseed phosphatides are treated separately (Chapters 7 and 13).

The unsaponifiables, characterized by insolubility in water after saponification of the oil, and amounting to 0.5–2 per cent (refs. 78–80 and Chapter 7), represent a mixture of several lipid classes, *viz*. sterols, tocopherols, hydrocarbons, higher aliphatic and terpene alcohols. In analytical treatments the material is in most cases isolated as described above by extractions preceded by saponification of the oil. Thus the isolated material is not necessarily obtained in its original form, since many of the components contain bonds which can have been split off by hydrolysis. Non-destructive enrichment procedures, *e.g.* fractional crystallization, have been applied to the isolation of unsaponifiables[81, 82].

The major lipid class of the unsaponifiable material is the sterols, constituting approximately 50 per cent. Chromatographic techniques are at present to be preferred for compositional studies of these components. Thus separation into classes is frequently achieved by thin-layer and column chromatography (Fig.8:5), and GLC is used for further separation within each class and for quantification.

Karleskind[83] studied separation conditions of rapeseed unsaponifiables on various thin-layer supports. He proposed a subsequent urea fractionation in order to achieve a separation of aliphatic and triterpene alcohols. Mordret[84] systematically investigated parameters for the separation of rapeseed unsaponifiables by column and thin-layer chromatography, *e.g.* support, solvent systems, recoveries etc. (Fig.8:5).

Using TLC, Fedeli *et al.*[85] found 0.59 per cent sterols and 0.03 per cent triterpene alcohols in rapeseed oil. By the same technique Rutkowski *et al.*[86] reported 61 per cent sterols, 13 per cent hydrocarbons and 12 per cent triterpene alcohols of total unsaponifiables. Fedeli[85], using silicic acid column with step-wise increase of the eluent power, established the composition as hydrocarbons (8.5–9%), squalene (4.1–4.4%), aliphatic alcohols (7.0–7.5%), triterpene alcohols (9.0–9.5%) and sterols (60.0–63.6%).

8.4.1 Sterols

By gas chromatographic methods it has been shown by several authors[81, 82, 85–88]

STEROL COMPOSITION OF RAPESEED OIL

	Sterol content in oil (per cent)	Composition (rel. per cent)				Ref.
		Cholesterol C27* (1 C=C)	Brassica sterol C28* (2 C=C)	Campesterol C28* (1 C=C)	β-Sitosterol C29* (1 C=C)	
Rapeseed oil (refined)	0.45		12.9	34.2	52.9	86
Rapeseed oil (hydrogenated)	0.46		10.3	32.8	56.9	
Rapeseed oil: B. napus Sv. Matador**		1.7	13.0	31.9	53.4	88
B. napus Sv. Regina II		3.8	10.1	25.6	60.6	

* Number of carbon atoms in molecular skeleton (number of ethylenic bonds).
** Swedish Seed Association, Svalöv, Sweden.

that rapeseed sterols consist of three major components, *i.e.* brassicasterol, campesterol and β-sitosterol.

A fourth minor sterol with retention time equal to that of cholesterol has also been found. This sterol, which earlier was thought only to exist in animal oils and fats, was however later tentatively identified in several oils of vegetable origin as cholesterol. Recently Capella et al.[88], using gas chromatography–mass-spectrometry presented definite proof of cholesterol in rapeseed of known origin (Fig.8:6).

Usually the unsaponifiables are prefractionated before the GLC analysis.

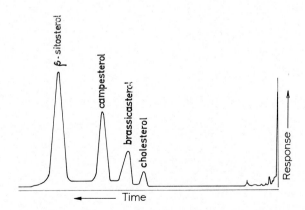

Fig.8:6. GLC separation of rapeseed sterols (*B. napus*, Sv. Regina II)[88].

Mordret[89] analyzed the sterols in the unsaponifiables on a non-polar column without any prefractionation. Seher *et al.*[90-92] have worked out various thin-layer chromatographic systems for the analysis of sterols.

8.4.2 *Hydrocarbons and alcohols*

The composition of the alcohol and hydrocarbon fractions is rather complex. The identities of the various components are for the most part tentatively based

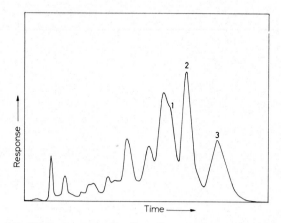

Fig.8:7. GLC of terpene fraction of rapeseed unsaponifiables[85].
Peak 1, (unresolved) β-Amyrin; Peak 2, Cycloartenol; Peak 3, 24-Methylene-cycloartenol.

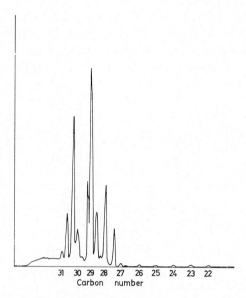

Fig.8:8. GLC separation of hydrocarbon fraction of rapeseed unsaponifiables[86].

188

on gas chromatographic retention data. Thirteen different terpene alcohols (Fig.8:7) have been detected in rapeseed oil[81, 82, 85], e.g. β-amyrin, cycloartanol, cycloartenol and 24-methyl cycloartenol.

Gas chromatographic analysis of the hydrocarbon fraction (Fig.8:8) has revealed at least 36 different compounds, viz. homologues of normal and branched paraffinic hydrocarbons with carbon numbers from 11 to 31[81, 82, 86].

8.4.3 Tocopherols

Another minor class of substances of great importance in several respects, the tocopherols, is included in the unsaponifiables, but they are functionally classified as lipovitamins. In rapeseed oils they amount to less than 0.1 per cent (Table 8:3).

TABLE 8:3

TOCOPHEROL CONTENTS OF PROCESSED RAPESEED OILS

Oil	Tocopherol ppm in oil	Composition in relative per cent			Ref.
		α-Toco-pherol	γ-Toco-pherol	Other tocopherols	
Rapeseed oil	691	32*	68*		108
Rapeseed oil	600**	—	—		104
Rapeseed oil	552**	—	—		104
Rapeseed oil (refined)	433	39	61		105
Rapeseed oil	560	27	73		109
Rapeseed oil crude (Swedish)	800	—	—		110
Rapeseed oil, lab. extr.	400–700	25–60	35–75	0–3 (δ)	111
Rapeseed oil ('Rüböl')	591	27	73		112
Hydr. (38/40), bleached***	750	32	68		106
Hydr. (38/40), deod.	630	34	66		106
Hydr. (36/38), deod.	700	38	62		106
Hydr. (32/34), deod.	650	34	66		106

* Reported as α + δ and β + γ + ε respectively.
** α, β + γ and δ were qualitatively identified, ascorbic acid (600) and pyrogallol (562) respectively as antioxidant in the saponification step.
*** Gas chromatographic data: β- and γ-tocopherol are not separated from each other (cf. Fig.8:9), but β-tocopherol is generally considered to be absent in most vegetable oils.

The procedure for their isolation is primarily based on the same scheme as referred to above for other minor constituents. Since they are a physiologically active group of substances, the literature on their determination is rather extensive[93–103].

The tocopherol contents in rapeseed oils reported in the literature vary depending on several different factors, e.g. variability between species, basic analytical errors, and degradation of the tocopherols during processing and storage. Quantitative accuracy of the multistep analysis is rather difficult to achieve due to degradation. In order to protect the tocopherols during the analytical procedure, antioxidants are usually added before saponification. In this connection Täufel[104] reported that ascorbic acid gave a higher yield than pyrogallol (Table 8:3). Herting et al.[105] mentioned the potential risk of iron in sodium sulphate used for drying the extract. Certainly there are many potential interactions which may seriously damage the analysis, and it is not very likely that the addition of any antioxidant can eliminate all kinds of losses. One way of getting around these problems is the use of an internal standard chemically closely related to the substance to be determined. δ-Tocopherol* is present in rapeseed oil only in trace amounts (Fig.8:9 and Table 8:3). If this tocopherol is allowed to accompany

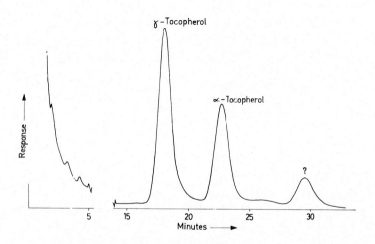

Fig.8:9. Gas chromatogram of tocopherol-fraction (TLC) of rapeseed oil unsaponifiables[106]. Column: 1/8″ × 2 m, stationary phase: OV-1 (ca. 2 per cent). Temperature: 240 °C. Injection temperature: 260 °C.

the sample as a standard throughout the analysis, i.e. also during the saponification[106], an idea of the losses involved may be obtained.

Losses during storage and processing are very dependent on the actual conditions[132]. Considerable losses during the hydrogenation process have been reported by Ward[107]. Herting et al.[105] on the other hand reported a minor decrease in the tocopherols during hydrogenation but the significance was said to be questionable (cf. Table 8:3).

* Nomenclature: ref. 93.

The natural colour of rapeseed oil originates from the presence of lipochromes, *i.e.* carotenoids and pigments such as chlorophylls and their degradation products pheophytins. Their constitutional chromophoric properties make their presence very apparent and easily detectable as a group by spectrophotometric means (Fig.8:10).

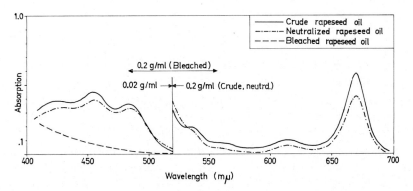

Fig.8:10. Visible spectra of rapeseed oil. Solvent: iso-octane. Cell: 1 cm.

Carotenoids absorb in the 400–500 nm range and chlorophylls in the 600–700 nm range. Minor absorption maxima at 535 nm and 610 nm in crude rapeseed oil have been reported by Box *et al.*[113] and can also be seen in Fig.8:10. The carotenoids, although rather similar with respect to their absorption properties, consist of many different compounds abundant in vegetable oils. Procedures for their differentiation by column and thin-layer chromatography and their spectroscopic properties have recently been reviewed by Srour[114]. The amount of carotenoids in rapeseed oil has been investigated by Benk *et al.*[115] using alumina columns for their separation. Values between 2.1–2.9 ppm carotenoids, of which 70–75 per cent consisted of β-carotene, have been reported. Box *et al.*[113] isolated the carotenoids by solvent extraction after saponification and subsequent partitioning between petroleum ether and 90 per cent methanol followed by chromatography on column and thin-layer. The authors discuss quantitative aspects of the procedure, *e.g.* the necessity of saponification in order to avoid too high values of carotenoids. They reported neo-lutein A and B, 23 ppm, as the major carotenoids, while carotene amounted to 1.8 ppm. Pheophytin A was the major component of the chlorophylls, 7.6 ppm, while chlorophylls A and B only amounted to 0.3 and 0.4 ppm respectively, calculated from measurements at four different wavelengths.

Niewiadomski *et al.*[116] determined the content of chlorophylls A and B and pheophytins A and B spectrophotometrically at four different wavelengths, *i.e.*

663.8 nm, 645.5 nm, 668 nm and 655.8 nm, respectively. From empirically constructed equations the amounts of the chlorophylls could then be individually calculated. The main component found was pheophytin A in various rapeseed oils, amounting to 18–26 ppm. The carotenoids were isolated from the rapeseed oil on an alumina column and spectrophotometrically measured at 450 nm. The content was calculated using β-carotene as a reference, since according to the authors it is the main component. Values between 7 and 4 ppm were reported[116].

8.6 SULPHUR

The content of sulphur in rapeseed oils originates mainly from the sulphur-containing glucosinolates in the rapeseed. The content of sulphur in crude oil is very much dependent upon the processing parameters, *i.e.* the pretreatment of the seeds in order to avoid splitting off the sulphur-containing aglycons (see Chapter 9).

The sulphur primarily affects the hydrogenation processing (see Chapter 10), and may also cause disturbances in other ways, *e.g.* by contributing to off-flavours. The relatively small amounts of sulphur present (crude rapeseed oil manufactured in Sweden normally contains between 5–15 ppm) necessitate the use of large samples in the analytical procedure in order to obtain sufficient sulphur for the

Fig.8:11. Apparatus for sulphur determination (combustion) according to Bladh[121]. From right to left: continuous sample feed unit (A), combustion chamber (B) and absorption tower (C).

determination. Baltes[117] has described a method based on reductive desulphurization and subsequent titration of the absorbed hydrogen sulphide. The method is rather time-consuming, and although it gives reproducible values if carefully performed, it seems that the method does not give the absolute sulphur content, since it has been observed[118] that the values are always lower than those obtained by combustion methods.

Modifications of the Schöniger combustion method to allow larger samples have been described in the literature[119, 120]. The apparatus pictured in Fig.8:11 was designed by Bladh[121] for the determination of sulphur in oils. The oil is continuously fed into the flask and the combustion gases are absorbed in an oxidative medium. The sulphate formed is subsequently titrated photometrically. The detection limit is theoretically unlimited. In practice a 2–5 g sample, which takes about 15 minutes for combustion, gives acceptable accuracy down to 1–2 ppm, and the whole analysis can be accomplished within 20 minutes.

8.7 MINERALS

The minerals in rapeseed are chiefly potassium, calcium, magnesium and phosphorus (*cf.* Table 7:21). They remain for the most part in the meal after extraction of the oil. Besides a minor contribution caused by mechanical contamination from the meal, the minerals found in the oil are related to phospholipids and similar compounds (Table 8:4).

TABLE 8:4

METAL CONTENTS OF RAPESEED OIL, DETERMINED BY ATOMIC ABSORPTION SPECTROMETRY

Rapeseed oil	Metal contents (ppm)					Ref.
	Ca	*Mg*	*Zn*	*Cu*	*Fe*	
Refined & bleached	1.0		1.15	0.15	0.55	127
Crude	38.8	19.2				
Refined	1.0	0.2				130
Degummed				≤0.05	1.2–3.1	
Neutralized				≤0.05	≤0.1	131

Determination of phosphorus is usually accomplished by classical colourimetric procedures (see Appendix II), while calcium and magnesium are preferably analyzed using atomic absorption spectroscopy.

The presence of heavy metals, of which iron and copper are of special interest, originates to a certain extent from natural sources but are mostly contaminations from external sources. Numerous methods for their determination are available. Even if colourimetric procedures can be applied directly to the oil[122, 123], which is generally more attractive than tedious ashing methods, they have limitations with respect to detection limits. Disturbances by co-complexation of other metals must also be emphasized. Contaminations from external sources, for example materials from mineral acids used in the extraction and wet-ashing procedures, and other chemicals, can be very hazardous for the analysis. Though it is probably true that activation analysis is generally the most accurate method, the expensive instrumental equipment makes it a rather exclusive method limited to a very few laboratories. The advent of atomic absorption spectroscopy (AA) has undoubtedly facilitated the analysis of metals in many respects and has been successfully applied in the oil and fat field[124]. List et al.[125] reported that atomic absorption spectrometry gave lower values in oil compared to colourimetric and activation analyses.

Without questioning the results there are probably no general differences between the methods referred to above and special circumstances may perhaps explain the deviations. Numerous other reports have demonstrated satisfactory recoveries and accuracy by AA[126-131].

REFERENCES

1   H. P. KAUFMANN, *Analyse der Fette und Fettprodukte*, Teil I–II, Springer-Verlag, Berlin/Göttingen/Heidelberg, 1958.
2   H. A. BOEKENOOGEN (Ed.), *Analysis and Characterization of Oils, Fats and Fat Products*, Interscience, New York/London/Sydney, Vol. I, 1964; Vol. II, 1968.
3   L. V. COCKS AND C. VAN REDE, *Laboratory Handbook for Oil and Fat Analysts*, Academic Press, New York/London, 1966.
4   T. P. HILDITCH AND P. N. WILLIAMS, *The Chemical Constitution of Natural Fats*, 4th Edn., Chapman & Hall, London, 1964.
5   T. P. HILDITCH, P. A. LAURENT AND M. L. MEARA, *J. Soc. Chem. Ind.*, 66 (1947) 19.
6   M. N. BALIGA AND T. P. HILDITCH, *J. Soc. Chem. Ind.*, 67 (1948) 258.
7   G. V. MARINETTI (Ed.), *Lipid Chromatographic Analysis*, Vol. 1, Marcel Dekker, Inc., New York, 1967.
8   R. A. STEIN, V. SLAWSON AND J. F. MEAD, in *Lipid Chromatographic Analysis*, Vol. 1, G. V. MARINETTI (Ed.), Marcel Dekker, Inc., New York, 1967.
9   E. C. HORNSHIP, A. CARMEN AND C. C. SWEELEY in *Progress in Chemistry of Fats and other Lipids*, Vol. 7, R. T. HOLMAN AND T. MALKIN (Eds.), Pergamon Press, Oxford/London/New York/Paris, 1963.
10   R. G. ACKMAN, *Lipids*, 2 (1967) 502.
11   B. TÖREGÅRD, unpublished data.
12   J. W. FARQUHAR, W. INSUER JR., P. ROSEN, W. STOFFELS AND E. H. AHRENS JR., *Nutrition Rev. (Suppl.)*, 17 (1959) 1.
13   H. P. BURCHFIELD AND ELEANOR E. STORRS (Eds.), *Biochemical Applications of Gas Chromatography*, Academic Press, New York/London (1962) p.549–554.
14   T. K. MIWA, K. L. MIKOLAJCZAK, F. R. EARLE AND I. A. WOLFF, *Anal. Chem.*, 32 (1960) 1739.
15   F. P. WOODFORD AND G. M. VAN GENT, *J. Lipid Res.*, 1 (1960) 188.
16   R. G. ACKMAN, *J. Am. Oil Chem. Soc.*, 40 (1963) 558.
17   R. A. LANDOWN AND S. R. LIPSKY, *Biochim. Biophys. Acta*, 47 (1961) 589.

18 R. G. Ackman and R. D. Burgher, *J. Chromatog.*, 11 (1963) 185.
19 R. G. Ackman, R. D. Burgher and P. M. Jangaard, *Can. J. Biochem. Physiol*, 41 (1963) 1627.
20 R. G. Ackman, *J. Am. Oil Chem. Soc.*, 40 (1963) 564.
21 R. G. Ackman and P. M. Jangaard, *J. Am. Oil Chem. Soc.*, 40 (1963) 744.
22 R. G. Ackman and R. D. Burgher, *J. Am. Oil Chem. Soc.*, 42 (1965) 38.
23 H. H. Hofstetter, N. Sen and R. T. Holman, *J. Am. Oil Chem. Soc.*, 42 (1965) 537.
24 J. K. Haken, *J. Chromatog.*, 23 (1966) 375; 26 (1967) 17; 39 (1969) 245.
25 G. R. Jamieson and E. H. Reid, *J. Chromatog.*, 20 (1965) 232; 39 (1969) 71.
26 M. K. Bhatty and B. M. Craig, *J. Am. Oil Chem. Soc.*, 41 (1964) 508.
27 D. F. Kuemmel, *J. Am. Oil Chem. Soc.*, 41 (1964) 667.
28 R. G. Ackman, *J. Am. Oil Chem. Soc.*, 43 (1966) 483.
29 J. L. Iverson, *J. Ass. Offic. Anal. Chem.*, 49 (1966) 332.
29a E. W. Haeffner, *Lipids*, 5 (1970) 489.
29b E. W. Haeffner, *Lipids*, 5 (1970) 430.
30 E. C. Horning, E. H. Ahrens Jr., S. R. Lipsky, F. H. Mattson, J. F. Mead, D. A. Turner and W. H. Goldwater, *J. Lipid Res.*, 5 (1964) 20.
31 J. Pokorny and J. Ruzicka, *Prumysl Postravin*, 15 (1964) 392; (CA 61 (1964) 5917c).
32 H. Grynberg, H. Szczepanska and M. Beldowicz, *Fette, Seifen, Anstrichmittel*, 66 (1964) 352.
33 A. M. Gad, S. Fiad, Z. E. Shoeb and M. M. Hassen, *Fette, Seifen, Anstrichmittel*, 67 (1965) 796.
34 J. Sliwiok and Z. Kwapniewski, *Mikrochem. Ichnoanal. Acta*, (1965) 657; (*C.A.* 63, 15115g).
35 F. D. Gunstone, *An introduction to the Chemistry and Biochemistry of Fatty Acids and their Glycerides* (2nd ed.,), Chapman & Hall, London, 1967.
36 G. Jurriens, *Analysis of Glycerides and Composition of Natural Oils and Fats* in *Analysis and Characterization of Oils, Fats and Fat Products*, Vol. 2, H. A. Boekenoogen (Ed.), Interscience, London/New York/Sydney, 1968.
37 G. Lakshminasayana, *J. Sci. Ind. Res. (India)*, 23 (1964) 506.
38 R. J. Vander Wal, *J. Am. Oil Chem. Soc.*, 37 (1960) 18.
39 M. H. Coleman and W. C. Fulton in *The Enzymes of Lipid Metabolism*, P. Desnuelle (Ed.), MacMillan, New York, 1961, p.127.
40 C. Litchfield, *Lipids*, 3 (1968) 170.
41 L. J. Morris, *J. Lipid Res.*, 7 (1966) 717.
42 B. de Vries, *Chem. Ind. (London)*, (1962) 1049.
43 L. J. Morris, *Chem. Ind. (London)*, (1962) 1238.
44 C. B. Basett, M. S. J. Dallas and F. B. Padley, *Chem. Ind. (London)*, (1962) 1050.
45 E. C. Nickell and O. S. Privett, *Separation Sci.*, 2 (3) (1967) 307.
46 H. P. Kaufmann and H. Wessels, *Fette, Seifen, Anstrichmittel*, 66 (1964) 13.
47 K. K. Carson and B. Serdarevoch, *Column Chromatography of Neutral Glycerides and Fatty Acids*, in *Lipid Chromatographic Analysis*, G. V. Marinetti (Ed.), Vol. 1, Marcel Dekker, New York, 1967, p.224.
48 F. B. Padley, *Chromatog. Rev.*, 8 (1967) 208.
49 A. Kuksis, *Gas Chromatography of Neutral Glycerides* in *Gas Chromatographic Analysis*, G. V. Marinetti (Ed.), Vol. 1, p.239.
50 C. Litchfield, R. D. Harlow and R. Reiser, *J. Am. Oil Chem. Soc.*, 42 (1965) 849.
51 R. D. Harlow, C. Litchfield and R. Reiser, *Lipids*, 1 (1966) 216.
52 B. Töregård, *5th Scandinavian Oil and Fat Meeting, 6–10 June 1969*, Tyringe, Sweden.
53 W. W. Christie, *J. Chromatog.*, 34 (1968) 405.
54 J. Barve *et al.*, Paper presented at the congress of the *International Society for Fat Research*, Rotterdam, 1968.
55 L. J. Morris and D. M. Wharry, *J. Chromatog.*, 20 (1965) 27.
56 B. de Vries and G. Jurriens, *Fette, Seifen, Anstrichmittel*, 65 (1963) 725.
57 F. D. Gunstone and F. B. Padley, *J. Am. Oil Chem. Soc.*, 42 (1965) 957.
58 R. A. Stein and V. Slawson, *Anal. Chem.*, 40 (1968) 2017.

59 C. B. Basett, M. S. J. Dallas and F. B. Padley, *Chem. Ind. (London)* (1962) 1050.
60 O. Renkonen and L. Rikkinen, *Acta Chem. Scand.*, 21 (1967) 2282.
61 T. P. Hilditch, P. A. Laurent and M. L. Meara, *J. Soc. Chem. Ind.*, 66 (1947) 1g.
62 H. P. Kaufmann and H. Wessels, *Fette, Seifen, Anstrichmittel*, 66 (1964) 81.
63 H. Grynberg, K. Ceglowska and H. Szczepanska, *Rev. Franc. Corps Gras*, 13 (1966) 595.
64 G. Jurriens and L. Schouten, *Rev. Franc. Corps Gras*, 12 (1965) 505.
65 M. R. Subbaram and C. G. Youngs, *J. Am. Oil Chem. Soc.*, 44 (1967) 425.
66 R. D. Harlow, C. Litchfield and R. Reiser, *Lipids*, 1 (1966) 216.
67 U. Persmark and B. Töregård, unpublished data.
68 F. H. Mattson and R. A. Volpenheim, *J. Biol. Chem.*, 236 (1961) 1891.
69 H. Grynberg and H. Szczepanska, *J. Am. Oil Chem. Soc.*, 43 (1966) 151.
70 H. Brockerhoff and M. Yurkowski, *J. Lipid Res.*, 7 (1966) 62.
71 L. Å. Appelqvist and R. J. Dowdell, *Arkiv Kemi*, 28 (1968) 539.
72 J. Marcinkiewicz and H. Niewiadomski, *Rev. Franc. Corps Gras*, 15 (1968) 511.
72a U. Persmark and K. A. Melin, Paper presented at the ISF–AOCS Congress, Chicago, 1970.
73 H. Brockerhoff, *J. Lipid Res.*, 6 (1965) 10.
74 S. B. Weiss, F. P. Kennedy and J. Y. Kiyash, *J. Biol. Chem.*, 40 (1960) 235.
75 H. Brockerhoff, R. J. Hoyle and N. Wolmark, *Federation Proc. (Abstracts)*, 24 (1965) 662.
76 C. Litchfield, *Lipids*, 3 (1969) 417.
77 W. Schlenk Jr., *J. Am. Oil Chem. Soc.*, 42 (1965) 945.
78 E. W. Eckey, *Vegetable Fats and Oils*, Reinhold Publ. Corp., New York (1954), p.435.
79 H. P. Kaufmann, J. Baltes, H. J. Heinz and P. Roevec, *Fette, Seifen, Anstrichmittel*, 52 (1950) 35.
80 L. Å. Appelqvist, *Acta Agr. Scand.*, 18 (1968) 3.
81 G. Jacini, E. Fedeli and A. Lanzani, *J. Ass. Offic. Anal. Chem.*, 50 (1967) 84.
82 E. Fedeli, *Rev. Franc. Corps Gras*, 15 (1968) 281.
83 A. Karleskind, *Chim. Anal. (Paris)*, 49 (1967) 86.
84 F. Mordret, *Rev. Franc. Corps Gras*, 16 (1969) 639.
85 E. Fedeli, A. Lanzani, P. Capella and G. Jacini, *J. Am. Oil Chem. Soc.*, 43 (1966) 254.
86 A. Rutkowski, G. Jacini, P. Capella and M. Cirimele, *Riv. Ital. Sostanze Grasse*, XLIII, (1966) 89.
87 A. Karleskind, F. Audiau and J. P. Wolff, *Rev. Franc. Corps Gras*, 13 (1966) 165.
88 P. Capella and G. Losi, *Industrie Agrarie*, VI (1968) 278.
89 F. Mordret, *Rev. Franc. Corps Gras*, 15 (1968) 675.
90 A. Seher and E. Homberg, *Fette, Seifen, Anstrichmittel*, 70 (1968) 481.
91 A. Seher and E. Homberg, Paper presented at DGF Tagung, Münster 1968. (*Fette, Seifen, Anstrichmittel*, 70 (1968) 843).
92 A. Seher, *Fette, Seifen, Anstrichmittel*, 71 (1969) 833.
93 H. T. Slower, L. M. Shelley and T. L. Burks, *J. Am. Oil Chem. Soc.*, 44 (1967) 161.
94 Analytical Methods Committee: The Determination of Tocopherols in Oils, Foods and Feeding Stuffs, *Analyst*, 84 (1959) 356.
95 M. Kofler, P. F. Sommer, H. R. Bolliger, B. Schmidli and M. Vecchi, in *Vitamins and Hormones*, Academic Press, N.Y., Vol. 20 (1967), p.407.
96 R. Strohecker and H. M. Henning, *Vitaminbestimmungen*, Verlag Chemie GmbH, Weinheim/Bergstr. (1963).
97 J. G. Bieri, *Lipid Chromatog. Anal.*, 7 (1969) 459.
98 A. Seher, *Microchim. Acta*, (1961) 308.
99 M. K. Gorina Rao, S. Venkob Rao and K. T. Achaya, *J. Sci. Food Agr.*, 16 (1965) 121.
100 M. Jaly, *Fette, Seifen, Anstrichmittel*, 69 (1967) 507.
101 P. W. Wilson, E. Kodicek and V. H. Both, *Biochem. J.*, 84 (1962) 524.
102 P. P. Nair and D. A. Turner, *J. Am. Oil Chem. Soc.*, 40 (1963) 353.
103 K. K. Carroll and D. C. Herting, *J. Am. Oil Chem. Soc.*, 41 (1964) 473.
104 K. Täufel and R. Serzisko, *Nahrung*, 6 (1962) 413.

105 D. C. HERTING AND E.-J. E. DRURY, *J. Nutr.*, 81 (1963) 335.

106 U. PERSMARK, B. TÖREGÅRD AND I. FRIBERG-JOHANSSON, unpublished data.

107 R. J. WARD, *Brit. J. Nutr.*, 12 (1958) 231.

108 F. BRO-RASMUSSEN AND W. HJARDE, *Acta Chem. Scand.*, 11 (1957) 34, 44.

109 J. GREEN, S. MARCINKIEWICZ AND P. R. WATT, *J. Sci. Food Agr.*, 5 (1955) 274.

110 E. OLIN, *Acta Polytechn. Scand. Met. Ser.*, No. 1 (1958) 56.

111 W. PRIORR, L. TOTH AND N. NOVAKAVIC, *Z. Lebensm. Untersuch.-Forsch.*, 138 (1968) 11.

112 A. NIEDERSTEBRUCH AND I. HINSCH, *Fette, Seifen, Anstrichmittel*, 69 (1967) 559.

113 J. A. G. BOX AND H. A. BOEKENOOGEN, *Fette, Seifen, Anstrichmittel*, 69 (1967) 724.

114 J. SROUR, *Rev. Franc. Corps Gras*, 16 (1969) 269.

115 E. BENK AND L. BRIXIUS, *Fette, Seifen, Anstrichmittel*, 67 (1965) 65.

116 H. NIEWIADOMSKI, I. BRATKOWSKA AND E. MOSSAKOWSKA, *J. Am. Oil Chem. Soc.*, 42 (1965) 731.

117 J. BALTES, *Fette, Seifen, Anstrichmittel*, 69 (1967) 512.

118 U. PERSMARK, unpublished data.

119 R. N. WEATLEY AND R. J. BARGES, *Hydrocarbon Process.*, 47 (1968) 133.

120 J. BELISLE, C. D. GREEN AND L. D. WINTER, *Anal. Chem.*, 40 (1968) 1006.

121 E. BLADH, personal communication.

122 T. H. NEWLOVE, The direct determination of trace metals in glyceride oils by spectro-photometric methods in *Laboratory Handbook for Oil and Fat Analysts*, L. V. COCKS AND C. VON REDE (Eds.), Academic Press, London/New York, 1966, p.346.

123 T. P. LABUZA AND M. KAREL, *J. Food Sci.*, 32 (1967) 572.

124 R. GUILLAUMIN, *Rev. Franc. Corps Gras*, 16 (1969) 497.

125 G. R. LIST, R. L. HOFFMANN, W. F. KWOLEK AND C. D. EVANS, *J. Am. Oil Chem. Soc.*, 45 (1968) 872.

126 G. R. ROGERS, *J. Ass. Offic. Anal. Chem.*, 51 (1968) 1042.

127 B. PICCOLO AND R. T. O'CONNOR, *J. Am. Oil Chem. Soc.*, 45 (1968) 789.

128 A. BURROWS, J. C. HEERDT AND J. B. WILLIS, *Anal. Chem.*, 37 (1965) 579.

129 D. R. WILLIAMS, *Spectrovision*, No. 19 (1968) 8, published by Unicam Instr. Ltd, Cambridge, England.

130 R. GUILLAUMIN AND N. DROUIN, *Rev. Franc. Corps Gras*, 13 (1966) 185.

131 U. PERSMARK AND L. JEPSSON, unpublished data.

132 A. RUTKOWSKI AND K. BABUCHOWSKI, in A. RUTKOWSKI (Ed.), *International Symposium for the Chemistry and Technology of Rapeseed Oil and other Cruciferae Oils, Warsaw, 1970*, p. 213.

CHAPTER 9

Manufacture of Rapeseed Oil and Meal

K. ANJOU

AB Karlshamns Oljefabriker, Karlshamn (Sweden)

CONTENTS

9.1 INTRODUCTION

The technological aspects of processing rapeseed for obtaining crude oil and meal will be dealt with in this chapter. The process involves the separation of oil from the oil-bearing material, which may be achieved by mechanical means (pressing), chemical means (solvent extraction), or a combination of both. The process also comprises the steps of seed cleaning, crushing, cooking, and oil degumming, and in the case of solvent extraction also desolventizing and solvent recovery.

The storage of crude oil and meal will also be dealt with in this chapter.

There is no principal difference between the extraction of rapeseed and that of other oil seeds. All extraction processes have certain objectives in common, *viz.* to obtain the oil undamaged and pure, to obtain the oil in as high a yield and as economically as possible and to produce an oil residue of the greatest possible value[1]. In processing rapeseed, equipment originally developed for other oil seed is generally used. However different oil seeds may demand different processing parameters. Thus with rapeseed, for example, the small seed size, the high oil content and the presence of glucosinolates influence the process, especially the operation of the preparatory equipment.

The oldest method of rapeseed oil extraction to obtain edible oil is the cold

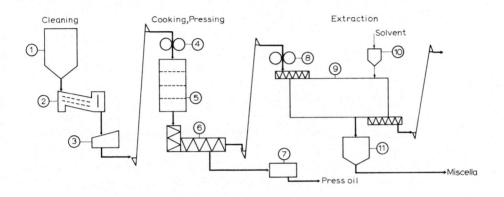

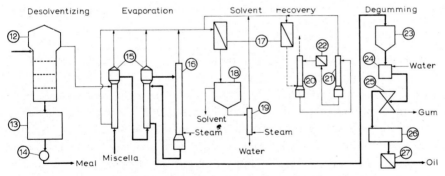

Fig.9:1. Flow diagram of a prepress extraction plant.

1. Silo	10. Solvent tank	19. Stripper
2. Cleaner	11. Miscella tank	20. Absorbator column
3. Stoner	12. Desolventizer	21. Stripping column
4. Crushing mill	13. Meal cooler	22. Heat exchanger
5. Cooker	14. Mill	23. Tank
6. Press	15. Long tubular evaporators	24. Mixer
7. Filter	16. Stripping column	25. Degumming separator
8. Flaking mill	17. Condensers	26. Oil dryer
9. Extractor	18. Separator tank	27. Oil cooler

References pp. 216–217

pressing operation, which gives an oil of mild flavour and is probably still used in some parts of the world[2-4].

The process of expeller pressing was then developed. With the use of high pressures, a good yield of oil was obtained that was mainly used as industrial oil. However, to produce high quality edible oil and meal, the pressures had to be lowered, but this also lowered the yield. In order to economize the process, especially for plants of high capacity, solvent extraction was adopted. Direct extraction as well as prepressing followed by solvent extraction of the press residue, have been tried. All three oil extraction methods are still used for rapeseed, but the prepressing process is the most common today.

In the following, the process, as it is used for rapeseed, will be dealt with step by step. In the last section the influences of the process on the products are discussed.

9.2 SEED CLEANING

The first step in the processing of rapeseed is cleaning to separate foreign material, such as sticks, stems, leaves, other seeds, sand and dirt, which is removed by screening and aspiration. Permanent- or electromagnets are used for the removal of iron objects.

Seed cleaning is an important operation in the preparation of rapeseed for extraction. Proper seed cleaning will increase capacity, reduce maintenance, and improve oil and meal quality[5]. Final cleaning of the seed is sometimes carried out before storing but usually it is done at the extraction plant just before processing.

9.3 DEHULLING

Rapeseed is presently not dehulled in commercial plants. The small size and the high oil content of the seeds make it rather difficult to dehull rapeseed efficiently, since the oil losses with the hulls would be too great. The hull content is about 15–20 per cent and its oil content is about 16 per cent (dry basis) giving losses of about 6–7 per cent of the oil[6, 7]. The hull content mainly depends on seed size, the smaller seeds having the higher hull content. Furthermore, the hulls may act as supporting material during the extraction providing better conditions and higher oil yields.

There is, however, a growing interest in dehulling rapeseed. This is primarily due to the high fibre content and dark colour of meal from undehulled seed, which lowers the feeding value of the meal. The glucosinolate content of the meal has, however, a greater influence on the meal quality (see Chapter 15). Extraction of dehulled seeds may also positively affect oil quality. Furthermore, there is a gain in the capacity of the extraction equipment. The whole problem of dehulling is of course dependent on the economic balance of yield and quality.

The normal way of dehulling oil seeds is first to crush the seeds in a mill. Then the hulls and the meats are separated by screening, air classification and gravity separation in several steps. This method may also be used for rapeseed. The problems are most likely to be found in the separation of the smaller fragments of hulls and meats. The high oil content of the seeds also causes the meats to adhere to the hulls. A usual technique for reducing this adherence is to dry the seeds slightly before dehulling, i.e. the moisture content is lowered 1–2 per cent-units[8].

Hull separation can be carried out before or after solvent extraction. Eapen et al.[9] have tried both alternatives in connection with detoxification of rapeseed in laboratory and semi-pilot plant experiments. Grinding wet-heat treated seeds in a stream of water resulted in squeezing out the meats intact from the hulls. After detoxification, drying, flaking and solvent extraction, the meat-hull mixture was air classified into a meat fraction and a hull fraction. They thus obtained 44 per cent of the crude material (moisture- and oil-free basis) as a white powder containing 60 per cent protein and a hull fraction containing 33 per cent protein.

9.4 CRUSHING

After cleaning, the rapeseed is crushed by roller mills to facilitate the following cooking and extraction. Usually one or a few pairs of rolls are used, which may be smooth or corrugated. The rolls in one pair can be run at equal or different speeds. Sometimes the crushing is done in a roll assembly consisting of five rolls placed one above the other, and also in this case both corrugated and smooth rolls are used. Too intense crushing will produce large amounts of fines, leading to difficulties in the extraction process and with the clarification of the oil.

Essential with rapeseed is that the seed must have been dried to a moisture content of 6–9 per cent before crushing. With too high a moisture content the myrosinase in the seeds will start hydrolyzing the glucosinolates when they are brought together during the crushing.

9.5 COOKING

Before pressing and extraction the crushed seed is subjected to a heat-treatment which is called conditioning or cooking. There are many reasons for the positive effect of cooking. The following objectives of cooking are generally considered[1, 10]: completion of the breakdown of oil cells, coagulation of the proteins to facilitate the separation of the oil, reduction of the affinity of the oil for the solid surfaces, insolubilization of phosphatides, increased fluidity of the oil by an increase in

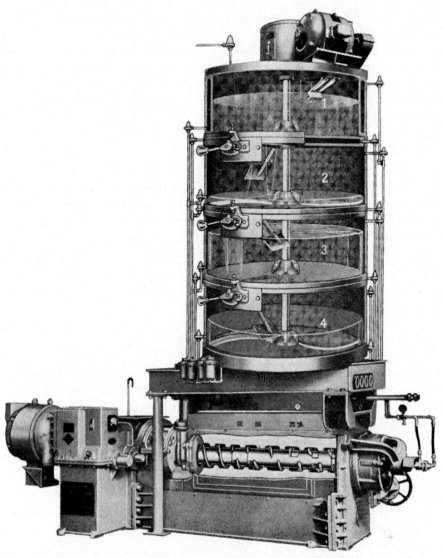

Fig.9:2. Cooker and screw press. (Courtesy of The French Oil Mill Machinery Co.)

temperature, destruction of molds and bacteria, inactivation of enzymes and drying of the seed to a suitable moisture content. The inactivation of myrosinase is the most important object of rapeseed cooking. This is essential for obtaining oil and meal of good quality and free of hydrolysis products from the glucosinolates (see Chapters 7, 10 and 15).

The cooking is generally carried out in 'stack cookers', which consist of a series of closed cylindrical steel kettles stacked one on the top of the other (see Fig.9:2). Each kettle is equipped with a sweep-type stirrer for agitating the crushed seeds during the cooking. The kettles are also jacketed for steam heating. Automatically operated gates provide a continuous flow of the seed down through the kettles. Moisture may be added through spray jets in the top kettle and removed through exhaust pipes in the lower kettles. Thus it is possible to control the moisture content of the cooking seed at each stage of operation.

The cooking conditions used are largely dependent on the subsequent extraction procedure and the meal quality desired. The primary operation parameters of rapeseed cooking are residence time, temperature and moisture. The residence time in the cooker is usually 30–60 min. The temperature may be different in the different kettles, usually rising down the cooker as the seed becomes dryer. Temperatures of 75–120 °C have been used, but usually the temperature is about 85–90 °C in the top kettle[11]. 80 °C is the minimum temperature for myrosinase inactivation. The highest temperatures are used when the seed is to be straight expeller pressed; for example 105–110 °C has been used followed by 5 min conditioning at 130–135 °C[12].

Since the moisture content of the seed affects the affinity between the meat and the oil, control of the moisture in the cooker is very important. Water or steam is sometimes added to the top kettle, but often there is no such addition in cooking rapeseed. The moisture content in the top kettle is 8–12 per cent and is reduced in the lower ones to 4–8 per cent. The lesser moisture the more powder is produced. For straight expeller pressing the seed is dried to a lower moisture content (4–5%) than when the seed is to be solvent extracted.

The method of cooking markedly determines the quality of both the oil and the oil-cake. Cooking is particularly important for the maintenance of a low level of oil refining losses[1]. The meal quality depends on the content and nutritional value of the protein and the content of goitrogenic substances. The protein of the meal is often extensively denatured in the cooker, which usually gives a better feeding value[6]. Too much heating, however, causes losses in amino acids, especially the basic ones. Lysine, which is an important essential amino acid, appears to be the most heat sensitive[12, 13] (see Chapter 15).

9.6 SCREW PRESSING

Extraction of rapeseed oil by means of screw presses (expeller presses) can be done as straight pressing or prepressing for subsequent solvent extraction. In the former case the oil content of the cake will be 4–7 per cent and in the latter 12–20 per cent[13].

The mechanical screw press has five essential elements: the main worm shaft and worms, the drainage barrel, the choke mechanism, the motor transmission and thrust bearings and the cooling system[14] (see Fig.9:2). The main worms shaft and worms are designed to exert a pressure of 1000–1400 kg/cm² [13] on the seed being processed and at the same time to convey the seed through and out of the pressure chamber. The choke mechanism permits a final adjustment of the pressure. The drainage barrel is made up of rectangular bars, which fit into a heavy barrel bar frame. The individual bars are separated by spacing clips which provide spaces of about 0.25 mm. This permits the drainage of oil from the seed being pressed and keeps the cake within the barrel, which also acts as a coarse filter medium. The screw press can be built up of a single horizontal shaft or a combination of a vertical and a horizontal shaft. A great deal of heat is produced during the pressing operation, especially when straight pressing is employed. In order to keep the temperature down, various cooling devices are required. The screw press may be equipped with water-cooled shafts and ribs in the bar cage or some of the expressed oil may be cooled and circulated over the exterior of the barrel.

Principally there is no difference between presses used for straight pressing and those used for prepressing. Ordinary high-pressure screw presses may be operated at low pressure and at an increased capacity, but especially designed machines are now normally used for prepressing.

The process of straight pressing has the advantage over solvent extraction in being simple and of lower investment cost. Disadvantages are, however, the high power consumption and the wear and tear of the machinery, giving high working costs. Another major drawback is that it is only possible to press down to a residual oil content of 4–7 per cent, giving high losses in yield[14]. There is also the effect of the high temperature generated during straight pressing causing an oil and particularly a meal of lower quality. The cooking conditions employed are also more severe when straight pressing is used.

The process of straight expeller pressing was the first method used for obtaining rapeseed oil and is even today applied in some cases. Now, however, the solvent extraction process is dominant, but it is most often combined with a prepressing step using screw presses. This utilizes the advantages of both techniques, and is especially useful in the case of seed with high oil content such as rapeseed.

After pressing, the oil contains fine meal particles. Before further treatment the oil is thus settled in a tank and filtered in filter presses with added filter aids[13]. The amount of fines, and thus the filtration rate, is highly dependent upon the cooking process. The fines may be returned to the cooker or, preferably, fed directly to the solvent extractor.

9.7 FLAKING

After straight expeller pressing the cake is cooled, moistured and milled to a meal. If a prepressing operation is used the crushed cakes are instead flaked for solvent extraction. This step is very important because the seed particles must be thin enough to be extracted readily, and yet large and coherent enough to form a mass through which the solvent will freely flow.

The flaking mill consists of one pair of smooth rolls by which a flake thickness of 0.2–0.3 mm is obtained. The moisture content of the seed material must not be too low, if thin coherent flakes are to be formed.

Sometimes when the oil content in the press cake is low (12–14%) the flaking is not so important and the cake is just milled[15]. This may also be the case if the flaking is done before the cooking.

9.8 SOLVENT EXTRACTION

Extraction of rapeseed with solvents constitutes the most efficient method of obtaining the oil from the oil bearing material. The advantage over mechanical pressing is greater for oil seeds with low oil content. Oil seeds high in oil content such as rapeseed are, however, also most economically solvent extracted, especially in high capacity plants. Generally prepressing is used to give a press cake containing 12–20 per cent oil, which has the most advantageous properties for solvent extraction. This process also gives the lowest residual oil content, about 0.5 per cent. However, direct solvent extraction of rapeseed is also used, and will be dealt with below.

In spite of the extensive application of solvent extraction of oilseeds, there is no universally accepted theory of the manner in which the basic process occurs. The rate of extraction is profoundly affected by the flake thickness and the flake area, the temperature, the solvent and moisture. Karnofsky[16] concluded from extraction studies that the bulk of the oil is extracted rapidly but the remainder of the oil only slowly and with increasing difficulty. Based on the porous nature of the flakes, two entirely different mechanisms have been advanced to explain solvent extraction of oils[17, 18]. These are the theory of molecular diffusion and the theory that the rate of extraction is determined by the rate of solution of undissolved oil. It is generally accepted that extraction takes place by both mechanisms simultaneously, but that dissolution of the difficulty soluble materials eventually determines the extraction rate.

The object of solvent extraction is to remove as much of the oil as possible from the meal with a minimum of solvent. This is most effectively achieved by continuous counter-current extraction. The principle is that the solvent should be passed in several stages through the travelling mass of the oil-bearing material. The characteristics of a good extractor are that it produces a miscella (solvent plus oil) of high oil content and with low content of fines, has relatively small dimensions, high capacity, reliability and safety of operation and has a low labour requirement[6].

Continuous extractors fall into two classes[5]: percolation extractors in which

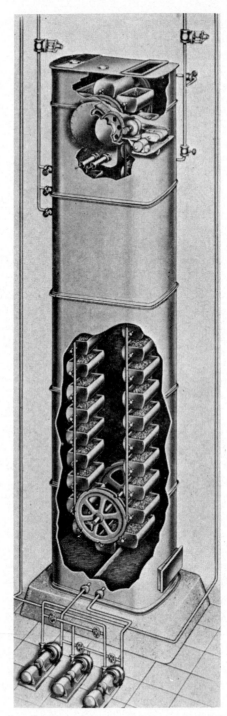

Fig.9:3. Vertical basket extractor. (Courtesy of The French Oil Mill Machinery Co.)

the solvent flows by gravity through a bed of solids, and immersion extractors in which the solids are immersed in the solvent or miscella. There are also extractors that comprise a combination of these two types. Modern extractors are of the percolation type, which gives higher miscella concentrations and allows use of the solid bed as a filter giving very small amounts of fines in the miscella. This is achieved by recirculation of miscella within the percolation extractor. In addition, the percolation type usually incorporates much better drainage of liquid from the spent solids. This fact and the increased miscella concentration decreases steam requirements for solvent recovery.

There are many modern percolation extractor designs, but the general principle is mainly the same. This may become clear from the following brief descriptions of some extractors.

The oldest type is the vertical basket extractor (Bollman extractor) still built in America and Europe, which consists of a vertical chamber in which a number of baskets with perforated bottoms are carried on a chain (see Fig.9:3). The baskets are moved vertically. The flakes are charged at the top and are concurrently extracted by miscella (half concentrated) on the descending side. On the ascending side the flakes are countercurrently extracted by fresh solvent, which is added just below the last drainage part. The meal is discharged at the top and transported away by a screw conveyor.

This type of extractor has the advantage of producing miscella free of fines because of the effective filtration in the concurrent extraction step. The large volume and height of the extractor is, however, an important drawback[19]. Modifications have thus been developed resulting in rectangular and horizontal basket extractors.

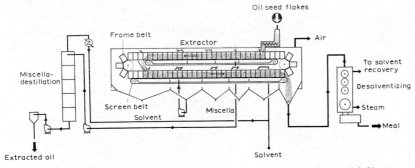

Fig.9:4. Lurgi frame belt solvent extractor. (Courtesy of Lurgi AG)

Lurgi has developed a frame belt extractor[20], composed of two screen belts supporting the frame baskets (see Fig.9:4). These have no bottoms and move at the same speed as the screen. The prepared seed material is delivered at one end of the upper part of the extractor, and at the other end the material falls down into the lower returning frame baskets. From the other end of this screen belt the meal is discharged. The solvent passes the seed material in several steps countercurrently. Under each step there is a hopper collecting the miscella. This is then partly pumped and sprayed over the material in the same step and partly pumped to the next step. The apparatus is divided into three zones—the washing zone, the extraction zone with circulating miscella, and the filtering zone in which the concentrated miscella is percolated through the bed until free from flakes. There are also drip-off zones between the steps, preventing the miscella of low concentration from diluting the more concentrated one. The shifting of the material when falling to the lower belt prevents any formation of channels and secures uniform de-oiling.

Another horizontal extractor is the De Smet endless belt type extractor[21]. The main differences as compared with the Lurgi extractor are that there is only one belt and no baskets.

Many constructions of extractors are based on the principle of turning sector-shaped compartments (baskets) in a horizontal plane around a vertical axis. The best known example is the Rotocel extractor manufactured by Blaw-Knox, Rosedowns and Krupp[1, 19, 21] (see Fig.9:5). This construction principle provides a simple, compact, and flexible extractor. As each compartment passes the feeder pipe, the seed material is deposited therein. As the filled compartment continues to revolve, it is sprayed at intervals mainly countercurrently with solvent from

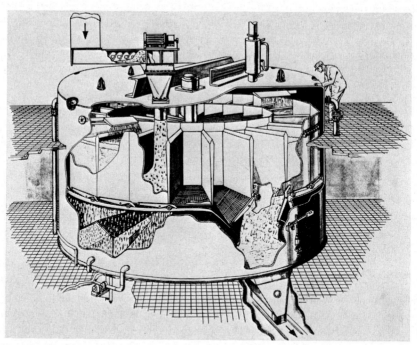

Fig.9:5. Rotocel solvent extractor. (Courtesy of Rosedown & Thompson Ltd. and Blaw Knox Co.)

overhead nozzles. The resulting miscella drains continuously through the mesh type floor of the compartment and collects in troughs. At the completion of the circuit the floor of the compartment opens, thus dumping the residual meal into a hopper. As soon as the floor is restored to its original position, the compartment receives another load of unextracted flakes, and the cycle is repeated. The whole extractor is entirely enclosed in a vapour tight cylindrical tank, and makes one complete revolution in about 90 min.

French has developed a similar extractor called the 'Stationary Basket Extractor', where the baskets are stationary and the filling spout, miscella distributors, miscella collection pans and the flake discharge hopper are rotating[21].

With the percolation type of extractor it is difficult to extract seeds of high oil content such as rapeseed directly. The volume reduction of the seed material and the tendency to produce fines causes improper packing of the bed and channelling, giving a lower yield of oil. Only if the extraction is performed in steps with repacking of the bed, as in the Lurgi extractor[21] or in the 'Carrousel Extractor' built in several floors from Extraktionstechnik[15], can direct extraction be used.

The filtration extraction ('Filtrex' from Wurster & Sanger) process was developed

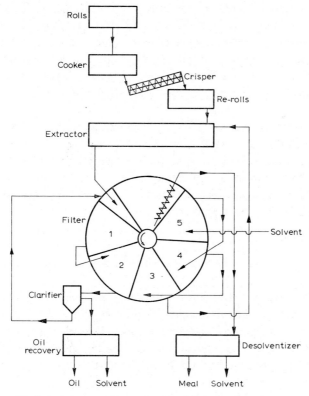

Fig.9:6. Schematic diagram of filtration extraction unit[22].

References pp. 216–217

for direct extraction of various oilseeds and is also used for rapeseed[22] (see Fig.9:6). The extraction is carried out in two steps, slurrying and filtration. Another distinctive feature of this process is that the cooking step is essential to the process. The cooked flakes are rapidly cooled in an evaporative step called 'crisping' and are then rerolled, producing a material with good extraction and filtration properties. The 'crisped' seed material is continuously fed together with one of the miscella filtrates to the slurry mixer, which is a mildly agitated horizontal tank. The slurry is kept there for 30 to 45 min at a temperature of 50–60 °C, and is then filtered on a rotating horizontal screen vacuum filter. The concentrated miscella drains through the filter leaving a 5 cm thick bed. The cake is counter-currently washed with solvent, but the most concentrated miscella is refiltered through the bed to reduce the content of fines. The meal, which only contains 25–30 per cent solvent, is removed continuously. The residual oil content, which can be obtained economically by using direct extraction methods, usually amounts to more than 1–2 per cent.

The most widely used solvent for extraction is commercial n-hexane, a petroleum fraction with a boiling range of 63–66 °C, especially refined for use in the vegetable oil extraction industry[1, 17]. Although other solvents are sometimes used for other oilseeds, only hexane is used for rapeseed. The high erucic acid content of rapeseed oil makes its solubility in polar solvents such as absolute ethanol or isopropanol less than that for other oils[23, 24]. Loury and Feng[25] reported that acetone, ethanol and isopropanol gave poor yields and that dichloroethane and trichloroethylene gave red-coloured oils. Cyclohexane gave the same results as n-hexane. Franzke et al.[26] found that the higher the aromatic content in the hexane, the darker was the extracted oil.

The solvent to meal solids ratio in modern extractors is about 1.1–1.3[17]. The concentration of oil in the miscella leaving the extractor is 20–35 per cent, somewhat dependent on the oil content of the extracted seed material[1, 21]. By extracting high oil content seeds in the Lurgi extractor the figures 45–50 per cent have been obtained[20]. The amount of solvent left in the meal after extraction in a percolation type extractor is about 30–35 per cent[8].

9.9 MEAL DESOLVENTIZING

After completed solvent extraction the extracted material is discharged from the extractor and conveyed to the desolventizer, where the hexane is removed by indirect heating and by direct steam injection. Besides a good desolventizing effect leaving a meal with not more than 0.01 per cent hexane, this step should give the meal an appropriate heat treatment, so that the feeding value of the meal will remain high.

The oldest equipment for desolventizing consists of a series of horizontal steam

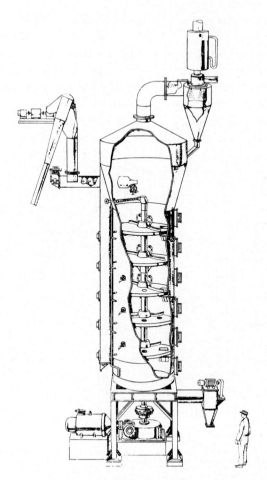

Fig.9:7. Desolventizer-toaster. (Courtesy of Rose, Downs & Thompson Ltd. and Blaw Knox Co.)

jacketed conveyer tubes. The wet meal is introduced at one end at the top and is moved downwards in zig-zag fashion. In the bottom tube sparging steam is sometimes introduced to evaporate the last traces of solvent. This desolventizing system is however rather inefficient and suffers from dust control problems and other problems of operation[5], but it usually gives a light coloured meal.

The most common desolventizer equipment today is the desolventizer-toaster, originally developed to give high quality soybean meal[19]. The equipment is similar to a cooker but is totally enclosed, and with a dust collector at the top (see Fig.9:7). Live steam is the main source of heat for the vapourization of the solvent from the meal, and it is introduced in the top stages. Solvent vapours replace the steam, which condenses and replaces the liquid solvent. The moisture content of the meal is then about 20 per cent. The toasting process taking place in the lower stages denatures the proteins and reduces the amount of glucosinolates in the rapeseed meal[6, 15] by the action of moisture and heat, in order to give the meal a high feeding value. In the toasting section the meal is also dried to a moisture content of 6–11 per cent. The temperature of the meal in the desolventizer-toaster is about 100–110 °C and the residence time about 30 min. The meal will usually be rather dark coloured after this process.

Another modern system is flash desolventizing, in which superheated solvent vapours are recycled over the meal to evaporate the solvent (see Fig.9:8). The

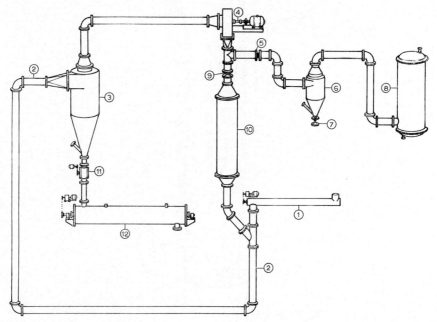

Fig.9:8. Flash-desolventizer. (Courtesy of Extraktionstechnik and Engineering Management Incorporated)

1. Feed conveyor	7. Scrubber discharge lock
2. Desolventizing tube	8. Solvent condenser
3. Flake separator	9. Recirculating control valve
4. Circulating blower	10. Vapour superheater
5. System pressure control valve	11. Separator discharge lock
6. Vapour scrubber	12. Flake stripper

desolventizing time can be reduced to a few minutes or even lower. A final steam stripping is required to take away the last traces of solvent from the meal. This is carried out in a special 'deodorizer' or in an ordinary desolventizer-toaster. The meal will have a light colour and the protein solubility can be easily regulated by this process[19].

After desolventizing the meal is immediately cooled in a meal cooler, where the final drying also takes place. The moisture content should be 9–10 per cent for good storage and transport quality[6]. Finally the meal is milled in a hammer mill.

Fines in the miscella are first removed in a filter press or in modern plants by recycling the miscella through the bed in the extractor. To keep the evaporation costs low the miscella should be as concentrated as possible.

The evaporation is ordinarily carried out in a long tubular evaporator where 90 per cent of the solvent is recovered. The oil is then passed through a stripping column which can be either a packed column or a bubble cap column. There the last traces of solvent are removed by applying sparging steam at the bottom of the column. To obtain good heat economy of the process the evaporation is divided into two stages and may be done under vacuum. The miscella in the first (second) stage is heated by the hot vapours from the desolventizer and the hot oil from the stripper is used for heating the second (first) stage evaporator. This system will lower the steam consumption to about 25 per cent[17]. The residual amount of solvent in the oil is less than 0.1 per cent. A separate problem is the recovery of solvent from the final discharge of vent gases. Refrigerated vent condensers have been used extensively. A new system of solvent recovery has been developed based on the principle of solvent absorption by a vegetable oil. This equipment is almost always installed as a part of new plants[18]. The main advantage is the simplicity of the equipment, especially since a complicated compressor for the cooling medium is unnecessary. Sometimes a water scrubber is used for solvent recovery[13]. A solvent loss as low as 0.06 per cent based on the raw material feed can be achieved in large plants[18].

9.11 DEGUMMING

The crude solvent extracted oil contains about 2 per cent 'lecithin' gums, which are usually removed by the degumming (desliming) process. If the gums are not removed, they will cause trouble by settling out in storage tanks, and they will tend to produce large refining losses[1].

The oil is treated with a small amount (4%) of water or a dilute water solution of acids or salts[1, 27-29] in order to precipitate the gums (mainly phosphatides, see Chapter 13). Often degumming with water at the extraction plant is followed by a degumming with phosphoric acid at the refining plant (see Chapter 10). The precipitated gums and the water are separated from the oil in continuous centrifuges, which are of the same type as those used in continuous refining. The advantage is especially that these centrifuges are hermetical and all contact with air can be avoided. This is important for preventing oxidation since the oil temperature is high (80 °C)[30]. If the removed gums are to be used for commercial lecithin production, they should be dried as soon as possible. This is carried out in vacuum with agitation or in a thin film evaporator at 70–110 °C, and the drying can be combined with hydrogen peroxide bleaching. Usually, however, the separated gums are transported back to the desolventizer-toaster, since rapeseed lecithin is of low value and soybean lecithin dominates the market. In the toaster the gums make the meal powder agglomerate more easily[6]. The degummed oil, which still contains about 0.5 per cent phosphatides, is mixed with the press oil and dried in vacuum at 80–90 °C. After cooling, the oil can be stored for a long time if kept cold.

9.12 STORAGE OF CRUDE OIL AND MEAL

The crude degummed oil may be stored for a longer or shorter time before refining. The longest time the oil can be stored depends mainly on its properties and stability but also on the way in which it is stored.

Rapeseed oil is usually stored in tanks. If these are large the temperature must be kept high enough to avoid crystallisation of the oil[30]. Rapeseed oil stored in this way at 25–30 °C will already after 3–4 months be highly oxidized. It is then better to store the oil in small tanks, in which it may solidify and then heat the oil to melt it, when it is to be used. The heating must be controlled closely in order to avoid local overheating and this is best done by indirect heating.

The amount of free fatty acids (FFA) and the degree of oxidation of the oil as measured by the peroxide value (PV) and benzidine value (BV), are the main properties which change during the storage. The amount of free fatty acids in crude oil of good quality should not be greater than 1.5 per cent in order to avoid high refining losses[31]. The amount is usually lower and the increase during the storage is also rather low, and thus there is usually no problem with free fatty acids.

Oxidative stability has the greatest significance for the storage of rapeseed oil. The linolenate content of the oil makes it prone to oxidation. The tocopherol content in rapeseed oil, however, makes the stability acceptable. It is important that the rapeseed is fully ripened and of good quality to give an oil of high stability[32–34]. The tocopherol content increases and chlorophyll (which is pro-oxidative) decreases in amount during ripening. Germinated and mouldy seeds also possess higher lipase activity and the oil produced is less stable.

Rapeseed meal is stored in bulk at a maximum of 10 per cent moisture in silos or metal tanks with conical bottoms. There is no problem in storing meal for a short time if properly done and with an appropriate percentage of moisture. Moisture adsorption due to high relative humidity causes heating, loss of ether-extractable fat, increase in free fatty acids, and slight decreases in protein and nonreducing sugars[35].

To obtain a rapeseed oil of high quality, it is always necessary to start with well ripened high quality seeds[34] (see Chapter 5). Furthermore it is essential to use optimal processing parameters throughout the production. The earlier the oil and meal get any damage the more serious it is for the end products. In other words, the basic quality of the oil is not improved, only more or less damaged[8].

One of the most important considerations with regard to quality is to have a continuous and rapid flow of material from one piece of equipment to another. In crushed and flaked seeds deterioration takes place rapidly, particularly if the material has a high moisture content or is of poor quality. It is important to avoid oxygen contact with the oil, to keep the temperatures low, to make the hold-up times short, to eliminate sources of catalysis, and to make the treatments in the dark[36].

There are two enzyme systems in rapeseed that may influence the oil and meal quality, *viz.* lipase and myrosinase. They must be inactivated at an early stage in the process, and this is done in the cooker.

The lipase activity of rapeseed may result in high free fatty acid values of the oil. The optimum temperature for lipase activity is 37–40 °C[37]. Lipase inactivation is highly dependent on moisture and temperature[37–40]. Inactivation starts at 60–70 °C and is rapid at 90–100 °C in an aqueous medium, and even more rapid in an acidic medium. In the dry state activity decreases only very slowly, and temperatures of 140–160 °C are needed for rapid inactivation.

The myrosinase in rapeseed catalyses the hydrolysis of the glucosinolates to give glucose, sulphate, isothiocyanates and (*l*)-5-vinyl-2-oxazolidinethione (see Chapter 15). The latter product is goitrogenic, thus yielding meal of low feeding value. They also act as a poison for hydrogenation catalysts if they are dissolved in the oil[41] (see Chapter 10). It is thus important to keep the glucosinolates intact by effectively inactivating myrosinase by proper cooking. Myrosinase and glucosinolates are brought together by the crushing of the seed, but the hydrolysis can only take place if the moisture content is high enough (13%). The enzyme has the greatest activity at 40–70 °C[42].

Inactivation of myrosinase starts at 70–80 °C if the moisture content is 6–10 per cent and is usually performed at 80–90 °C in the cooker[13, 43]. Eapen *et al.*[44] have performed various laboratory heat treatments in order to evaluate their effect on myrosinase inactivation. Dry-heat treatment (30 min, 105 °C) proved unsatisfactory, but steam blanching (5 min), microwave heating (3 min) or immersion of the seed in boiling water (1.5 min) were effective. If the moisture content of the seed material is high and the temperature rise is not quick enough, glucosinolates may be hydrolyzed before the myrosinase is inactivated[22].

Heat treatments during the process may also have a negative effect on oil and meal quality. If cooking temperatures above 110 °C are employed the extracted oil may be difficult to hydrogenate, which has been attributed to chemical breakdown of the glucosinolates to give oil-soluble sulphur-containing compounds[22]. Microwave heating (1.5 min) and steam blanching (5 min) have resulted in dark-coloured oils, while dry cooking (30 min, 105 °C) and immersion in boiling water (1.5 min) gave lighter-coloured oils[44]. Rapeseed stocks with green seeds often yield dark coloured oil, but an immersion of the seed in boiling 0.5 per cent sodium hydroxide has been found to give lower colour values[44].

The colour of the meal is also affected by the heat treatments, especially that in the desolventizer-toaster. High temperature and moisture favour nonenzymatic browning reactions (Maillard-reactions) between reducing sugars and amino acids. At the same time the feeding value may be lowered by a decrease in the amount of lysine, which is the most sensitive amino acid in this respect[13] (see Chapter 15). This decrease in lysine is also the reason that the straight expeller pressed meal is of lower nutritional value than solvent extracted meal[45, 46].

A small lowering in oxidation stability of the oil may take place during the extraction process, together with a slight increase in the peroxide value[47] and the benzidine value[48]. Heat treatments of the seeds and exposure of the oil to light and heat result in higher peroxide values[49].

The fatty acid composition of the oil can also be affected by the extraction technique used. The fatty acid composition in the neutral oil is different from that in the phospholipids (see Chapter 7). These are most difficult to extract and are thus obtained only in the solvent extracted oil. By using repeated extraction of white mustard with different solvents Iverson[50] found that the last extracted oil was different from the first oil. Zeman and Pokorny[51] also found that the oil remaining in meal extracted with hexane contained a higher amount of poly-unsaturated fatty acids and a lower content of erucic acid.

REFERENCES

1 D. Swern, *Bailey's Industrial Oil and Fat Products*, 3rd ed., Interscience Publ., New York, 1964.
2 E. W. Eckey, *Vegetable Fats and Oils*, Reinhold Publ. Corp., New York, 1954, p.433.
3 H. P. Kaufmann, *Neuzeitliche Technologie der Fette und Fettprodukte*, Vol. I, Aschendorffsche Verlagsbuchhandlung, Münster, 1956, p.40.
4 H. J. Lips, N. H. Grace and E. M. Hamilton, *Can. J. Res.*, 26 (1948) 360.
5 R. P. Hutchins, *J. Am. Oil Chem. Soc.*, 33 (1956) 457.
6 A. Rutkowski and H. Kozlowska, *Sruta Rzepakowa, Wydawnictwo Przemyslu Lekkiego i Spozywczego*, Warszawa, 1967.
7 C. G. Youngs, *Fats and Oils in Canada, Semi-annual Review*, 2:2 (1967) 39.
8 R. P. Hutchins, *J. Am. Oil Chem. Soc.*, 45 (1968) 624 A.
9 K. F. Eapen, N. W. Tape and R. P. A. Sims, *J. Am. Oil Chem. Soc.*, 46 (1969) 52.
10 A. M. Altschul, *Processed Plant Protein Foodstuffs*, Academic Press, New York, 1958.
11 W. L. Goble, *Oil Mill. Gaz.*, 72 (1968) 66.
12 D. R. Clandinin and E. W. Tajcnar, *Poultry Sci.*, 40 (1961) 291.

13 C. G. Youngs, *Rapeseed Meal for Livestock and Poultry, A Review*, Can. Dep. Agric., Ottawa, 1965, p.24.

14 J. W. Dunning, *J. Am. Oil Chem. Soc.*, 33 (1956) 462.

15 W. Depmer, *Int. Symp. Chem. Technol. Rapeseed Oil Cruciferae Oils*, Warsaw, 1970, p. 45.

16 G. Karnofsky, *J. Am. Oil Chem. Soc.*, 26 (1949) 564.

17 K. S. Murti, *Indian Oilseeds J.*, 9 (1965) 292.

18 K. W. Becker, *J. Am. Oil Chem. Soc.*, 41 (1964) 8.

19 K. W. Becker, *Chem. Eng. Progr. Symp. Ser.*, 64 (1968) 60.

20 A. Bhattacharya, *Chem. Age, India*, 17 (1966) 300.

21 H. P. Kaufmann, *Neuzeitliche Technologie der Fette und Fettprodukte*, Vol. IV, Aschendorffsche Verlagsbuchhandlung, Münster, 1965.

22 J. R. Reynolds and C. G. Youngs, *J. Am. Oil Chem. Soc.*, 41 (1964) 63.

23 R. K. Rao and L. K. Arnold, *J. Am. Oil Chem. Soc.*, 33 (1956) 389.

24 R. K. Rao and L. K. Arnold, *J. Am. Oil Chem. Soc.*, 34 (1957) 401.

25 M. Loury and H. Feng, *Rev. Franç. Corps. Gras*, 5 (1958) 83.

26 C. Franzke, G. Heder and B. Klingberg, *Die Nahrung*, 11 (1967) 277.

27 C. Defromont, R. Guillaumin, D. Delahaye and N. Drouhin, *Rev. Franç. Corps Gras*, 9 (1962) 486.

28 R. Guillaumin and M. Boulot, *Rev. Franç. Corps Gras*, 7 (1960) 506.

29 B. F. Teasdale, *Rev. Franç. Corps Gras*, 15 (1968) 3.

30 E. Olin, *Acta Polytech. Scand. Chem. Met. Ser.*, 1 (1958) 56.

31 E. Olin, *Sveriges Utsädesförening Tidskr.*, 67 (1957) 91.

32 A. Rutkowski, Z. Makus and A. Stefaniuk, *Roczniki Technol. i Chem. Zywnosci*, 2 (1957) 91.

33 A. Rutkowski and Z. Makus, *Oleagineux*, 13 (1958) 203.

34 A. Rutkowski and Z. Makus, *Fette, Seifen, Anstrichmittel*, 61 (1959) 532.

35 C. Defremont and D. Delahaye, *Rev. Franç. Corps Gras*, 8 (1961) 359.

36 J. Dahlén, *Livsmedelsteknik*, 1967, p.245.

37 A. Rutkowski, Z. Franczek and W. Korzeniowski, *Oleagineux*, 18 (1963) 701.

38 J. Pokorny, G. Janicek and H. Zwain, *Prumysl Potravin*, 15 (1964) 217.

39 H. Zwain, J. Pokorny and G. Janicek, *Ind. Aliment. Agr.*, 80 (1963) 607.

40 L. Jiraskova, J. Pokorny and G. Janicek, *Sb. Vys. Skol. Chem.-Techn. Praze Potravin.*, E 20 (1968) 79.

41 K. Babuchowski and A. Rutkowski, *Seifen Öle Fette Wachse*, 95 (1969) 27.

42 G. C. Mustakas, L. D. Kirk and E. L. Griffin, Jr., *J. Am. Oil Chem. Soc.*, 39 (1962) 372.

43 L.-Å. Appelqvist and E. Josefsson, *J. Sci. Food Agr.*, 18 (1967) 510.

44 K. E. Eapen, N. W. Tape and R. P. A. Sims, *J. Am. Oil Chem. Soc.*, 45 (1968) 194.

45 D. R. Clandinin, *Poultry Sci.*, 46 (1967) 1596.

46 K. J. Goering, O. O. Thomas, D. R. Beardsley and W. A. Curran, Jr., *J. Nutr.*, 72 (1960) 210.

47 A. Rutkowski, *Fette, Seifen, Anstrichmittel*, 61 (1959) 1216.

48 U. Holm, K. Ekbom and G. Wode, *J. Am. Oil Chem. Soc.*, 34 (1957) 606.

49 L.-Å. Appelqvist, *J. Am. Oil Chem. Soc.*, 44 (1967) 206.

50 L. J. Iverson, *J. Am. Offic. Anal. Chem.*, 49 (1966) 332.

51 I. Zeman and J. Pokorny, *Prumysl Potravin.*, 15 (1964) 289.

CHAPTER 10

Processing of Rapeseed Oil

R. WETTSTRÖM

AB Karlshamns Oljefabriker, Karlshamn (Sweden)

CONTENTS

10.1 INTRODUCTION

Rapeseed oil is mainly processed in the same way as other vegetable oils such as soyabean oil, although some differences exist due to its rather high content of erucic acid. This may be changed in the near future since varieties with low or negligible erucic acid content have been developed by plant breeders (p.116). One aim of the various processing steps is to remove substances which can give unpleasant flavour to the oil. The natural impurities consist of free fatty acids, mono- and diglycerides, phosphatides, steroids, vitamins, hydrocarbon pigments, proteins, sulphur compounds and oxidation products from the triglycerides. The crude oil must be of good quality, since an improperly pretreated and stored crude oil needs more elaborate refining, and in spite of a more intense and expensive treatment, such an oil is also less stable than is normal for an oil originally of good quality. First of all the extraction must be done carefully, since incorrect heat treatment can give dark oils, especially if the seed stock contains green seeds[1]. Furthermore, if the enzyme myrosinase has reacted with the glucosinolates (p.149), some oil-soluble sulphur compounds may be obtained. Even a good quality crude rapeseed oil contains about 10 ppm sulphur.

Besides being responsible for flavour disturbances, the sulphur compounds are severe catalyst poisons in the hydrogenation step. In the hydrogenation the melting point of the oil is increased and products with varying consistency can be obtained for various purposes, such as the production of margarine and shortening. The processing of rapeseed oil normally consists of degumming and refining, bleaching, hydrogenation and post-bleaching (if desired), and deodorization. A general review of rapeseed oil processing has previously been published by Teasdale[2].

10.2 DEGUMMING AND ALKALI REFINING

Lecithins and other gums are often removed at the end of the extraction process to avoid tank residues (Chapter 9). Nevertheless, there are still some mucilaginous components in the oil, mainly phosphatides, which can be measured by the amount of phosphorus present[3]. The presence of phosphatides will increase the risk for losses due to the formation of emulsions during the alkali treatment, and decomposition to coloured products may occur during the deodorization. Various methods for degumming the oil have been suggested[4], for example treatment with sulphuric acid, hydrochloric acid and also nitric acid[5-7]. Although an effective degumming agent, nitric acid is found to give a dark-coloured deodorized oil with bad flavour[8]. For edible fats it is more advisable to treat the oil with 0.02–0.1 per cent of phosphoric acid (85 per cent technical grade) directly followed by alkali treatment according to the Alfa Laval Short-Mix process[9], or to treat the oil at a temperature of 60–90 °C with 0.25–0.3 per cent by volume of 85 per cent phosphoric acid for 5–30 minutes under vacuum and then separate the precipitates by centrifugation according to the Zenith process[10]. One advantage with these two methods is the removal of calcium and magnesium ions by the phosphoric acid, which decreases the soap content in the refined oil[11].

10.2.1 Neutralization

Normally the process of neutralization is carried out with sodium hydroxide of various concentrations, but also with other alkaline solutions, e.g. sodium carbonate (Clayton–De Laval process). However, the latter process has the disadvantage that carbon dioxide is formed (ref. 4, p.86). The sodium hydroxide combines with the free fatty acids to form sodium soaps, but if used in excess it will also react with the triglycerides. A stronger alkaline solution (4 M) will increase this risk, while the use of more dilute alkali increases emulsion formation. Together with the reaction time these factors must be borne in mind in order to secure low neutral fat losses. The efficiency of the refining can be expressed in various ways as shown in Table 10:1.

TABLE 10:1

DEFINITION OF DIFFERENT EFFICIENCIES

$$\text{Refining factor} = RF = \frac{\text{Per cent total refining loss}}{\text{Per cent fatty acid content of crude oil}}$$

$$\text{Fatty acid factor} = \frac{\text{Per cent total fatty matter in soap-stock of crude oil}}{\text{Per cent fatty acid content of crude oil}}$$

$$\text{Refining efficiency} = \frac{\text{Yield of neutral oil} \times 100}{\text{Neutral oil content of crude oil}}$$

When different factories or processes are to be compared, it is important to know whether or not the total refining loss or the yield of neutral oil includes the bleaching step. To obtain a rapeseed oil of high quality it is often necessary to perform another treatment with alkali, using either a weak solution of sodium hydroxide or a mixture of sodium hydroxide and calcium carbonate. After the alkaline treatment the rapeseed oil has to be washed thoroughly with soft water[11], since the calcium and magnesium ions in the hard water can be dissolved in the oil phase. It is practical to use either water from the condensers or desalted water from ion exchangers for this purpose. Immediately after washing, the oil must be dried under vacuum to avoid an increase in free fatty acids due to hydrolysis. A refined and dried rapeseed oil may be stored for a few days before bleaching, although it is of course preferable to make this storage as short as possible. Closed tanks in which the oil is protected by nitrogen are recommended for storage in order to avoid contact with air during the processing, especially when the oil is warm. These precautions become more necessary after the oil has been refined since some substances with antioxidative effect are removed during the processing[12]. The addition of antioxidants to rapeseed oil has no effect when the oil is properly stored, and is only of limited value when stored under improper conditions[13]. The method of refining greatly influences the oxidative stability[9, 11, 14].

(a) Batch process

In batch refining, good results have been obtained with Swedish rapeseed oil under the following conditions[15]: the degummed oil is heated to 70 °C and 1 volume per cent of a solution containing 10 per cent acid (*e.g.* oxalic, citric or phosphoric acid) is added under stirring. The agitation is continued for one hour, whereupon sodium carbonate (25 per cent solution) is added in an amount of 0.3 volume per cent, and the agitation is continued for another hour while the temperature is increased to 80 °C. The oil is then neutralized with 3.5 M sodium hydroxide (20 °Be). After one hour of separation the oil is washed with 10 volume per cent of water and the mixture is left for two hours' sedimentation. The separated soap phase is drained and the oil is heated to 95 °C. The oil is re-refined with 3 per cent 1.5 M sodium hydroxide (6 °Be) under agitation for three hours. In order to break the emulsion, 0.25 per cent sodium chloride is added and stirring is continued for another hour. The oil is washed three times with 10 per cent of water. Finally, 0.05 per cent of a citric acid solution in ethanol (1 : 1) is added in order to split the remaining soaps. This also increases the bleaching effect, probably due to the fact that iron and other complexing metals form chelates with citric acid.

In order to avoid the emulsion losses, suggestions have been made to use electrolytes or alcoholic alkaline solution. In the Marchon process the use of sodium xylene sulphonate in the alkali has proved to decrease the refining losses[16]. From one to three per cent of the hydrotrope is used for each unit of acid value in the crude oil. A 40 weight per cent hydrotrope solution is mixed with the normally required amount of lye and is added to the heated oil (70 °–90 °C) with vigorous agitation. Compared to normal batch refining the settling of the soap stock is considerably promoted by the high specific gravity and the low viscosity of the hydrotrope solution with the dissolved soap. After separation the soap stock is split with sulphuric acid and the fatty acids separate. The hydrotrope solution has to be concentrated by evaporation to 40 per cent before recycling to the process. The sodium sulphate obtained by splitting the soaps accumulates and the excess will crystallize, leaving a deposit which must be removed by filtration. The advantage with the Marchon process is more evident for batch refining. The improvement of continuous refining may be too small to pay the extra capital and processing costs.

(b) Continuous process

Emulsion in the soap stock causes high refining losses, even if the emulsion layer is separated as such and afterwards split by treatment with sodium chloride. Better results have been achieved with continuous processing using separators, *e.g.* Alfa Laval, Sharples and Westfalia[4, 17]. With the continuous system a uniform quality is obtained compared to that from batchwise neutralizing.

Alfa Laval Short-Mix equipment is especially well designed for the refining of

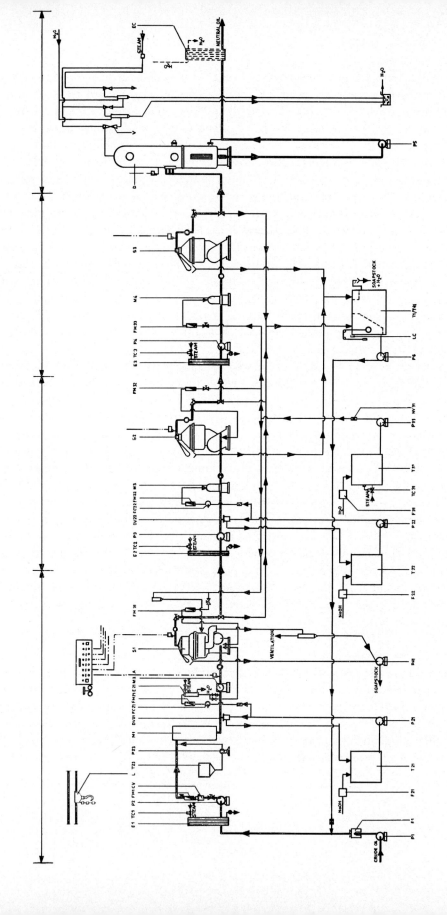

TABLE 10:2

ALFA LAVAL SHORT-MIX COMMERCIAL OPERATING DATA[9]

Crude oil	Location of refinery	% FFA in crude oil	Wesson loss %	Refining loss %	Refining factor	% FFA in refined oil
Rapeseed	Germany	1.70	2.00	3.23*	1.90	0.06
	Germany	1.20	1.50	2.22*	1.85	0.04
	Sweden	2.20	2.80	3.95*	1.80	0.07
Soyabean	Germany	0.52	0.90	1.15*	2.22	0.06
	Netherlands	0.54	0.94	0.98	1.82	0.05
	Netherlands	8.60	—	12.10	1.40	0.04

* Including re-refining.

Fig.10:1. Alfa Laval Short-Mix Refining Plant.
Continuous, all-hermetic and self cleaning.

Centrifugal Separators:
S1   Hermetic self cleaning separator
S2, S3   Hermetic separator

Pumps
P1   Crude oil pump
P2, P3, P4   Oil feed pump
P5   Refined oil pump
P6   Slop oil pump
P21, P22   Lye pump
P23   Phosphoric acid pump
P31   Water pump
P41   Soapstock pump

Tanks:
T21, T22   Lye tank
T23   Phosphoric acid tank
T31   Water tank
T1/T41   Slop tank/catch basin

Mixers:
M1   Degumming mixer
M2   Neutralizing mixer
M3   Re-refining mixer
M4   Water-washing mixer

Strainers:
F1   Crude oil strainer
F21, F22   Lye strainer
F31   Water strainer

Heat Exchangers:
E1, E2, E3   Oil heater
EC   Oil cooler
E21   Lye heater

Flow Meters:
FM1   Oil flow meter
FM21, FM22   Lye flow meter
FM31, FM32   ⎰ Water flow meter
FM33,   ⎱

Vacuum Dryer:
D   Vacuum dryer
V   Steam ejector

Auxiliary Equipment:
A   Alarm system
L   Lifting tackle

Controllers:
CV   Constant oil-flow controller
DV21, DV22   Lye-dosing ratio controller
FC21, FC22   Lye-flow control valve
LC   Oil level controller
TC1, TC2, TC3 Oil temperature controller
TC31   Water temperature controller
NV 31   Non-return valve for water

Fig.10:2. Alfa Laval Hermetic Self-Cleaning Separator, Type SRPX 213HGV-14H.
Capacities between approx. 1000 and 6000 l/h (200–1300 igph).

rapeseed oil and gives an oil of high quality with low losses as shown in Table 10:2[9, 18].

In the most recent design (Fig.10:1), called the Alfa Laval All Hermetic Self Cleaning Short Mix, a new type of separator is used (Fig.10:2). At regular intervals the sludge is discharged from the bowl, which reduces the frequency with which the bowl must be washed.

The crude oil is treated with 0.1–0.3 volume per cent of phosphoric acid in a mixer and then neutralized in the next mixer with 3.5 M sodium hydroxide (20 °Be), somewhat in excess of the stoichiometrically calculated amount. The soap phase is removed in the first separator and the oil is treated once more in a mixer with about 1–3 per cent of 2 M sodium hydroxide (12 °Be) before entering the second separator.

Although phosphoric acid pretreatment is preferred, it is also possible to use a mixture of 10 per cent sodium carbonate together with the sodium hydroxide in the re-refining step. Due to a decrease in the content of lecithin this will increase the decolourizing and oxidative stability[11]. Finally, in a third step the oil is washed with 10 per cent of water, separated and dried in a vacuum drying tower[18].

Rapeseed oil refined in batch or continuously according to Alfa Laval was found by Rutkowski and co-workers[19] to give oil with the same quality, while

224

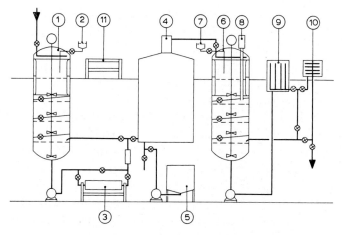

Fig.10:3. The Zenith Process

1. Phosphoric acid treatment vessel
2. Phosphoric acid dosage
3. Centrifuge for sludge
4. Neutralizing vessel
5. Soapstock draining tank
6. Bleaching vessel
7. Citric acid dosage
8. Bleaching earth dosage
9. Bleaching filter
10. Polishing filter
11. Panel board

Möller[20] reported a greater increase in the peroxide value of the oil directly after batch refining compared to hermetic continuous refining.

The short-mix process counteracts the saponification which occurs at higher lye concentrations by a short contact time (30–40 sec). Another method to decrease the refining loss is to neutralize with a weak lye, which is used in the Zenith Process[10, 21–23]. Fig.10:3 shows a refining installation designed for rapeseed oil and other oils containing lecithin.

By pretreatment with 0.15–0.3 per cent of phosphoric acid (85 per cent) the impurities will give a precipitate, partly due to denaturation of phosphatides. The acid also reacts with magnesium in the chlorophylls, thus decreasing the green colour of the oil. The precipitate formed is removed by a separator. The pretreated oil is continuously introduced at the lower end of a vertical column filled with 0.3 M sodium hydroxide. In the bottom of the neutralizing column the oil stream is distributed into small droplets which rise to the top of the column. Above the caustic lye an oil layer is built up and continuously discharged to the next step. The lye can be used until the concentration has decreased to about 0.1 M, and then has to be renewed. The neutralized oil contains about 10–100 ppm of soap, the last traces of which would not be eliminated by a normal washing with hot water. Instead a solution of citric acid in water is added in an amount corresponding to 0.05 weight per cent of citric acid. The oil is then dried in the so called C-unit, where free fatty acids and sodium citrate are formed. The amount of free fatty acids formed is negligible and in any case is removed in the deodorizing step, while the citrate is adsorbed on the bleaching earth. For rapeseed oil a refining factor of 1.8, a fatty acid factor of 1.5 and a refining efficiency of 99.0 per cent are reported[23].

In the Zenith Process the crude oil is semi-continuously supplied to the plant while the bleached oil is continuously discharged. The lye can easily be renewed by using two neutralizing vessels. Other systems for weak lye treatment are described in which both the oil and the lye are continuously supplied[24].

A system for bleaching is also included in the Zenith Process for refining. The system is designed with a vacuum vessel holding three or more trays, and the whole system is kept under vacuum. The oil is sucked into the upper tray through a spraying device for de-aerating and drying the oil. In the first tray the oil is mixed with a solution of citric acid during the drying, and is then pumped to the second tray where bleaching earth is added by a dosage arrangement. The lowermost tray serves as a buffer from which the fat is continuously pumped to the filter.

10.3 BLEACHING

Bleaching of the neutralized oil is performed in order to remove coloured substances and small amounts of soaps.

Even a well refined and water-washed oil may still contain 10–100 ppm of soap. It is especially important that the bleaching step remove traces of soap if the oil is to be hydrogenated. A low soap content is favoured by vacuum bleaching (ref. 25, p.780), but the adsorbed soap partly inactivates the bleaching earth and will thus decrease the decolourizing effect. The coloured compounds in rapeseed oil are among others the green pigments chlorophyll A and B as well as their products of decomposition, pheophytins A and B, which have been determined by Niewiadomski and co-workers[26]. Seed stocks containing as little as 3 per cent unripened seeds normally give noticeably dark oils[1]. Such dark coloured oils

TABLE 10:3

PHOSPHATIDE, TOCOPHEROL AND IRON CONTENTS OF RAPESEED OIL ACCORDING TO RUTKOWSKI[28]

	Phosphatide ppm	Tocopherol ppm	Iron µg/100 g
Pressed oil	101–166	442–589	36–65
Oil in miscella	166–261	452–656	217–519
Mixed crude oil	97–140	417–585	126–384
Neutralized oil	17–34	326–521	101–217
Bleached oil	10–23	249–423	30–87

require bleaching with a larger amount of bleaching earth in order to give an acceptable edible oil. It is not yet known whether this is due to decomposition products of the colour pigments themselves or if there are other colourless products which increase as the disagreeable flavour increases. However, Täufel and co-workers[27] have shown that chlorophyll has a prooxidative effect on unsaturated fats exposed to light. An acid-activated earth is more advantageous because the pigments are unstable under acidic conditions (ref. 25, p.772).

There is some risk that antioxidants are removed during the bleaching. Rutkowski[28] has found decreasing amounts of tocopherols and phosphatides (Table 10:3), together with a decreased oxidative stability for rapeseed oil according to the AOM test[29] after each processing step and also after bleaching at 70–80 °C. On the contrary Täufel and co-workers[30] report an increasing stability of bleached oil compared to neutralized oil, but they also found the best stability for crude oil.

The choice of bleaching earth is a question of economy where decolourizing ability, filterability and oil retention have to be compared to the price. A non-activated earth may be economically feasible for a rapeseed oil of good quality with light colour but as soon as the colour of the crude oil increases an activated earth will become more economical[8]. The addition of activated carbon did not increase the bleaching power[31]. The advantage of using only one quality of

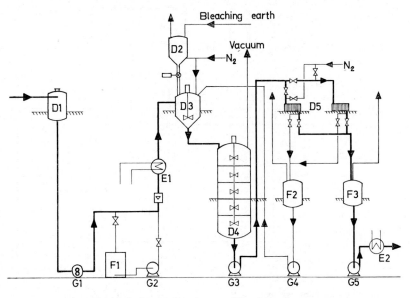

Fig.10:4. Lurgi Continuous Bleaching System

D1	Drying tank		E1	Heater
D2	Clay hopper		E2	Cooler
D3	Mixer			
D4	Bleaching vessel		F1	Holding tank
D5	Filter press		F3	Holding tank
			G1-G5	Pumps

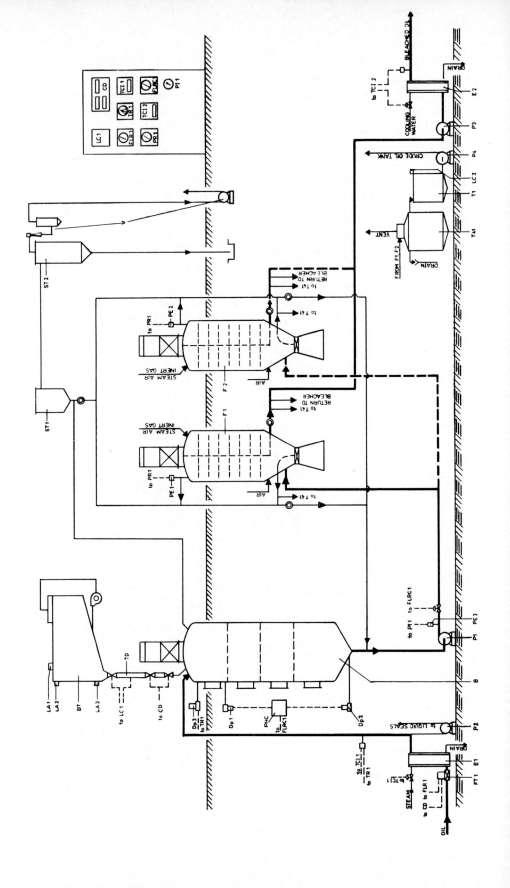

bleaching earth in a factory is however so evident that normally an activated earth is used for all kinds of oils. The amount of earth must be adjusted to the quality of the oil to minimize the cost of earth and oil retention. The formation of free fatty acids increases with the amount of bleaching earth[31, 32].

Vacuum bleaching has proved to be advantageous, especially for the removal of soaps, but also to protect the oil against oxidation. Nonetheless, it has been found that the bleaching earth should not be too well dried, since an earth completely dried at high temperature is inactive. According to Kögler[33], there is an optimal water content for bleaching earth. For bleaching at 95–98 °C under atmospheric pressure the optimal water contents for soyabean and cottonseed oil were 8–12 per cent and 9–16 per cent, respectively. Further increase in water content decreased the bleaching power. In spite of this, an earth with 24 per cent water was reported[33] to give better decolourizing than a fully dried earth.

A good idea of the amount of bleaching earth required for a certain oil can be obtained by testing on a bench scale[34, 35], but it is always advisable to follow the results in the plant.

Fig.10:5. Alfa Laval Auto-Bleach Process

Bleacher
B   Bleacher

Filter
F1, F2   Filter

Pumps
P1   Feed pump
P2   Service pump
P3   Feed pump
P4   Slop pump

Instrumentation
TCI 1, TCI 2   Temperature controller
TR1   Temperature
FLRC1   Flow controller
PHC   Pneumatic computer
Dp1-2-3   Pressure transmitter
FT1   Flow transmitter
CD   Digital control device

Tanks
BT   Bleaching earth tank
ST1   Oil trap
ST2   Steam condenser
T41   Catch basin
T1   Slop tank

Heat Exchanger
E1   Oil heater
E2   Oil cooler

Dosing Equipment
TD   Dry dosing equipment

FLR1   Flow recorder
PE 1-2-3-4   Pressure transmitter
PI1   Pressure indicator
PR1   Pressure recorder
LC 1-2   Level controller
LA 1-2-3   Level transmitter

Auxiliary Equipment
V   Vacuum equipment

Open bleaching tanks are still used, but most factories endeavour to bleach under vacuum. Such equipment should be airtight, especially below the oil level, and furnished with agitation in order to provide intimate contact between the oil and the bleaching earth. There are also various systems for continuous bleaching[4, 36-38], *e.g.* the Lurgi System (Fig.10:4).

The dried oil and the deaerated bleaching earth are dosaged with agitation in a mixing tank. The mixture runs through a bleaching reactor with a certain residence time and is then filtered.

Together with the Zenith process for refining, a system for bleaching has also been described (see p.226).

An entirely automatized bleaching system is the Alfa-Laval Auto-Bleach process (Fig.10:5). The bleaching plant comprises three main sections: bleaching reactor with automatic control device, dosing equipment for bleaching earth, and a filtering unit. The oil is heated to a suitable temperature in the heater E1, and goes to the upper compartment of bleacher B, where deaerated bleaching earth is added continuously. The amount of bleaching earth and the oil flow are ratio controlled, which ensures a constant percentage of bleaching earth in the oil. A program controls the residence time independent of the capacity, and the only change in the bleacher at different throughputs is some variation in the level in the compartments.

From the bottom of the bleacher the oil is pumped to one of the filters by means of pump P1. The filtered oil is then cooled in the heat exchanger E2.

The filter unit consists of two alternatively working Funda filters[39], each one containing a number of horizontal filter plates on a common shaft. When one filter has to be emptied, the oil in this filter is pumped by means of pump P1 to the second filter. The bleaching earth on the plates of the first filter is deoiled and discarded. When changing from one oil to another, all oil, even the residual volume, can be completely filtered in one filter, thus avoiding mixing between two different oils.

(a) Filtration

The bleached oil is sometimes filtered through open filter presses, which are manually emptied and cleaned, but several new types with automatic or mechanical discharge have been presented. At the same time it is still desirable to protect the oil from exposure to the air. A large amount of soap in the refined oil can disturb the filtration, and it has been reported[40] that a badly degummed rapeseed oil can decrease the filtration capacity to 1/5 of that normally obtained. The capacity can be increased by pre-coating with a slurry of filter aid in a clean oil or just by having the right type of filter aid[41]. The oil retention in the filter cake may be rather high, about 25 weight per cent, even after blowing with steam.

Various methods have been suggested to counteract retention losses, for example extraction of the cake in the filter itself or in a special plant[4, 17, 25, 42]. More attractive is extraction with hot water (95 °C) directly in the filter, the so-called Thomson process. For the Funda type filter about 10–12 per cent oil is then left in the cake[39].

10.4 HYDROGENATION

The previously described processes, alkali refining and bleaching as well as deodorization (see below), attempt to clean the oil without changing its properties. In hydrogenation there is a change in the composition of the fatty acids. The amount of linolenic acid in rapeseed oil is reduced, which increases its stability against oxidation. Hydrogenation also increases the melting point more or less depending on how many double bonds are reduced. Fat hardening is still an empirical process, but a great deal of systematic work has been done which is of great help in the estimation of suitable process parameters, although most of the work has been done for other oils than rapeseed oil. Dutton[43] has summarized research results up to 1968.

10.4.1 Selectivity and isomerisation

In the triglycerides, linolenic, linoleic and oleic acid are hydrogenated at different rates. A simplified reaction model makes it possible to define selectivity[25].

$$\text{Linolenic} \xrightarrow{k_1} \text{Linoleic} \xrightarrow{k_2} \text{Oleic} \xrightarrow{k_3} \text{Stearic}$$

The linolenic selectivity (S_{L1}) is the ratio between k_1 and k_2, $S_{L1} = k_1/k_2$, and the linoleic selectivity is the ratio between k_2 and k_3, $S_{L2} = k_2/k_3$.

In this model the concentration of hydrogen at the surface of the catalyst is not considered. The rate constants will thus be influenced not only by the temperature and catalysts but also by those process parameters which influence the concentration of hydrogen at the catalyst surface. The importance of mass transport processes has been demonstrated by Coenen and his co-workers[44, 45], who have presented a comprehensive discussion of various selectivity concepts[45], in which they also consider the triglycerides themselves and the effect of pore diffusion. Attempts have recently been made to mathematically formulate a rate equation including the external mass transport steps[46].

A low concentration of hydrogen on the surface of the catalyst favours selective hydrogenation, and can be obtained by low hydrogen pressure, a high concentration of catalyst and high temperatures (above 180 °C). At the same time a low concentration of hydrogen promotes conjugation, which causes migration of double bonds and the formation of *trans* isomers. Under these conditions the conjugation of trienoic acids can cause polymerization and cyclization[47]. The amount of aromatic fatty acids (AFA), about 0.1 per cent, found by Coenen and co-workers[47] for rapeseed oil hydrogenated at 200 °C and with shortage of hydrogen, is not found in commercial hydrogenated rapeseed oil processed in Sweden (AFA below 0.01 per cent), and probably appears only under conditions such that extremely high amounts of *trans* acids are produced.

The conjugation and *trans* isomerization of especially linolenic acid will after oxidation give a special objectionable flavour to the hydrogenated rapeseed oil as it does for all other oils containing linolenic acid. This flavour disappears when the oil is deodorized but after long storage the hydrogenation flavour will appear again. Chang and co-workers[48] characterized some of these flavours in hydrogenated soyabean oil as saturated aldehydes and methyl ketones.

Keppler and co-workers[49] separated the volatile substances from hydrogenated linseed oil by gas-liquid chromatography and could show the presence of 6-*trans*-nonenal and identify this as the principal carrier of hydrogenation flavour.

However, the selectivity concept, as defined above, relates to the changes in the amounts of trienoic, dienoic, and monoenic acids and not to the formation of *trans* isomers or changes in physical properties such as melting point, dilatation curves etc., but these properties are of course influenced by the selectivity. A low linoleic selectivity means a larger amount of stearic acid at a certain iodine value, which increases the melting point. On the other hand the content of *trans* isomers influences the dilatation to a greater extent than the selectivity ratios[71] (Fig.10:6). In rapeseed oil it is also possible that the erucic acid and oleic acid in the triglycerides hydrogenate at different rates. Hilditch and Paul[50] reported as early as 1935 that erucic acid and oleic acid in the triglycerides hydrogenate at about

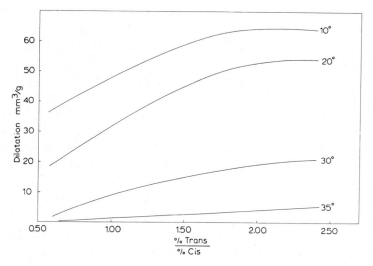

Fig.10:6. Dilatation *vs. Trans:Cis*-content for rapeseed oils hydrogenated under different conditions to an iodine value of 75 (ref. 65).

the same rate, but as methyl esters there was a preference for methyl oleate. Their conclusions were confirmed by Coenen and co-workers[51]. Jakubowski[52a] stated that in rapeseed oil erucic acid hydrogenates more slowly than oleic, while Koman[52b] found the reverse rate order. In any case the difference in hydrogenation rates which is often found between rapeseed oil and soyabean oil is probably due to the sulphur compounds in the former.

In the hydrogenation process there are three phases involved, each of which has to be taken into account.

10.4.2 Hydrogen

There are several ways of producing hydrogen[25] but today the choice normally is between the hydrocarbon reforming process and electrolysis of water. Electrolytic hydrogen is of high purity, normally more than 99.8 per cent, and no purification is required. In the reforming process desulphurized hydrocarbons are used, and the hydrogen produced contains about 0.05 per cent methane, 0.3 per cent carbon monoxide and 0.1 per cent carbon dioxide after scrubbing with ethanolamine. Passage through a methanation catalyst increases the purity to about 99.5 per cent hydrogen with less than 10 ppm carbon oxide[53].

10.4.3 Oil

The isothiocyanates and oxazolidinethiones in rapeseed oil are not only objectionable for refined oil but also function as poisons for the catalysts. Rutkowski and co-worker[54] have found that a catalyst is deactivated in proportion to the amount of sulphur present in the oil. An amount of 0.016 ppm sulphur had an almost negligible influence on the hydrogenation, while 5 ppm reduced the catalytic activity by about 50 per cent. They also found an addition of 6 ppm phosphorus as rapeseed lecithin harmful for the hydrogenation, and that this effect dominated sulphur poisoning in rapeseed oil. The decomposition products of the sulphur compounds extracted together with the oil are not fully removed by alkali refining or bleaching. One alternative is to treat the oil with a small amount of catalyst at 150–200 °C together with a slow flow of hydrogen in order not to reduce the double bonds. This method has been recommended for fish oil[55]. The best solution is to pretreat the seeds so that no sulphur compounds are extracted with the oil[1]. Reynolds and Youngs[56] describe a technique for rapeseed which gives an oil that needs about 0.03–0.08 per cent nickel for hydrogenation[2]. A comparable figure is found for Swedish rapeseed oil. Deodorization will only be helpful if the sulphur compounds are volatile, but in that case they are also removed if the bleaching is done under vacuum.

10.4.4 Catalyst

In most cases a nickel catalyst, such as nickel formate reduced in oil at about 250 °C or dry-reduced nickel carbonate on a support, may be used. One advantage with the supported, dry-reduced catalyst is its better filterability. A normal analysis of such a catalyst gives 20–25 per cent nickel, 10–15 per cent silica and 60–65 per cent hydrogenated fat, while the wet-reduced catalyst normally ranges from 22 to 30 per cent nickel in hydrogenated fat. One of the most important demands upon a catalyst is uniformity between different lots. The catalysts can be compared to a standard catalyst by hydrogenation under standardized conditions. Cocks and Van Rede[57] describe a method in which they add 0.07 per cent nickel to the oil and compare the decrease in refractive index for hydrogenation of a standard sesame oil for half an hour. The activity is expressed as the relative refractive index decrease:

$$\text{Relative refractive index decrease} = \frac{\text{Refr. index decrease with sample}}{\text{Refr. index decrease with standard}} \times 100$$

Still more important than good activity is that the catalyst gives a hardened product of constant quality. The simplest way to check the quality of the product is to compare the dilatation curve for the same type of oil hydrogenated to the same iodine value with the sample and the standard catalyst or to compare the fatty acid composition and *trans* acid content. Catalysts based on other metals have hitherto been of little practical interest. Platinum and palladium have been suggested[58–60] but seem to offer no advantages over nickel. Copper catalysts, however, give a better linolenic acid selectivity[61, 62], but due to their low activity the hydrogenation has to be performed under a higher pressure, 5–10 atm. Traces of copper in the hydrogenated oil are objectionable and would demand a special production line for copper hydrogenated products, equipped with an effective post-bleaching process.

10.4.5 Equipment and systems

For batch hydrogenation there are reactors with and without recirculation of hydrogen, both of which types are used for the hydrogenation of rapeseed oil. The so called 'dead-end' system (*i.e.* without recirculation) sometimes requires purging during the reaction. A normal batch in Europe is 10,000–15,000 kg, while in the U.S.A. a normal charge for soyabean oil is about 30,000 kg.

As continuous processes for refining and bleaching have become more common, interest in continuous hydrogenation has been aroused. From Russia, for example, work concerning stationary catalysts has been reported[63–65]. However, the only tests on rapeseed oil seem to be have been carried out only on a bench scale[66]. During recent years two commercial systems have been introduced, one (Lurgi) working with constant hydrogen consumption and another (BUSS) working at constant pressure with normal powder catalyst.

(a) The Lurgi system for continuous hydrogenation[67] (Fig.10:7)
The installation consists of one or more pressure reactors, divided by a series of plates into a number of reaction chambers, which can be cold or heated for adjustment to the desired temperature.

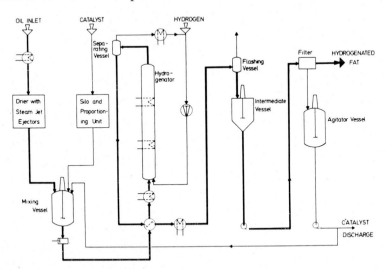

Fig.10:7. Lurgi System for Continuous Hydrogenation.

The refined and bleached oil is mixed with new or reused catalyst and fed to the bottom chamber of the first tower by a volumetric pump. The hydrogen is introduced at a constant flow to the feed line, from which it bubbles through the oil and catalyst mixture in the reactor. From the drop chamber in the last reactor the oil phase and the unused hydrogen flow through a cooler into a separating vessel, where the liquid fat and catalyst separate from the hydrogen gas. By the gas circulation pump the unused hydrogen is returned to the first reaction chamber. Due to the gas circulation no agitation is necessary. In relation to the volume of the liquid phase there is a small gaseous volume, so small changes in reaction rate will immediately increase or decrease the gaseous pressure and thus neutralize the change. The decrease in the iodine value is directly affected by the flow rates of hydrogen and oil.

(b) Buss system for continuous hydrogenation[68] (Fig.10:8)
The equipment is composed of four or more reactors in cascade. In each reactor the contact between oil, catalyst and hydrogen is carried out in a mixing nozzle, *13*, by rapid circulation of the oil and catalyst mixture. The refined and bleached oil is dried under vacuum and fed to the reactor. The four to six chambers have

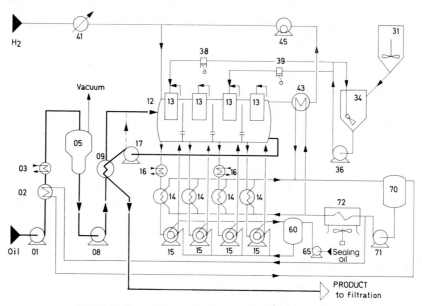

Fig.10:8. Buss System for Continuous Hydrogenation.

Reactor		*Heat exchangers*	
12	4 to 6 apartment reactor	14	Oil cooler
13	Mixing nozzle		

Pumps

Flow meters

01	Crude oil pump	41	Hydrogen flow meter (FIRC)
08	Oil feed pump		
15	Circulation pump with sealing oil	*Tanks*	
17	Oil pump	31	Catalyst tank
36	Catalyst suspension pump		
45	Gas circulation pump		
38, 39	Membrane pump for catalyst dosage		

the same gaseous space but the oil is separated by walls between the chambers. By an overflow the oil is transported through the reactors. The hydrogen is directly fed to the reactor through a flowmeter, *41*, or the hydrogen can also be circulated from the last to the first reactor by a ring pump, *45*. A suspension of catalyst in oil is fed in a fixed amount by a pump, *38*, to the first chamber and to the third chamber by another pump *39*, which is controlled by the hydrogen flowmeter, thus making the addition of hydrogen constant.

For cascade reactors such as Lurgi and Buss it is necessary to have a sufficient number of steps, so the residence time distribution will not be too unfavourable. Changes in reaction time by alterations in temperature, pressure and amount of catalyst will probably influence the end product in the continuous processes in a manner similar to that observed in batch hydrogenation. The advantage of such reactors would then be a more uniform product, a better heat economy and perhaps less operation maintenance.

10.4.6 Practical aspects of the hydrogenation of rapeseed oil

In order to reduce the iodine value of 1000 kg of oil by one unit, 39.4 moles of hydrogen are required, which corresponds to 79 g or 0.883 m³ at 0 °C and 760 torr absolute pressure. There are no published data on the heat of reaction in the case of rapeseed oil, but the corresponding values for peanut oil and soyabean oil are 24.5 and 26.1 kcal/mole and double bond respectively[69].

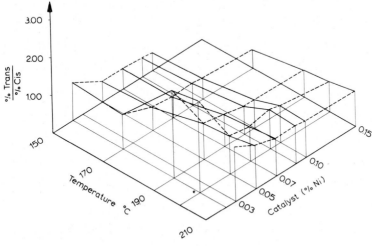

Iodine number: 75

Fig.10:9. *Trans:Cis*-content *vs.* temperature and nickel concentration for rapeseed oil hydrogenated to an iodine value 75 with new catalyst[71].

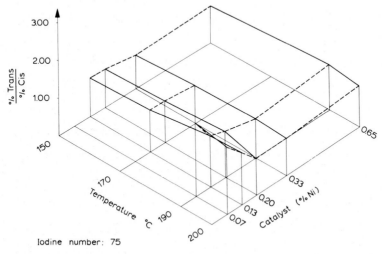

Iodine number: 75

Fig.10:10. *Trans:Cis*-content *vs.* temperature and nickel concentration for rapeseed oil hydrogenated to an iodine value 75 with reused catalyst[71].

References pp. 246–248

TABLE 10:4

ANALYSIS OF RAPESEED OIL HYDROGENATED WITH 0.07 PER CENT UNSUPPORTED NICKEL AT 185° C[71]

Time min	Iodine value (Hanus)	Trans isomers wt. %	Conjugated* Dienes wt. %	Conjugated* Trienes wt. %	Non-conjugated** Dienes wt. %	Non-conjugated** Trienes wt. %	Fatty Acid Weight Percentage Composition*** 16:0	16:1	18:0	18:1	18:2	18:3	20:0	20:1	20:2	22:0	22:1	22:2	24:0	24:1	Polyunsaturated**** fatty acid, %
0	105.7	2.7	0.26	0.00	13.8	9.7	3.0	0.3	0.8	12.2	13.9	10.3	0.0	8.6	0.5	0.0	47.7	1.1	0.4	1.2	26.0
3	103.4	3.0	0.30	0.00	12.0	8.6	3.0	0.4	0.9	13.4	13.8	9.8	0.0	9.5	0.6	0.0	47.5	0.3	0.4	1.3	23.4
6	99.9	6.2	0.51	0.00	13.0	7.3	2.9	0.3	0.8	14.6	12.9	7.3	0.0	8.7	0.4	0.0	49.3	1.1	0.3	1.2	19.8
9	98.3	8.4	0.65	0.00	9.7	5.4	3.1	0.3	1.0	16.5	13.0	7.0	0.0	8.6	0.5	0.0	48.0	0.6	0.3	1.1	17.3
12	95.6	10.2	0.66	0.00	8.3	3.5	2.9	0.3	0.9	17.5	12.3	5.4	0.0	8.7	0.4	0.3	48.9	0.8	0.3	1.3	13.7
15	92.7	13.5	0.65	0.00	6.7	3.2	3.4	0.4	1.2	19.2	11.7	5.1	0.0	8.5	0.6	0.5	46.9	0.9	0.2	1.3	11.0
25	85.5	24.1	0.28	0.00	3.6	1.3	3.1	0.4	1.3	24.4	9.4	1.7	1.0	8.1	0.0	0.8	48.4	0.0	0.2	1.2	4.2
35	77.7	35.6	0.00	0.00	1.2	0.0	3.1	0.5	2.6	27.6	6.3	0.3	1.0	8.1	0.0	2.0	47.0	0.0	0.2	1.2	0.7
45	71.3	41.5	0.00	0.00	0.0	0.0	3.0	0.4	4.5	27.1	4.7	1.0	1.8	7.3	0.0	3.7	45.2	0.0	0.4	1.1	0.2
60	62.1	43.8	0.00	0.00	0.0	0.0	3.2	0.4	7.8	25.1	2.9	1.0	2.3	6.8	0.0	9.3	40.8	0.0	0.7	0.9	0.0
75	53.9	42.0	0.00	0.00	0.0	0.0	3.1	0.3	11.7	22.4	1.9	0.5	3.5	5.6	0.0	15.4	33.9	0.0	0.7	1.0	0.0
90	46.4	38.0	0.00	0.00	0.0	0.0	3.1	0.5	15.8	18.0	1.2	0.5	4.6	4.5	0.0	21.2	29.1	0.0	0.9	0.8	0.0
105	40.3	32.9	0.00	0.00	0.0	0.0	3.1	0.6	17.2	14.1	1.7	0.0	5.0	4.1	0.0	23.5	29.0	0.0	0.9	0.8	0.0
120	30.0	26.5	0.00	0.00	0.0	0.0	3.2	0.4	20.0	14.1	1.1	0.4	5.8	3.7	0.0	28.8	20.8	0.0	0.9	1.2	0.0

* Determined by ultraviolet spectroscopy according to Mehlenbacher[76]
** The reported values are measured according to Mehlenbacher[76] multiplied with a factor 1.2.
*** Determined by gas-liquid chromatography.
**** Determined by ultraviolet spectroscopy after oxidation with lipoxidase[77].
Estimation of cis-methylene-interrupted polyenoic acids according to MacGee[78].

The alkali refined and bleached oil is heated under vacuum in the reactor to about 140 °C and then a predetermined amount of catalyst is added. At about 150 °C the hydrogen is supplied and the temperature rises to about 180 °–200 °C. At this temperature the reaction is continued until the desired iodine value and melting point have been reached. By altering the amount of catalyst, temperature, pressure and agitation, a fat with the same melting point but with different plasticity can be produced[70]. The influence of the temperature and amount of catalyst on the building of *trans* isomers is shown in Fig.10:9 for unused catalyst, and in Fig.10:10 for reused catalyst[71]. Hydrogenated fat for margarine or shortening is normally described by the amount of solids[72], dilatation[73] or solid fat index (S.F.I.)[74] at various temperatures. Normally there are certain differences in the products of different companies with regard to dilatation due to variations in hydrogenation techniques and crude oils. In common is, however, the desire to increase the stability of the oil by eliminating linolenic acid. Jakubowski[75] has

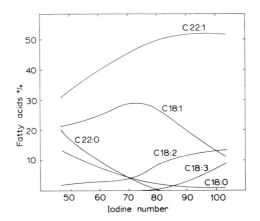

Fig.10:11. Fatty acid content of rapeseed oil hydrogenated under standard conditions[71].

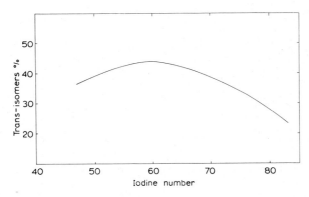

Fig.10:12. *Trans* content *vs.* iodine number for hydrogenated rapeseed oil[71].

reported that industrially hydrogenated Polish rapeseed oil contains less than 1 per cent linolenic acid at an iodine value of 85. The amount of octadecenoic acid found consists of iso-linoleic acid to a great extent. According to Jakubowski[75] a hydrogenated rapeseed oil of iodine value 70 (m.p. 30 °C) contains only about 2 per cent *cis,cis*-9,12-octadecadienoic acid.

Laboratory hydrogenation of a normal blend of Swedish rapeseed oil showed that no conjugated trienes, but almost 1 per cent conjugated dienes could be detected at the beginning of the hydrogenation (Table 10:4). In the same table the gas chromatographic analyses and the amounts of polyunsaturated fatty acids after conjugation with alkali and lipoxidase are given[71].

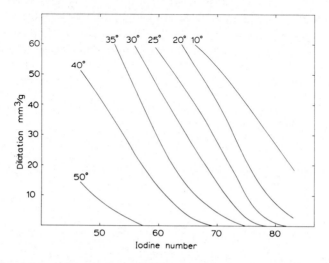

Fig.10:13. Dilatation *vs.* iodine number at different temperatures for rapeseed oil hydrogenated under standard conditions[71].

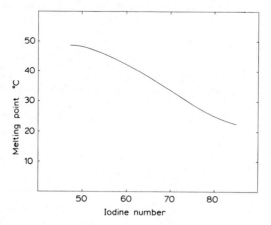

Fig.10:14. Melting point (AOCS) *vs.* iodine number for hydrogenated rapeseed oil[71].

TABLE 10:5

DILATATION FOR DIFFERENT VARIETIES OF RAPESEED OIL AND MUSTARD SEED OIL, HYDROGENATED IN THE LABORATORY UNDER THE SAME CONDITIONS WITH A COMMERCIAL CATALYST[71]

Variety	Rapeseed oil m.p. 34°C Dilatation						Rapeseed oil m.p. 38°C Dilatation						Rapeseed oil m.p. 40°C Dilatation					
	J.V.	10°C	20°C	30°C	35°C	40°C	J.V.	10°C	20°C	30°C	35°C	40°C	J.V.	10°C	20°C	30°C	35°C	40°C
Brassica napus, winter type	71.5	49.5	35.0	10.0	3.0	0.0	67.5	53.0	49.0	21.5	10.0	2.0	63.8	63.0	58.5	34.5	20.0	8.0
Brassica campestris, winter	71.3	48.0	33.0	10.0	4.0	0.0	67.5	57.5	47.0	20.5	10.5	2.0	63.5	62.0	58.0	34.0	20.5	8.0
Brassica napus, summer type	72.5	49.0	34.0	10.0	2.5	0.0	68.0	59.0	50.0	23.0	11.0	2.0	64.0	65.0	61.0	36.0	21.0	8.0
Sinapis alba	72.5	46.5	31.0	10.0	3.5	0.0	68.5	54.5	43.5	19.5	9.0	2.0	64.4	62.5	56.5	33.0	19.5	8.0
Brassica campestris, summer	74.5	49.5	32.0	10.0	3.0	0.0	70.5	58.0	45.5	21.0	9.5	2.0	66.0	64.0	57.5	34.0	20.0	8.0
Brassica napus, winter type	72.5	42.5	31.0	10.0	3.0	0.0	69.5	56.5	44.5	20.0	9.0	2.0	65.3	63.5	57.0	33.0	19.0	8.0
Brassica napus, winter type	74.0	48.5	30.5	10.0	4.5	0.1	70.5	56.0	41.0	17.5	8.5	2.0	66.0	63.5	53.0	29.0	16.5	8.0

In Fig.10:11 the fatty acid composition as determined by gas chromatography is given for another blend of Swedish rapeseed oil. The total *trans* content[79] increases during the hydrogenation up to an iodine value of 60, and will then decrease with decreasing iodine value (Fig.10:12).

An oil with the indicated composition and *trans* content gives the dilatation and melting point curves shown in Fig.10:13 and 10:14.

Rapeseed oil contains oils from various cultivated rapeseed species (see Chapter 7). A comparison between different Swedish varieties has shown that there is only a small change in the dilatation for a given variety compared to an average (see Table 10:5).

10.4.7 Controlling the hydrogenation process

A common way of following and controlling the degree of hydrogenation is to determine the refractive index and melting point. There are different methods for the determination of melting points, but normally a rapid method such as slip-point[53] is used. Some factories combine slip-point with congeal-point[80], which provides a notion of the dilatation. A good control of catalyst quality, the amount charged and the hydrogenation temperature together with a proper pre-treatment of the oil will give a product which meets the demand of uniform dilatation, assuming that the hydrogenation is stopped at the right iodine value. Modified more rapid methods for iodine value, dilatation or solid fat index can be used.

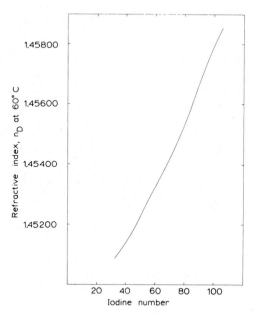

Fig.10:15. Refractive index at 60 °C *vs.* iodine number for hydrogenated rapeseed oil[71].

242

By continuously determining the amount of hydrogen the degree of hydrogenation can be followed, but the leakage in the seal for the agitator shaft has to be considered and this method is often combined with iodine value or more easily refractive index. There is a good correlation between refractive index and iodine value (Fig.10:15).

10.4.8 Postbleaching

When the hydrogenation has reached the desired end point the oil is cooled and the catalyst removed by filtration. In many places this is still done in open filter presses, but as in the case of the filtration of bleaching earth, there is successive change to closed automatic filters. The catalyst has a tendency to clog the filter cloths, and must therefore be mixed with filter aids. Supported catalysts are often found to be more easily filtered than formate catalysts. A great deal of the producer's efforts are devoted to increasing filterability.

After the catalyst is filtered off the oil contains mainly nickel as soaps[81]. To remove 5–10 mg nickel per kg of oil, one per cent Fullers' earth was needed, and the addition of citric or lactic acid considerably improved the efficiency. 100 g acid per 1000 kg of oil decreased the need of bleaching earth to 0.5 per cent for an oil containing 5 mg nickel per kg oil[81]. According to Drew Chem. Corp.[82] the best efficiency is obtained with one kg of a 25 per cent solution of citric acid in water to 10,000 kg hydrogenated fat added to the fat after the bleaching earth. Alkaline refining after hydrogenation seems to be unnecessary if the postbleaching is carried out as described with citric acid.

10.5 DEODORIZATION

Whether the oil has been neutralized and bleached or hydrogenated the fat contains minor amounts of volatile compounds with a certain flavour. Especially unhydrogenated rapeseed oil may have a strong seedness flavour, but also other undefined flavours may exist, *e.g.* a pungent flavour arising from glucosinolates and their decomposition products (ref. 4, p.156). In the deodorization, which is a steam distillation, these volatile compounds are removed together with the free fatty acids (about 0.05 per cent) obtained in the bleaching and hydrogenation steps. At the same time, the colour is improved and peroxides are decomposed partly to volatile products.

It is difficult to give general rules for rapeseed oil in addition to those for other oils[4, 17, 25, 42, 83] due to variations in equipment from factory to factory. Normally rapeseed oil needs a more efficient deodorization than other vegetable oils. On the other hand, if the temperature is too high the natural antioxidants, such as tocopherols, may be distilled or decomposed. From laboratory tests it was con-

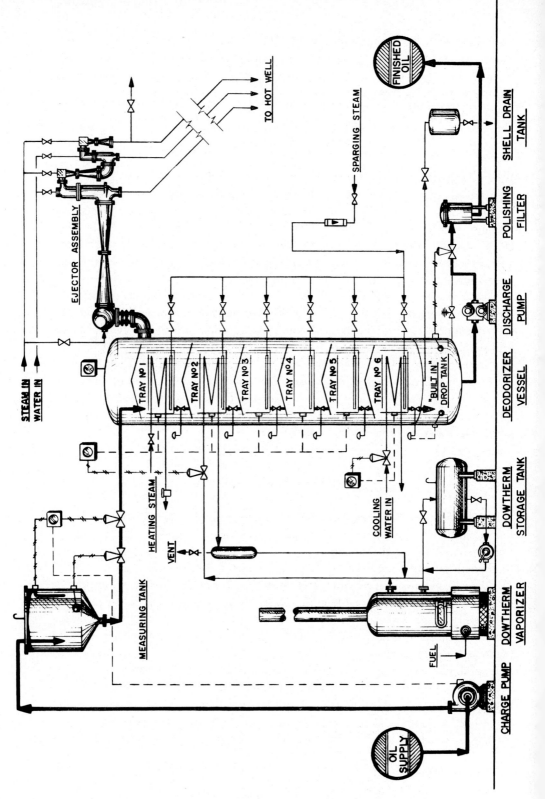

Fig.10:16. Six Tray Semi-Continuous Deodorizer.
Courtesy Votator Division Chemetron Corp.

cluded[84] that for rapeseed as well as for soyabean oil there is no formation of conjugated dienes at 210 °C, but that they appear in amounts up to one per cent when deodorization is carried out at 280 °C. The content of *trans* isomers increases from 2–3 per cent at 210 °C to about 10 per cent at 280 °C. The experiments were carried out at 5 torr with a decreasing residence time, t, as the temperature was increased following the expression $t =$ const. $\times 1/P_T$, where P_T is the vapour pressure of stearic acid at temperature T. A shorter deodorization time seemed to be more favourable for rapeseed oil than for soyabean oil.

The stability of the deodorized oil with regard to oxidation and flavour as well as colour depends on the storage temperature and on whether it is protected by nitrogen or not. However, earlier steps in the process influence the stability. A minor increase in the peroxide value of the neutralized oil will decrease the flavour stability of the deodorized oil[85].

According to Harada and co-workers[86] rapeseed oil from pressure milling was stable, while that from solvent extraction showed some colour reversion. For soyabean oil it is stated[87] that the colour reversion is dependent on the amount of 'tocored' in the crude oil. During the processing the 'tocored' seems to transform to a colourless compound which remains in the deodorized oil and will transform back to the red 'tocored' in the reverted oil.

Rapeseed oil for margarine may be deodorized together with the other oil

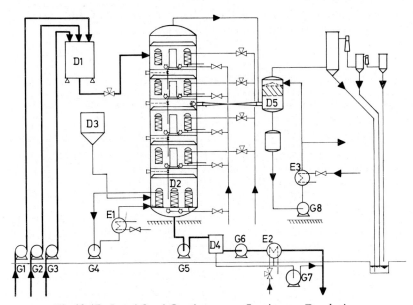

Fig.10:17. Lurgi Semi-Continuous or Continuous Deodorizer.

D1	Scale	E1, E2, E3	Cooler
D2	Deodorizer		
D3	Tank for citric acid	G1, G2, G3	Pump for bleached oil
D4	Filter	G5	Pump for deodorized oil

components or may be separately deodorized. A batch deodorization normally takes about eight hours with heating and cooling and reaches a temperature of about 195–200 °C. Before the deodorization, 0.01 per cent citric acid may be added to the oil and the same amount after the deodorization in the cooling step (ref. 4, p.203).

A great step forward was achieved when Bailey[88] introduced the Votator principle (Fig.10:16), which made it possible to deodorize rapeseed oil at a higher temperature (230 °C) for a shorter time (*i.e.* a cycle time of 27 minutes, giving an effective processing time of about two hours). A great number of equivalent designs of the same principle have appeared, among them the Lurgi plant (Fig. 10:17) which can operate either continuously or semi-continuously.

REFERENCES

1   K. E. Eapen, N. W. Tape and R. P. A. Sims, *J. Am. Oil Chem. Soc.*, 45 (1968) 194.
2   B. F. Teasdale, *Rev. Franc. Gorps Gras*, 15 (1) (1968) 3.
3   *AOCS Official and Tentative Methods* Ca 12-55; *DGF-Einheitsmethoden* C IV 4 (61).
4   A. J. C. Andersen, *Refining of Oils and Fats for Edible Purposes*, Pergamon Press Ltd, London, 1962.
5   R. Guillaumin and M. Boulot, *Rev. Franc. Corps Gras*, 7 (1960) 506.
6   C. Detrommt, R. Guillaumin, D. Delahaye and N. Drouhin, *Rev. Franc. Corps Gras*, 9 (1962) 486.
7   R. Guillaumin and N. Drouin, *Rev. Franc. Corps Gras*, 13 (1966) 185.
8   L. Wictorin and C. Svensson, Private communication.
9   B. Braae, *Chem. Ind. (London)*, (1958) 1152.
10 L. O. Bergman, *Canad. Patent* 664, 968 (1963).
11 B. Braae, U. Brimberg and M. Nyman, *J. Am. Oil Chem. Soc.*, 34 (1957) 293.
12 A. Rutkowski, *Fette, Seifen, Anstrichmittel*, 61 (1969) 1216.
13 A. Stawowczyk, *Acta Polon. Pharm.*, 17 (1960) 229; *C.A.*, 55 (1961) 12891 d.
14 B. Braae and M. Nyman, *Communication at '2:a Nordiska Symposiet för Fetthärskning'* in Helsingör, Sep. 1957.
15 AB Karlshamns Oljefabriker.
16 A. Koebner and T. Thornton, *Brit. Patent* 1,002,974 (1965).
17 H. P. Kaufmann, *Neuzeitliche Technologie der Fette und Fettprodukte*, 4. Lieferung, Aschendorffsche Verlagsbuchhandlung, Münster Westl. 1965, pp.694–701.
18 B. Braae, *Fette und Seifen*, 55 (1953) 859.
19 A. Rutkowski, H. Kozlowska and M. Dlugosz, *Roczniki Technol. Chem. Zywnosci*, 10 (1964) 155; *C.A.* 61 (1964) 7241 b.
20 P. Möller, *Communication at '2:a Nordiska Symposiet för Fetthärskning'* in Helsingör, Sept. 1957.
21 L. O. Bergman, *Canad. Patent* 664, 969 (1963).
22 L. O. Bergman, *Canad. Patent* 665, 192 (1963).
23 L. O. Bergman and A. Johnsson, *Fette, Seifen, Anstrichmittel*, 66 (1964) 203.
24 J. J. Hepburn, *Brit. Patent* 1,090,114 (1967).
25 F. A. Norris in D. Swern (Ed.), *Bailey's Industrial Oil and Fat Products*, Interscience Publ., New York, 1964.
26 H. Niewiadonski, I. Bratkowska and E. Mossakowska, *J. Am. Oil Chem. Soc.*, 42 (1965) 731.
27 K. Täufel, Cl. Franzke and G. Heder, *Fette, Seifen, Anstrichmittel*, 61 (1959) 1225.
28 A. Rutkowski, *Fette, Seifen, Anstrichmittel*, 61 (1959) 1216.

29 A. Rutkowski and Z. Makus, *Roczniki Technol. Chem. Zywnosci*, 2 (1957) 79; *C.A.* 52 (1958) 13132 e.
30 K. Täufel, H. Freude, Cl. Franzke and G. Heder, *Nahrung*, 5 (1961) 646.
31 N. H. Grace, *Can. J. Res.*, 26 (1948) 349.
32 H. P. Kögler, *Fette, Seifen, Anstrichmittel*, 70 (1968) 394.
33 H. P. Kögler, *Fette, Seifen, Anstrichmittel*, 65 (1963) 940.
34 H. Pardun, E. Kroll and O. Weber, *Fette, Seifen, Anstrichmittel*, 70 (1968) 531, 643.
35 A. Greentree, *J. Am. Oil Chem. Soc.*, 43 (1966) 4A.
36 (a) R. R. King, S. E. Pack and F. W. Wharton, *U.S. Patent* 2,428,082 (1947) 389.
    (b) R. R. King and F. W. Wharton, *J. Am. Oil Chem. Soc.*, 26 (1949).
37 W. A. Singleton and C. E. McMicheal, *J. Am. Oil Chem. Soc.*, 32 (1955) 1.
38 A. A. Robinson, *U.S. Patent* 2,483,710 (1949); *C.A.* 44 (1950) 356 b.
39 H. Müller, *Fette, Seifen, Anstrichmittel*, 66 (1964) 438.
40 H. Wilke, *Fette, Seifen, Anstrichmittel*, 65 (1963) 470.
41 R. J. Zilli, *Communication at the A.O.C.S. Spring Meeting 1969*.
42 R. Lüde, *Die Raffination von Fetten und Fetten Ölen*, Theodor Steinkopf, Dresden and Leipzig (1962).
43 H. J. Dutton, *Progress in the Chem. of Fats and other Lipids*, 9:3 (1968) 351.
44 J. W. E. Coenen, *Actes du Deuxième Congrès International du Catalyse*, Paris, 1960, 2705.
45 J. W. E. Coenen, H. Boerma, B. G. Linsen and B. de Vries, *Proc. The Third Congress on Catalysis*, Ed. Sachtler, Schuit and Zwietering, North Holland Publ. Comp., Amsterdam, 1965, p.1387.
46 M. Pihl and N. Schöön, *Acta Polytech. Scand. Chem. Met. Ser.*, to be published.
47 J. W. E. Coenen, Th. Wieske, R. S. Cross and H. Rinke, *J. Am. Oil Chem. Soc.*, 44 (1967) 344.
48 S. S. Chang, Y. Masuda, B. D. Mookherjee and A. Silveira, Jr., *J. Am. Oil Chem. Soc.*, 40 (1963) 721.
49 J. G. Keppler, J. A. Schols, W. H. Feenstra and P. W. Meijboom, *J. Am. Oil Chem. Soc.*, 42 (1965) 246.
50 T. P. Hilditch and H. Paul, *J. Soc. Chem. Ind.*, 54 (1935) 331.
51 J. W. E. Coenen and H. Boerma, *Fette, Seifen, Anstrichmittel*, 70 (1968) 8.
52 A. Jakubowski, I. Sobierajska and K. Modzelewska, *Rev. Franc. Corps. Gras*, 9 (1962) 678.
52a V. Koman, *Prumysl Potravin*, 15 (1964) 394; *C.A.* 61 (1964) 13541 e.
53 B. J. Mayland, R. L. Harvin and C. R. Trimarke, *J. Am. Oil Chem. Soc.*, 41 (1964) 26.
54 K. Babuchowski and A. Rutkowski, *Seifen-Oele-Fette-Wachse*, 95 (1969) 27.
55 J. Baltes, *Fette, Seifen, Anstrichmittel*, 69 (1967) 512.
56 J. R. Reynolds and C. G. Youngs, *J. Am. Oil Chem. Soc.*, 41 (1964) 63.
57 L. V. Cocks and C. van Rede, *Laboratory Handbook for Oil and Fat Analysts*, Academic Press, London/New York (1966).
58 M. Zajcew, *J. Am. Oil Chem. Soc.*, 39 (1962) 301.
59 A. E. Johnston, D. Macmillan, H. J. Dutton and J. C. Cowan, *J. Am. Oil Chem. Soc.*, 39 (1962) 273.
60 C. R. Scholfield, R. O. Butterfield, V. L. Davison and E. P. Jones, *J. Am. Oil Chem. Soc.*, 41 (1964) 615.
61 C. Okkerse, A. de Jonge, J. W. E. Coenen and A. Rozendaal, *J. Am. Oil Chem. Soc.*, 44 (1967) 152.
62 (a) S. Koritala and H. J. Dutton, *J. Am. Oil Chem. Soc.*, 43 (1966) 86.
    (b) S. Koritala and H. J. Dutton, *J. Am. Oil Chem. Soc.*, 46 (1969) 556.
63 V. I. Shlyakhov, D. V. Sokolskii, F. G. Golodov and V. K. Orlov, *Gidrirovanie Zhirov, Sakharov Furtorula*, (1967) 117; *C.A.* 70 (1969) 18942 p.
64 D. V. Sokolskii, V. I. Komarov and K. A. Zhubanov, *Maslo. Zhir. Prom.*, 34 (1968) 42; *C.A.* 70 (1969) 5343 u.
65 F. M. Kantsepolskaya, A. I. Glushenkova and A. L. Markman, *Maslo. Zhir. Prom.*, 35 (1969) 16; *C.A.* 71 (1969) 1406 q.
66 I. Kaganowicz, *Tluszcze Jadalne*, 11 (1967) 258; *C.A.* 68 (1968) 88388 v.
67 T. Voeste and H. J. Schmidt, *U.S. Patent* 3,444,221, (1969).

68 G. LEUTERITZ, *Fette, Seifen, Anstrichmittel*, 71 (1969) 441.
69 H. P. KAUFMANN, *Studien auf dem Fettgebiet*, Verlag Chemie, Berlin 1935.
70 K. NAKAZAWA, S. MITSUNAGA AND K. TADA, *Abura Kagaku*, 5 (1965) 292; *C.A.* 51 (1957) 12511 e.
71 R. B. WETTSTRÖM, unpublished results.
72 W. D. POHLE AND R. L. GREGORY, *J. Am. Oil Chem. Soc.*, 44 (1967) 397.
73 *DGF-Einheitsmethoden* C-IV 3e (57).
74 *AOCS Official and Tentative Methods* Cd 10-57.
75 A. JAKUBOSKI AND Z. KOWZAN, *Rev. Franc. Corps Gras*, 11 (1964) 67.
76 V. C. MEHLENBACHER, *The Analysis of Fats and Oils*, The Garrard Press, Publ. Champaign, Ill., 1960, pp.522–531.
77 I.-B. WALLINDER, private communication.
78 J. MACGEE, *Anal. Chem.*, 31 (1959) 298.
79 *AOCS Official and Tentative Methods* Cd 14-61.
80 *AOCS Official and Tentative Methods* Cd 14-59.
81 H. SZEMRAJ AND A. JAKUBOWSKI, *Pr. Inst. Lab. Badaw. przem. spozyw.*, 18 (1968) 103.
82 Drew Chemical Corp., *Technical Bulletin*.
83 O. B. EIOFEEVA AND I. P. LEVSH, *Tr. Tashkentsk Politekhn. Inst.*, 22 (1963) 89; *C.A.* 64 (1963) 3863 b.
84 S. ANDLID AND L. WICTORIN, Private communication.
85 B. BRAAE AND M. NYMAN, *Riv. Ital. Sostanze Grasse*, 43 (1966) 133; *C.A.* 65 (1966) 2908.
86 I. HARADA, Y. SARANTI AND M. ISHIKAWA, *Nippon Nogei Kagaku Kaishi*, 34 (1960) 545, *C.A.* 54 (1960) 25900 i.
87 M. KOMODA, N. ONUKI AND I. HARADA, *Agr. Biol. Chem.*, 31 (1967) 461.
88 A. E. BAILEY, *J. Am. Oil Chem., Soc.*, 26 (1949) 166.

CHAPTER 11

Edible Products from Rapeseed Oil

Ü. RIINER and E. HONKANEN

AB Karlshamns Oljefabriker, Karlshamn (Sweden)

CONTENTS

11.1 INTRODUCTION

It is still a rather common misconception that rapeseed oil is mainly an inedible oil. The McGraw-Hill Encyclopedia of Science and Technology[1] reports 1960: 'Rapeseed is used to some extent for edible purposes and for soap making. It is also a source of oil, which is used chiefly in mixtures with mineral oils for lubrication, or alone for tempering steel plates. The refined oil is known as colza oil'. Although statistics covering the disposal of the harvested seeds from the different *Brassica* varieties are not readily available from the major producing countries, China and India, it seems justified to state that the major part of the approximately 2,000,000 metric tons of rapeseed oil extracted annually from the seeds at the present time is utilized for human consumption (see Chapter 2). This corresponded to *ca.* 7 per cent of the world supply of liquid edible oils in 1970.

The utilization of rapeseed oil before 1940 has been covered in Chapter 1. The purpose of this chapter is to review the use of rapeseed oil as food after 1940 with emphasis on quality factors and food technological aspects that are specific for the *Brassica* oils. As mentioned, literature is not readily available on the edible uses of rapeseed oil in China and India and therefore mainly information from Europe and North America is included. For general quality factors and specific problems with other vegetable and animal oils and fats reference is made to other monographs[2-8]. In the monographs by Bailey[2] and Eckey[3] limited aspects of rapeseed oil utilization are covered.

In the literature the designation rapeseed oil cannot always be led back to a specific variety or strain. Where unspecified the designation rapeseed oil will be used in this review for the seed oils from *B. napus* (rape) and *B. campestris* (turnip rape), as well as for other Cruciferae seed oils with related compositions.

11.1.1 Edible products from rapeseed oil after 1940

In Chapter 1 the historical background for rapeseed cultivation and utilization is reviewed and the situation is revealed until 1940. During the years 1940–1945, when large regions of the world with a temperate climate were blocked from importing oils and fats due to World War II, considerable growing of rapeseed for edible oil production began. Beare[9] has reviewed the use and Anderson[10] the cultivation of rapeseed oil in wartime Europe. During the same period seeds of rape were brought to Canada and experimental farming was started[11]. From Chapter 2 it may be seen that a steady increase in the world production of rapeseed has taken place. The factors that have influenced this will not be analyzed here. However, two basic requirements seem to be necessary for successful growing and utilization for edible purposes: The specific technology that is required to achieve a high quality oil (see Chapters 5, 9 and 10) must be known and applied, and in the second place the oil must be regarded as safe for human consumption by the authorities that have the food legislative power in their hands (see Chapter 14 for the nutritive value of rapeseed oil). In some countries the utilization of rapeseed oil for edible purposes is still prohibited, for example in Portugal[12].

Statistics covering the types of foods, prepared industrially or otherwise, in which rapeseed oil is used in various countries are not available from normal

TABLE 11:1

THE UTILIZATION OF RAPESEED OIL IN CANADA[13]

Year	Margarine oil	Per cent as Shortening oil	Salad oil
1967	36	38	26
1968	28	39	33

literature sources. Information from some countries can, however, be indicative: In Canada in 1967 and 1968 rapeseed oil was the second vegetable oil after soyabean oil[13]. It accounted for 19 per cent (101,742 tons) in 1967 and 21 per cent (116,692 tons) in 1968 of the total production of deodorized oils. The portions used in margarine, shortening and as salad oil are presented in Table 11:1.

Also in the United Kingdom utilization of rapeseed oil has increased rapidly in recent years[14] from very small quantities in 1961 to approximately 50,000 tons in 1968. 24 per cent of this was used in margarine and 12 per cent in shortening, accounted for as compound cooking fat. The rest must have been used in the food industry or as a table oil. In Sweden a voluntary agreement between the margarine industry and the state has existed under which the industry buys a quantity of rapeseed oil that corresponds to the margarine production (*cf.* Chapter 2). Also within the European Economic Community the principal use of rapeseed oil is considered to be in margarine[15].

In the following, the main areas of use of rapeseed oil for edible products will be discussed separately. First, however, some general factors affecting the quality of rapeseed oil will be commented upon.

11.2 QUALITY FACTORS

Rapeseed oil and related Cruciferae seed oils are classified in Bailey's Industrial Oil and Fat Products[2] as erucic acid oils but could as well have been included under the title linolenic acid oils together with *e.g.* soyabean oil. These two compositional factors, *i.e.* the presence of glycerides containing erucic and linolenic acids, thus constitute the major difference between rapeseed oil and other fats and oils of importance in food uses. There are other specific minor components in rapeseed oil (see Chapter 7) that distinguish this oil from others. One such example is the presence of the sulphurous compounds, glucosinolates, in the seeds. These are, however, eliminated to a great extent by proper processing (see Chapter 10) and can therefore perhaps be disregarded when end uses of the oil are discussed. The influence of other seed components such as moisture, chlorophyll, enzymes, free fatty acids, phospholipids etc., that can affect the oil quality and/or processing yields markedly, are covered in Chapter 10. The lack of appropriate technological know-how and proper facilities for cultivation, seed handling and processing have at times certainly prevented the production of high quality edible rapeseed oil. At the present time these prerequisites seem to be available (see Chapters 5 and 10).

11.2.1. Flavour and flavour stability

A properly produced rapeseed oil is odourless and tasteless. The presence of

7–11 per cent linolenic acid in the glycerides makes it however more susceptible to autoxidation than *e.g.* groundnut oil and places it in the same category as soyabean oil. The linolenic acid as a source of the off-flavour of these oils has been studied by Moser *et al.*[16] and others[17, 18]. The rate of autoxidation and development of off-flavours depend as for other fats and oils on a number of factors which may be summarized as the resultant effect of glyceride composition, prooxidants, antioxidants and the conditions of handling and storage[19]. The comprehensive work that has been performed on the flavour and flavour stability of the related soyabean oil has recently been reviewed by Cowan[20], and can be used as a guide when discussing rapeseed oil stability. Holm *et al.*[21] reported on the type of off-flavour in rapeseed oil and other oils by stating: 'Thus rapeseed and soya oils with aldehyde contents over 2 had a noticeable oily, green off-taste after a couple of days while cottonseed oil and peanut oil formed a nutty, metallic, or bitter taste only at higher aldehyde contents'. The type of off-flavour that develops can thus be characterized as similar to that found in soyabean oil, and therefore comparisons can be fruitful between these two oils.

In the work of Moser *et al.*[16] the flavour characteristics and the oxidative stability of crambe seed, mustard seed, rapeseed and soyabean oils are subjected to further study with the purpose of establishing similarities or dissimilarities among these linolenate-containing oils. The oils were produced in the laboratory and the flavour evaluations were made by a test panel of 20 trained judges, and analyses of variance and F-tests were used to test the means on a 10-point scale. The description of the flavour of freshly deodorized oils was dominated by the term 'buttery', with 'nutty' or 'beany' flavours present in minor amounts. After 4 days storage at 60 °C the terms 'rancid', 'beany', 'painty', 'grassy' were given to oils when they did not contain citric acid. The presence of citric acid markedly protected the oils and there was no general difference between soyabean oil and the samples of Cruciferae seed oils. However, when samples with added citric acid were exposed to light for two hours the soyabean oil was given the same descriptive terms as the oil without citric acid, while the Cruciferae oils exhibited a significant drop in the flavour scores and developed a definite 'rubbery' flavour

TABLE 11:2

THE EFFECT OF CITRIC ACID ON CRUCIFERAE OILS EXPOSED TO LIGHT FOR 2 HOURS[16]

Oil	Without citric acid		With citric acid	
	Score	Description	Score	Description
Crambe seed	7.2	Buttery, beany	5.7	Rubbery, grassy, buttery
Mustard seed	4.6	Rancid, buttery, grassy	2.1	Rubbery, rancid, grassy
Rapeseed	6.9	Buttery, grassy	4.8	Rubbery, buttery
Soyabean	6.1	Grassy, buttery, rancid	6.5	Grassy, buttery, rancid

that was often accompanied by a 'garlic' or 'onion-like' flavour (Table 11:2). The authors assume that this effect may be due to isothiocyanate compounds introduced into the oils from the meal during extraction and not completely removed by deodorization. Whatever the cause, the compounds in question must exhibit specific photochemical reactions in the presence of citric acid. In any case it is a remarkable and specific effect caused by minor components in the Cruciferae seed oils which, as stated by the authors, 'project a new problem for study'. Moser et al.[16] also report results on industrially produced Swedish and Canadian rapeseed oils obtained in alkali refined and bleached form and deodorized in the laboratory. These results are reproduced in Table 11:3 in order to indicate the

TABLE 11:3

THE EFFECT OF CITRIC ACID ON THE KEEPING QUALITY OF SWEDISH AND CANADIAN RAPESEED OIL[16].

| Treatment | Swedish | | | | Canadian | |
| | Continuous screw-pressed | | Extracted | | Extracted | |
	without citric acid	with citric acid	without citric acid	with citric acid	without citric acid	with citric acid
Initial flavour	7.8	8.4	8.0	8.1	7.7	7.4
4 Days at 60 °C	4.2	7.0	4.9	6.9	4.5	6.1
2 hours' light exposure	6.8	4.8	5.8	2.9	4.0	5.4
AOM-value*	20.0	9.0	15.1	4.6	22.4	21.4

* Peroxide values after 8 hours under AOM conditions.

extrapolation of the results with the laboratory produced samples to products obtained from large scale operation with the processing praxis prevailing before 1965.

The positive effect of citric acid on the flavour stability when the oils were kept in darkness is demonstrated in all three samples as well as the deleterious effect on the rapeseed oil produced in Sweden when exposed to the standardized illumination for two hours. The Canadian oil, however, does exhibit a higher flavour score with citric acid than without, being the only exception to the off-flavour inducing effect of citric acid in the presence of light. The authors remark, however, that the characteristic 'rubbery' flavour was found in all samples but evidently less pronounced in the Canadian oil.

In Fig.11:1 the peroxide values after 8 hours' treatment under the AOM test conditions[22] are reported for the laboratory samples of soyabean oil and the Cruciferae oils. It may be noted that the effect of citric acid treatment is most marked for the soyabean oil investigated. The oxidative stability is indicated to

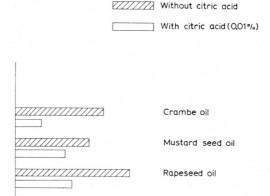

Fig.11:1. The effect of citric acid on the AOM-values of soyabean oil and Cruciferae oils[16].

be less for the soyabean oil than for the rapeseed oil before treatment with citric acid but superior after treatment with citric acid. These stability tests, performed at *ca.* 100 °C, are in good agreement with the flavour evaluations except for the stability to light.

Comparison of these data with other reports on the flavour and flavour stability of rapeseed oil in the literature is not very easy to make since different methods of evaluation have been used. A recent monograph (ref. 2, p.258) reports among the analyses of typical commercial samples of salad and cooking oils the keeping quality for soyabean oil as "AOM stability, 8 hours", which is the time needed for the sample to reach the peroxide value 125 under the AOM test conditions. Taking into consideration the fact that the peroxide value in Fig.11:1 after 8 hours for a citric acid-treated soyabean oil is found to be ca 4 and that Cowan[20] has reported values as low as 3.3 mekv/kg under the same experimental conditions for so-called protected soyabean oils, this value may be suspected to be representative of a processing technology that is obsolete. As a comparison, 19 lots of rapeseed oil produced in Sweden 1966/1967 were followed[23] with AOM tests. The 8-hour peroxide values for citric acid-treated oils were between 9.2 and 11.5, in good agreement with Fig.11:1, and the AOM stability was 22–23 hours. The addition of 0.01 per cent ascorbyl palmitate reduced the 8-hour peroxide value to 0.8 and increased the AOM stability to 33 hours. These results indicate, as can be expected, that the kinetics of the autoxidation of soyabean oil and rapeseed oil are different. Three reasons for this deserve to be pointed out here:

(a) The fatty acid composition. Frequently rapeseed oils contain 8–10 per cent linolenic acid and 14–15 per cent linoleic acid (see Chapter 7, p.135) while soyabean oils contain 6–9 per cent linolenic acid and *ca.* 50 per cent linoleic acid. If other

conditions are assumed to be equal the soyabean oil reacts faster with oxygen than the rapeseed oil due to the larger amount of dienoic compounds[24, 25].

(b) *The glyceride composition.* Raghuveer and Hammond[26] have shown that the position of fatty acids in the triacylglycerol molecules influences the rate of autoxidation in the liquid state. When more than one reactive group is present in the molecule the rate is faster than when the groups are located on different molecules. Randomization of mixtures of trilinoleoylglycerol and tridecanoyl-glycerol decreased the rate of autoxidation while randomization of cocoa butter, Borneo tallow, corn oil and soyabean oil caused an increase in the rate of aut-oxidation. As shown by Mattson and Volpenhein[27] and others[28] the dienes and trienes of rapeseed oil are preferentially located in the 2-position of the triacyl-glycerols which should be advantageous from the above point of view. On the other hand the position of linolenic acid in soyabean oil is nearly random (see references in [28]), whereas the dominant linoleic acid has a slight preference for the 2-position. The above arguments are indicated in the paper of Appelqvist and Dowdell[28] and in a review by Appelqvist[29].

(c) *The composition of tocopherols.* The content of different tocopherols in rape-seed oil is reported in Chapter 7, p.146, where it appears that both the content and composition are rather favourable from the point of view of autoxidation. In contrast the most advantageous concentration of tocopherols in soyabean oil is lower than the natural concentration[30], and therefore processing is directed towards a reduction of the tocopherols present[20].

In several studies Holm *et al.*[21] and Holm[31, 32] have established the importance of the degree of pre-oxidation, *i.e.* the content of primary and secondary oxidation products in crude oil on the flavour stability of deodorized rapeseed oil.

The non-volatile carbonyl compounds that are present in the oils as a result of the decomposition of primarily formed hydroperoxides, and are not eliminated during the deodorization, were measured by the benzidine test[33] and the results correlated with the flavour stability of the oil. These results emphasize the importance of a protected storage and handling of rapeseed oil, quite in accord-ance with the principles established during years of research and development work on soyabean oil[20]. It is therefore evident that comparisons between flavour and flavour stabilities of different oils must be made with great consideration of the pretreatment of the samples. A deficiency in the carefully performed work of Moser *et al.*[16] is that the degree of preoxidation of the crude oils is not documented.

The degree of preoxidation is of course not the only factor that influences shelf life. Previously it was noted (p.254) that the addition of ascorbyl palmitate had a markedly retarding effect on the rate of autoxidation of rapeseed oil. Antioxidant effects have further been reported on the addition of propyl gallate[34], butylated hydroxyanisole (BHA), butylated hydroxytoluene (BHT), nordihydroguaiaretic acid and mixtures of BHA and BHT[35]. Sedlácek[36, 37] compared the antioxidative efficiency of 0.2 per cent ascorbyl palmitate, 0.02 per cent BHA and 0.01 per cent propyl gallate, respectively, in rapeseed oil and sunflower oil by storing samples at 28 °C for 35 weeks and at 2–4 °C for 52 weeks. The samples containing 0.2 per cent ascorbyl palmitate were most effectively protected and were given the flavour score 6 on a 10 to 0 point scale after 30 weeks at 28 °C compared to 2 for the blank. At the earlier stages of storage the off-flavour was characterized as harsh and in addition a somewhat bitter off-flavour was noticed in samples containing ascorbyl palmitate. In these experiments the rapeseed oil reacted slower with oxygen than the sunflower oil. This work was also extended to shortenings manufactured from rapeseed oil and sunflower oil, giving essentially the same results[38]. The effect of BHA, BHT, propyl, lauryl, dodecyl and cetyl gallates and some mixtures with these compounds and citric acid on the autoxidation of hydrogenated rapeseed oil has also been evaluated[39]. As a generalization of the reported work with antioxidants in rapeseed oil and hydrogenated rapeseed oil it can be stated that BHA and BHT have a rather small retarding effect on the rate of autoxidation while the addition of 0.2 per cent ascorbyl palmitate markedly reduces it. When choosing antioxidants, specific off-flavours and heat resistance of the compounds as well as local regulations (*cf.* Appendix III) have to be considered. These factors are not specific to rapeseed oil and will not be commented on further. For information on the heat resistance of antioxidants reference is made to the literature[40, 41].

The stabilization of rapeseed oil and other oils by a combined polymerization and deodorization process has been reported[42] but seems to be out of practice at the present time.

Regarding the flavour stability of the different types of Cruciferae oils no general comparisons can be made because there is an insufficient control of the many factors that influence the experimental results. Differences in fatty acid compositions can be indicative but by no means conclusive. Overall measurements of the rates of oxygen consumption at forced or actual conditions can prove more useful in practice than any single analytical method. In the work of Ohlson[43] such a method, based on measurements of oxygen concentration in samples treated in different ways as a function of time, was used to indicate differences between the samples of winter and summer species of *Brassica napus* and *Brassica campestris* used. The winter type of *B. campestris* was found to have the slowest rate of autoxidation.

In conclusion the flavour and flavour stability of properly processed, protected and stabilized rapeseed oil is completely satisfactory for an oil to be used in today's food industry. There are several reports showing that rapeseed oil is superior to oils such as soyabean and sunflower. There are also reports in which rapeseed oil was found to be inferior to soyabean oil, *e.g.* in the presence of citric acid and light. This only stresses the well-known point that many factors influence the flavour of vegetable oils and fats, some of which are general for all and some of which are specific for a certain oil.

Among the physical properties the viscosity and the behaviour on solidification and melting can be of specific interest for the use of rapeseed oil for edible products.

(a) Viscosity

The viscosity of an oil is of significance not only for the chemical engineer in the design of pumps, piping and other pieces of equipment, but also for the food technologist as it affects the consistency of foods. Due to its composition the viscosity of rapeseed oil is higher than that of other vegetable oils used commercially at the present time (ref. 2, p.90 and ref. 3, p.84). The most complete compilation of viscosity data for oils and fats is still perhaps found in the works of Kaufmann and Funke[44, 45]. At 20 °C the dynamic viscosity of rapeseed oil of usual composition is around 90 centipoises compared to about 60 centipoises for soyabean oil and to 77 centipoises for olive oil. The authors reported viscosities of 87.9–90.4 centipoises for rapeseed oil ranging in saponification numbers between 175.5 and 173.6.

The dependence of the viscosity of rapeseed oil on temperature was shown to follow the Walther formula[46] for mineral oils,

$$\log \log(\Gamma + 0.8) = m(\log T_1 - \log T) + \log \log(\Gamma_1 + 0.8),$$

where Γ is the kinematic viscosity in centistokes and T the absolute temperature in °K. With $m = 3.075$ good agreement between calculated and observed values was found in the temperature range 20–150 °C.

The calculation of the viscosities of rapeseed oil mixtures with other oils is tested by mixing rapeseed oil and palm kernel oil and applying the formula used by Kendall and Monroe[47]. Perhaps a satisfying degree of accuracy for food technological purposes may be reached by using the same methods that are applied to mineral oils[48], *i.e.* linear interpolation of the function $\log \log(\Gamma + 0.8)$ when this is plotted against the volume fraction of one of the two oils.

Kinematic viscosities at 65–95 °C have been determined by Tremazi *et al.*[49] for a hydrogenated rapeseed oil, iodine value 1.5, containing *ca.* 60 per cent behenic acid, and it is stated that hydrogenated rapeseed oil is probably the highest melting and most viscous fat which may be incorporated in food products.

(b) Melting and solidification

The high content of erucic acid in rapeseed oil would imply a high melting point of this oil. The melting point of the most stable polymorphic form of trierucylglycerol is *ca.* 30 °C[50]. It has, however, been shown[27] that erucic and eicosenoic acids are preferentially situated in the 1- and 3-positions of the glycerol residue, thus leading to mixtures of molecules containing oleic, linoleic and linolenic acids in the 2-position. These compounds have not yet been characterized, but melting points between 4 °C and 5.8 °C have been reported for rapeseed oils from West Pakistan, Sweden and Canada[49]. Riiner[51] investigated a number of Cruciferae seed oils with X-ray and thermal methods and found generally the same polymorphic behaviour for oils containing between 8 per cent and 63 per cent of erucic acid. It was concluded that the requirement for a Cruciferae oil to remain clear at 0 °C is that the erucic acid content is below *ca.* 39 per cent for oils from *Brassica napus* and below *ca.* 43 per cent for oils from *Sinapis alba*. The behaviour on melting and crystallization was shown to vary regularly with the erucic acid content (Fig.11:2) and thus the phase behaviour of a certain oil can be predicted

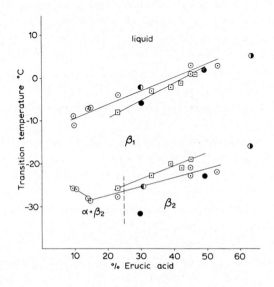

Fig.11:2. Phase transitions in Cruciferae oils. Erucic acid content > 8%. (ref. 51). ○, *Brassica napus*; ☐, *Sinapis alba*; ●, *Brassica campestris*; ◗, *Crambe abyssinica*; ◖, *Conringium orientalis*.

from its composition. This can perhaps be of value when oils for foods that are stored and used at low temperatures are under consideration.

Riiner and Melin[52] have reported on the solidification characteristics of hydrogenated rapeseed oil. The initially formed α-phases were shown to have a longer lifetime in rapeseed oil than in other hydrogenated vegetable oils, which can influence the consistency and processing conditions of margarine and shortenings. Other specific characteristics of the solid phases of hydrogenated rapeseed oil are found in the patent literature and will be pointed out in the following parts of this review.

In the following specific uses of rapeseed will be reviewed. Once again it is stressed that the quality of an oil, as well as the prepared food containing the oil, varies with the processing of the raw materials, hygienic factors, and the quality criteria applied. Therefore the selection of material cannot be made on a completely objective basis nor can the evaluation of results be expressed in internationally generalizable terms. It is, however, still believed that reviewing the literature on rapeseed oil can give guidance for the selection of oils and fats for use as food.

11.3.1 Salad and cooking oils

It is natural to begin with the use of rapeseed oil as such for cooking, frying, food dressings, mayonnaises and as a table oil. Taking into consideration the fact that the major part of rapeseed oil at present is produced and used in Asia, where the food industry is rather undeveloped, it is perhaps true that the larger part of the world production is used directly in cooking.

Without doubt the rapid growth of rapeseed oil utilization in Canada has brought about or been promoted by the careful evaluation of the use of rapeseed oil in foods. From other sources such evaluations are hardly available. The early work was done 20 years ago by Lips et al.[53] who compared samples of rapeseed oil and mustard seed oil with corn oil in mayonnaise, pastry and for frying doughnuts. Disregarding the processing parameters included in this work, the flavour of the Cruciferae oils was found comparable with that of corn oil in cold uses (oil, mayonnaise) but inferior in hot uses (frying). The autoxidation susceptibility was higher for Cruciferae oils than for corn oil. Because technological progress has taken place during the past 20 years it can be of interest to compare the work of Lips et al. with two recent evaluations[54, 55]. In the first[54] samples of rapeseed, soyabean, sunflower, groundnut and corn oils, obtained from commercial manufacturers, were compared in cold and hot uses. A panel of 14 persons did not find any significant difference between salad dressings prepared from these oils. French fries fried in corn oil were preferred to those fried in rapeseed oil or soyabean oil, but no significant differences were established between French fries prepared in rapeseed, soyabean, sunflower or groundnut oil. The number of panelists was 8 in these evaluations. It was noticed that the corn oil and the groundnut oil had the strongest flavours before frying. Despite the passing of 20 years even in this evaluation corn oil was found more acceptable for frying than rapeseed oil. In the more extensive evaluation by Vaisey and Shaykewich[55] two brands of rapeseed oil were found comparable in frying performance to corn, sunflower and soyabean oils. In this case doughnuts were fried at 175 °C twelve successive times, each time the total frying time being approximately 10 minutes. Neither the odour scoring of the oils used nor the flavour scoring of doughnuts distinguished the oils from each other. The smoke point determinations indicate that all the oils were of comparable quality from the beginning and that the breakdown of the soyabean oil after 12 fryings had perhaps advanced more than that of the other oils. These two contradictory results indicate the relativity of the measurements but also seem to permit the conclusion that a rapeseed oil which compares favourably with other vegetable oils in hot uses can be manufactured. In an 8 days' accelerated storage test at 60 °C (Schaal Oven Test) one brand of rapeseed oil remained unchanged in odour and was in this respect equal to corn oil and soyabean oil. The other brand was superior to sunflower oil though less stable than corn oil. One zero-erucic acid oil that also was evaluated was the least stable of the oils examined.

For readers interested in the chemistry of flavour compounds, reference is made to the work of Hrdlicka and Pokorny[56] who identified some of the volatile compounds formed in rapeseed oil on cooking and frying.

The expression zero-erucic rapeseed oil refers to seed oils from *Brassica* strains obtained by recent plant breeding work in Canada (*cf.* Chapter 6). Downey *et al.*[57] give a summary of this work and report that the oil is essentially free from erucic acid and is named Canbra oil, a contraction of the words Canada and Brassica. The single flavour evaluation reported above cannot be taken as evidence for a statement that Canbra oil is inferior to ordinary rapeseed oil in flavour stability. As the melting point of Canbra oil is below −10 °C[51] it is, however, possible to hydrogenate it in order to increase the oxidative stability without too large an increase in the content of solids at low temperatures. Teasdale[58] claims that the reduction of the linolenic acid content to 3 per cent improves the flavour stability only slightly whereas a reduction to levels below 1 per cent gives a substantial improvement. Winterization of hydrogenated Canbra oils with 0.4–0.9 per cent linolenic acid at 7–8 °C was reported to give salad oil yields of approximately 95 per cent. This high yield is compared to yields from rapeseed oil with 25–40 per cent erucic acid and to yields from soyabean oil. At 1 per cent residual linolenic

TABLE 11:4

THE RELATIONSHIP BETWEEN LINOLENIC ACID CONTENT AND STABILITY OF HYDROGENATED CANBRA AND SOYABEAN OILS[59]

Oil	Linolenic acid per cent	AOM stability hours	Schaal Oven Test days
Canbra oil	3.2	35	18
Canbra oil	2.9	50	21
Soyabean oil	2.1	25	14

acid these yields are of the order of 75 per cent. Description of the manufacture of hydrogenated cooking and frying oils from Canbra oil without winterization is also found in the patent literature[59]. The stabilities obtained are illustrated in Table 11:4.

Salad and food dressings often contain liquid oil (40–50 per cent) dispersed in an aqueous phase containing salt, acetic acid, egg yolk and other ingredients. In mayonnaise the oil content is often 70–80 per cent. As the products are preferentially stored under refrigeration the oil used should not crystallize, otherwise the emulsion will break. From the results of the study by Riiner[51] it can be concluded that Cruciferae oils with erucic acid contents above 40–45 per cent can be hazardous to use in the industrial production of food dressings and mayonnaise distributed under refrigeration conditions. There is, however, in the patent literature[60] a specific application of hydrogenated rapeseed oil in the manufacture of food dressings with an oil phase in fact containing crystalline phases. Solid fats (2–8 per cent) are melted with liquid oil (92–98 per cent), mixed with the aqueous phase, and chilled. The solid fat portion, with iodine value ≤ 12, consists of a blend of a so-called β-tending fat and hydrogenated rapeseed oil in proportions ranging from 1:4 to 4:1. The expression "β-tending" refers to the polymorphic form that is attained after tempering[61].

11.3.2 Margarine

As indicated in Section 11.1.1 an important use of rapeseed oil in Europe and Canada is in margarine formulations. A standard formula in Germany during World War II was reported to be[9]:

10 per cent animal fats,
10 per cent hydrogenated rapeseed oil, melting point 40–42 °C,
38 per cent hydrogenated rapeseed oil, melting point 30–32 °C,
42 per cent rapeseed oil.

Under the prevailing extreme conditions the product quality must have been of secondary importance, after product quantity. At the present time, it is reported that rapeseed oil or hydrogenated rapeseed oil is seldom used in margarine in West Germany[62]*. The reason for this is said to be that the products in question give the margarine a harsh mouth-feel and that this is caused by specific phosphatides in the oil. This should apply particularly to oil from rapeseeds grown in West Germany. In other localities, in any case, this characterization of margarine containing rapeseed oil has not been given the same emphasis. It must be borne in mind, however, that consumer acceptance as well as the factors that influence oil quality vary considerably from country to country. One such important quality factor is the requirement on the shelf-life of the product with respect to

---

* *Cf.* however ref. 16.

TABLE 11:5

HYDROGENATED RAPESEED OILS FOR MARGARINE IN CANADA, 1967[63]

Type	I.V.	Fatty acid composition %										Trans Acids %	S.F.I.* °C			
		16:0	16:1	18:0	18:1	18:2	18:3	20:0	20:1	22:0	22:1		10	21.1	26.7	33.3
Soft																
A	93.3	2.5	0.3	2.1	42.7	14.1	2.9	1.0	10.8	0.4	23.2	17.6	3.3	0.8	—	—
B	87.5	2.5	0.3	2.9	47.0	10.9	1.0	1.0	10.8	0.8	22.2	29.9	6.4	2.0	—	—
Hard																
C	70.7	3.6	0.3	8.9	38.0	3.8	—	1.2	8.6	1.7	33.9	—	58	40	33	15
D	64.2	2.7	0.2	14.0	44.5	2.9	—	2.7	8.1	4.8	20.1	—	59	45	37	19

* Official and Tentative Methods of the A.O.C.S., Chicago 1967, Cd 10-57.

time, temperature and packaging. A consequence of this is *e.g.* that in some countries liquid rapeseed oil and soyabean oil are accepted for margarine manufacturing whereas only hydrogenated forms of these oils are considered appropriate in others.

In a recent monograph[63] on margarine it is pointed out that hydrogenated rapeseed oil is preferentially used in two forms, one with a melting point *ca.* 34 °C (iodine value 72) and the other with a melting point of *ca.* 42 °C (iodine value 60). This statement seems, however, to be only locally valid. A great range of hydrogenated fats can be manufactured from rapeseed oils, as from other oils, ranging from practically liquid to very hard. For the market conditions in Canada in 1967, Teasdale[63] reported that margarines with a butter-like consistency are produced by blending two types of hydrogenated rapeseed oil, one soft and one hard. Examples of soft and hard components are given in Table 11:5. The components A, B and D are produced from the oil from *B. campestris* and C from *B. napus*. By blending *e.g.* 57 per cent B and 43 per cent C or 65 per cent B and 35 per cent D butter-like margarines are produced, the fat blends having the following SFI values*: 10 °C 26–28; 21.1 °C 15–16.5; 33 °C 3–4. Soft margarines, spreadable from the refrigerator, were reported to be composed of three to six fat components, of which two can be chosen from either group in Table 11:5.

Zalewski and Kummerow[64] have reported a systematic investigation on the suitability of rapeseed oil in two-component margarine oil blends, containing one hydrogenated and one non-hydrogenated component. By dilatometric and chemical methods margarines in the United States of America were characterized and

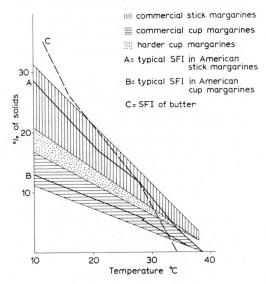

Fig.11:3. Solid fat index range for American stick and cup margarines[64].

---

* SFI = solid fat index.

divided into two groups, one group called stick margarines and the other cup margarines.

The SFI values *vs.* temperatures are plotted in Fig.11:3. The hydrogenated fats in the investigation which fulfilled the SFI curve requirement for stick and cup margarines in blends with various portions of liquid oils were the following three:

A, selectively hydrogenated soyabean oil (I.V. = 68, *trans* fatty acids = 63 per cent),

B, unselectively hydrogenated rapeseed oil (I.V. = 65, *trans* fatty acids = 48.5 per cent),

C, selectively hydrogenated rapeseed oil (I.V. = 68, *trans* fatty acids = 58 per cent).

The soyabean hardstock A which contained a high solids content at 38 °C was more suitable for a cup type margarine but both types could be prepared by blending with rapeseed oil or corn oil. Corn oil blends showed less change in solids content at a temperature range of 21–27 °C but a steeper SFI curve in the range 27–38 °C, which ensured a lower solids content at 37 °C. Rapeseed oil hardstock B was suitable for cup type margarine blends but not suitable for the stick type due to the high solids content at 37 °C. The rapeseed hardstock C allowed for both proper cup type and stick type margarines. It was further shown that the hardstocks B and C gave the same SFI curve irrespective of which liquid oil was used in the two-component mixtures. The oils used were cottonseed, corn, rapeseed, safflowerseed and soyabean oil. This indicates that the solubilities of the two hydrogenated oils are either negligible or follow the laws of ideal solubility. It is indicated that the presence of erucic acid in hydrogenated rapeseed oil favourably influences the crystallization and storage of margarine so that the graininess of the margarine due to β' to β-transformations is not to be expected. When hydrogenated soyabean oils of certain types are used this can occur[65].

Zalewski and Kummerow[64] also demonstrated the effect of transesterification on blends of hydrogenated rapeseed oil and liquid oils. As could be expected the SFI curves after transesterification were dependent on the liquid oil used although they were independent of the kind of oil before the transesterification. It was noticed that the differences between the rearranged blends of various liquid oils and the same hydrogenated rapeseed oil could be attributed to the amount of saturated fatty acids in the oils rather than to the degree of unsaturation. About twice as much hardstock was required to ensure the proper SFI values after transesterification. If one assumes that the hardstock is more expensive than the liquid oil it can be concluded that the SFI curve criteria alone cannot justify this treatment in commercial processing.

Transesterification of hydrogenated rapeseed oil with fats containing short chain fatty acids can, however, be justified under certain circumstances. Rudischer[66] reported that a substantial improvement in organoleptic properties could be attained by the transesterification of hydrogenated rapeseed oil blends with coconut oil whereas no significant improvement could be established by treating the hydrogenated rapeseed oil alone. Seiden[67] further emphasizes this point by showing that transesterified blends of hydrogenated rapeseed oil and coconut or palm kernel oils preferentially in combination with liquid oils, can be favourably used for the production of margarine with improved oiling-off properties and good eating quality. Another application of transesterification is the reduction of the melting points of high melting fats by the rearrangement of blends of the fat in question and a liquid oil. Thus high melting animal tallows could be made useful for margarine production. Although this application of liquid oils is by no means specific for Cruciferae oils, the properties of the resulting fat depend on the composition of the oil used. The principle is illustrated by the following example from the literature[68]: A mixture of 35 per cent hydrogenated beef tallow and 65 per cent rapeseed oil had a drop point 48.3–49.7 °C before the rearrangement and 30.9–33.5 °C after the rearrangement. Work with non-hydrogenated beef tallow[69] and hydrogenated lard[70] in mixtures with rapeseed oil has also been reported.

A specific use of highly hydrogenated rapeseed oil, *i.e.* with iodine values below 30, and preferentially below 10, in retarding the oiling-off of margarines is also claimed by Seiden[71]. Small quantities of hydrogenated rapeseed oil (0.2–2.5 per cent) in the formulae are reported to reduce the oiling-off tendency considerably, without too much adverse influence on the eating quality.

In the patent literature a specific hydrogenation process[72] is also described, the purpose of which is among other things to enable the production of an improved type of very soft margarine by using mixtures of hydrogenated rapeseed oil according to the patent along with liquid oils.

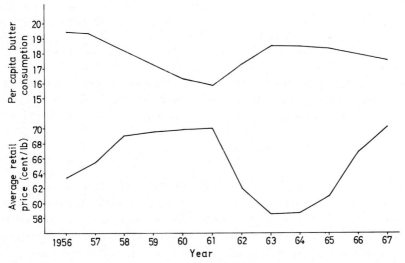

Fig.11:4. The dependence of per capita consumption of butter on retail price[74].

Reports on flavour evaluations of margarines are scarce in the literature. In a Canadian evaluation[54] rapeseed oil margarines were compared favourably with soyabean oil margarines.

In this connection the importance of price as a buying motive for margarine, one of the major outlets for edible rapeseed oil, can be illustrated by a recent study from Canada[73]. The results indicated that 50 per cent of the consumers asked bought margarine on a price basis, 40 per cent for health reasons and 10 per cent because they preferred the flavour. Only 2 per cent thought that they bought butter on price basis, 49 per cent for health reasons and 49 per cent for the flavour. This could be taken as an indication that the margarine manufacturers should concentrate development work on improving the flavour, but until such progress is made the price is the main competitive factor between butter and margarine in Canada. This is further illustrated in Fig.11:4 by the dependence of per capita consumption of butter on the average retail price during a 12 year period[74].

Rapeseed oil prices in Western Europe 1964–1967 are compared with those of some other vegetable oils and fats in Table 11:6. Soyabean oil and rapeseed oil

TABLE 11:6

THE PRICES OF SOME COMMON FATS AND OILS IN WESTERN EUROPE (POUNDS PER TON)[75]

Oil	1964		1965		1966		1967	
	£	s	£	s	£	s	£	s
Rapeseed oil	91	10	95	0	88	15	75	18
Soyabean oil	83	10	98	0	95	0	79	15
Coconut oil	113	5	117	0	107	10	104	0
Groundnut oil	107	5	130	5	113	0	117	5
Sunflower seed oil	105	15	122	10	108	15	93	0
Cottonseed oil	94	15	101	5	100	0	97	10

have been the cheapest oils on the market. The average price for rapeseed oil has been lower during three of the four years, thus constituting an important motive for use in margarine, provided that other requirements are fulfilled.

11.3.3 Shortenings or compound cooking fats

The definition of different types of edible fat products is not completely clear. The word shortening was originally used in American literature for fats employed in baking, making the bread shorter in texture. It has since then been transferred to fat products in general constituted by 100 per cent or close to 100 per cent fats and oils, used not only in baking but also in domestic and institutional cooking, as well as in the rapidly increasing different uses in the food industry. In the United

Kingdom[75] the expression compound cooking fat covers at least a part of the connotations of the American counterpart. Confusion between generic names is avoided when uses and properties are defined more precisely.

As pointed out previously systematic evaluations of rapeseed oil and hydrogenated rapeseed oil in edible products were performed in Canada when rapeseed growing was started. The work on hydrogenated rapeseed and mustard seed oils in baking and frying applications[76], made with samples produced on a pilot-plant scale, showed that these were as satisfactory as the commercial reference material for the preparation of pastry and doughnuts. The acceptability tests were made by a 24 member panel and supported by chemical and physical measurements of several properties of the samples. For the doughnut preparations the shortenings were used both in the dough and as frying fats. No significant differences were established between commercial and pilot-plant samples. Odour and flavour of pastry made with the experimental fats were rated as slightly preferable to those of pastry made with the commercial shortening. Baking value tests, performed with a standard biscuit batter, showed that plasticizing, i.e. controlled crystallization and agitation of the fats, was required to achieve volumes comparable to commercial shortening when the fats were used in non-melted form. Similar experiments were reported with hydrogenated oils from screened weed seed, mainly from Cruciferae varieties[77]. At the present time nearly 50,000 tons of rapeseed oil is used per annum in shortenings in Canada[13]. A more recent evaluation of shortenings containing rapeseed oil made with commercial products in Canada[54] revealed no differences between these and other types when used for the preparation of cakes, but gave a less tender and flaky pastry than lard. Compared to other vegetable shortenings, no difference was found.

In shortenings, as well as in margarines, a variety of hydrogenated products manufactured from rapeseed oil can be used. This situation is not specific for rapeseed oil alone (see Chapter 10). The early experiments in Canada, discussed above, were made on a restricted number of samples with the purpose of establishing the general performance characteristics of these samples. There are, however, some specific properties of hydrogenated Cruciferae oils that are desirable in some of the food preparations where shortenings have traditionally been used. It should be mentioned here that the tradition in some countries is to almost exclusively use margarines in baking. The composition and properties of the fat phase can of course be the same whether a dispersed water phase is present or not. Therefore the following discussion can also be valid for the fat phases of e.g. bakery margarines in some countries.

Tremazi et al.[49] compared the temperature ranges within which hydrogenated rapeseed oils and cottonseed oils could be expected to be plastic. It was stated that at comparable iodine values hydrogenated rapeseed oil shows a broader range of plasticity than hydrogenated cottonseed oil. This can be a consequence of the presence of the triacylglycerols containing erucic and eicosenoic acids, which

TABLE 11:7

RESULTS OF YELLOW LAYER CAKE TEST WITH 5% SUSPENSIONS HYDROGENATED FATS IN COTTONSEED OIL[80]

Type of hydrogenated fat	I.V.	Without emulsifier			With 1.2% Myverol 18:00 emulsifier*		
		Batter vol., cc/g	Cake vol., cc	Texture of cake	Batter vol., cc/g	Cake vol., cc	Texture of cake
None	—	0.98	1000	Hard	0.90	1000	Hard
Soyabean oil	1	0.96	1000	Med. hard	0.96	1060	Hard
Cottonseed oil	1	0.94	1080	Med. hard	1.03	1155	Med. hard
Palm oil	2	1.05	1080	Med. hard	0.93	1155	Med. hard
Herring oil	2	0.89	1055	Med. hard	—	—	—
Lard	7	0.94	1100	Med. hard	0.85	1000	Med. hard
Rapeseed oil	1	1.01	1045	Med. soft	1.31	1165	Soft
Mustard seed oil	1	1.15	1120	Soft	1.38	1210	Soft
Plastic shortening	—	1.23	1210	Soft	1.23	1210	Soft

* Distilled saturated monoacylglycerol.

upon hydrogenation, in part, are transformed to triacylglycerols containing behenic and arachidic acids. The action of these is then the same as that of high melting triacylglycerols intentionally added as plasticizers to the fat blends, a procedure sometimes used in the manufacturing of shortenings (ref. 2, p. 300). This effect is further elaborated by adding highly hydrogenated rapeseed oil as such as a plasticizer to fat blends intended to be used for shortening[78].

Besides being useful to provide an increased range of temperature at which the products remain plastic, hydrogenated rapeseed oil is also reported to have favourable aerating properties in baking applications where this is desirable, *e.g.* in cake batter mixing operations where fat and sugar are first beaten[79]. This effect was studied in detail by Linteris and Thompson[80] in their work with fluid shortening development. For comparison some results obtained by a standard yellow layer cake performance test with 5 per cent suspensions of a number of highly hydrogenated fats and oils in cottonseed oil are given in Table 11:7.

It may be seen that even without the addition of emulsifier, which is common practice in aerating shortenings and margarines, the Cruciferae oils hydrogenated to low iodine values gave fair batter and cake volumes. When emulsifier was added, these 5 per cent suspensions were completely satisfactory, in contrast to the other fats investigated. The authors further conclude that fractions made from rapeseed and mustard seed oils and certain peanut oil fractions possess the properties desired for the production of a fluid shortening having aerating properties. The triacylglycerols indicated to be especially effective in this respect were of the behenoyldistearoylglycerol and behenoylstearoylpalmitoylglycerol types. The hydrogenated rapeseed oil, iodine value 1, was, however, almost as effective as the triacylglycerol fractions used. It is interesting to note that no correlation was found between the polymorphic forms of the solid phases and their aeration properties in the systems investigated by Linteris and Thompson[80]. It has been a rather common opinion that fats in the β' form in general have a better creaming ability than fats in the β form (ref. 2, p.362). In this work the opposite was clearly demonstrated: When hydrogenated rapeseed oil was precipitated from liquid cottonseed oil at 21 °C, the solid phases were in the β' form and when precipitated at 29 °C they were in the β form. However, no difference in aeration properties was found.

An additional requirement of a fluid shortening is that the viscosity (fluidity) remains unchanged on storage at different temperatures and that the product remains homogeneous. Dispersions of hydrogenated rapeseed oil, iodine value 1, were stable between 16 °C and 37 °C for shorter periods but thickened upon cyclic tempering and prolonged storage. This solidification tendency was assumed to be caused by a leaching out of components initially present in solid solution and their redeposition as long interlocked crystals. This effect could be minimized by using blends of hydrogenated rapeseed oil (IV = 1) and hydrogenated mustard seed oil (IV = 30), perhaps by the formation of more stable solid solutions.

Settling of solid phases and the formation of oil layers on the surface was inhibited by the addition of finely divided (particle size 1–10 micron) hydrogenated soyabean stearine.

Another route to fluid shortening containing, among other fats, hydrogenated rapeseed oil in minor quantities (0.5–4.0 per cent) is described by Matsui *et al.*[81]. In this case the triacylglycerols do not seem to play a decisive role for the determination of the physical properties. Instead combinations of triacylglycerols with surface active lipids are used to achieve functionality.

11.3.4. Other food uses

Confectionary fats, and substitutes for cocoa butter, have received considerable attention in the literature both from fundamental and applied points of view. The fats can be divided into two groups, *i.e.* fats miscible with cocoa butter and fats immiscible with cocoa butter. The former group is preferentially prepared from fat sources containing triacylglycerols similar in composition to cocoa butter and the other group traditionally from lauric acid fats. It is evident that the average chain length of rapeseed oil fatty acids is too long to permit the isolation of compounds structurally similar to 2-oleoylpalmitoylstearoylglycerol, the main component of cocoa butter. The other group of fats is used together with defatted cocoa powder, thus imitating the properties of chocolate without the requirement that the fat must be miscible with cocoa butter.

Using combinations of hydrogenation and fractional crystallization, it has been reported that the physical properties such as hardness and rate of melting of cocoa butter can be imitated with rapeseed oil[82, 83]. The oil is first hydrogenated under conditions of high *trans* fatty acid formation and then subjected to fractionation in solvents. It is claimed that chocolate with good bloom resistance, hardness and mouth-feel can be produced. In another procedure for the production of hard butters[84] chiefly from lauric acid fats, *e.g.* coconut and palm kernel oils, by the vacuum distillation of transesterified blends, it is reported that the use of up to 15 per cent hydrogenated rapeseed oil, iodine value less than 5, is suitable.

In a new hydrogenation procedure[72] which is not specifically concerned with Cruciferae oils, it is pointed out that blends suitable as ice cream coating fats can be prepared containing 50–80 per cent hydrogenated rapeseed oil.

A specific use of monoacylglycerols derived from hydrogenated rapeseed oil, iodine value less than 8, is claimed by Bedenk[84].

11.4 FUTURE USES

It can be concluded from this review, covering at least a part of the literature on edible uses of Cruciferae seed oils, that they compete favourably with other vegetable oils and animal fats for use as food. The balance between different oils depends in part on differences in quality, but also on other technological and economical factors, that have nothing to do with the quality of the oils itself. There are, however, also some specific physical properties of hydrogenated rapeseed oil, not found in other vegetable oils, that are or will be utilized in fluid or plastic fatty products. These properties are due to the presence of longer fatty acids in the glyceride molecules. On the other hand the high content of erucic acid-containing triacylglycerols is in some applications a limitation. It seems however that rapeseed oil in its present form can uphold a place among the raw materials for the food industry of tomorrow with its increasing number of diversified products or product lines.

No doubt plant breeding efforts will lead to new Cruciferae varieties with altered oil compositions—considerable advances have already been made *(cf.* Chap. 6)—that can make the crop more competitive and fill the requirements of food manufacturers regarding functionality and shelf-life, thus offering a greater variation in the properties.

REFERENCES

1 *McGraw-Hill Encyclopedia of Science and Technology*, McGraw-Hill Book Company Inc., New York, Vol. 11, 1960 p.341.
2 D. SWERN Ed., *Bailey's Industrial Oil and Fat Products*, Interscience Publishers Inc., New York, 1964, p.218.
3 E. W. ECKEY, *Vegetable Fats and Oils*, Reinhold Publishing Corporation, New York 1954, p.434.
4 K. A. WILLIAMS, *Oils, Fats and Fatty Foods*, J. & A. Churchill Ltd., London, 1966.
5 K. S. MARKLEY, Ed., *Soyabeans and Soyabean Products*, Interscience Publishers Inc., New York, Vol. I, 1950 and Vol. II, 1951.
6 A. E. BAILEY, *Cottonseed and Cottonseed Products*, Interscience Publishers Inc., New York, 1948.
7 M. E. STANSBY, Ed., *Fish Oils, Their Chemistry, Technology, Stability, Nutritional Properties, and Uses*, The AVI Publishing Company Inc., Westport, 1967.
8 G. B. MARTINENGHI, *Tecnologia Chimica Industriale Degli Oli, Grassi e Derivati*, Editore Ulrico Hoepli, Milano, 1963.
9 J. L. BEARE, *Food. Manuf.*, 32 (1957) 378.
10 G. ANDERSSON, *Oléagineux*, 16 (1961) 767.
11 *Special Crops, Publication 187*, Manitoba Dept. of Agriculture and Immigration, Winnipeg, 1944.
12 *IFMA, The Situation in the Margarine Industry from a viewpoint of legislation*, The Hague, 1967, I-Por-6.
13 *Rapeseed oil Facts and Recipes*, Rapeseed Association of Canada, Publication No. 6, 1969.
14 *Tropical Products Quarterly*, 10 (No. 1) (1969) 100.
15 D. ELZ, *Oilseed Product Needs of the European Economic Community*, 1970, U.S. Department of Agriculture, 1967, p.153.
16 H. A. MOSER, C. D. EVANS, G. MUSTAKAS AND J. C. COWAN, *J. Am. Oil Chem. Soc.*, 42 (1965) 811.
17 H. J. DUTTON, C. D. EVANS AND J. C. COWAN, *Trans Am. Ass. Cer. Chem.*, 11 (1953) 116.
18 G. HOFFMANN, *J. Am. Oil Chem. Soc.*, 38 (1966) 1.

19   W. O. LUNDBERG, Ed., *Autoxidation and Antioxidants*, Interscience Publishers, New York, Vol. I, 1961, Vol. II, 1962.
20   J. C. COWAN, *J. Am. Oil Chem. Soc.*, 43 (1966) 300 A.
21   U. HOLM, K. EKBOM AND G. WODE, *J. Am. Oil Chem. Soc.*, 34 (1957) 606.
22   E. M. SALLEE, Ed., *Official and Tentative Methods*, A.O.C.S. Tentative Method Cd 12-57, Chicago, 1965.
23   U. RIINER, unpublished data.
24   R. T. HOLMAN AND O. ELMER, *J. Am. Oil Chem. Soc.*, 24 (1947) 127.
25   G. S. FISHER, W. G. BICKFORD AND F. G. DOLLEAR, *J. Am. Oil Chem. Soc.*, 24 (1947) 379.
26   K. G. RAGHUVEER AND E. G. HAMMOND, *J. Am. Oil Chem. Soc.*, 44 (1967) 239.
27   F. H. MATTSON AND R. A. VOLPENHEIN, *J. Biol. Chem.*, 236 (1961) 1891.
28   L.-Å. APPELQVIST AND R. J. DOWDELL, *Arkiv Kemi*, 28 (1968) 539.
29   L.-Å. APPELQVIST, *Acta Univ. Lund, Sectio II*, No. 7 (1968).
30   E. N. FRANKEL, P. M. COONEY, H. A. MOSER, J. C. COWAN AND C. D. EVANS, *Fette, Seifen, Anstrichmittel*, 61 (1959) 1036.
31   U. HOLM, *Rapport fra 3. Nordiska Fettharskningssymposium*, Oslo, 1962, p.94 (in Swedish); *C.A.* 59 (1963) 5696 b.
32   U. HOLM, *Livsmedelsteknik*, 6 (4) (1964) 146. In Swedish.
33   I.U.P.A.C., *Standard Methods for the Analysis of Oils, Fats and Soaps*, Butterworths Scientific Publications, London, 1966, II, D, 15.
34   C. PIETRZYK, *Roczniki Technol. Chem. Zywnosci*, 9 (1962) 29. In Polish; *CA* 61 (1964) 7242 h.
35   C. PIETRZYK, *Roczniki Technol. Chem. Zywnosci*, 9 (1962) 81. In Polish; *CA* 61 (1964) 8535 a.
36   B. A. J. SEDLÁČEK, *Nahrung*, 12 (1968) 721.
37   B. A. J. SEDLÁČEK, *Nahrung*, 12 (1968) 727.
38   B. A. J. SEDLÁČEK, *Fette, Seifen, Anstrichmittel*, 70 (1968) 795.
39   Z. GOLUCKI, *Acta Polon. Pharm.*, 22 (1965) 255. In Polish; *C.A.* 63 (1965) 18497 d.
40   T. E. FURIA, Ed., *Handbook of Food Additives*, The Chemical Rubber Co., Ohio, 1968, p.209.
41   N. M. EMANUEL AND YU. N. LYASKOVSKAYA, *The Inhibition of Fat Oxidation Processes*, Pergamon Press, Oxford, 1967, p.276.
42   H. J. LIPS, A. L. PROMISLOW AND N. H. GRACE, *J. Am. Oil Chem. Soc.*, 30 (1953) 213.
43   R. OHLSON in A. RUTKOWSKI (Ed.), *International Symposium for the Chemistry and Technology of Rapeseed Oil and other Cruciferae Oils, Warsaw, 1970*, p.279.
44   H. P. KAUFMANN AND S. FUNKE, *Fette und Seifen*, 45 (1938) 255.
45   H. P. KAUFMANN, *Analyse der Fette und Fettprodukte, I:Allgemeiner Teil*, Springer-Verlag, Berlin, 1958, p.925.
46   C. WALTHER, *Erdöl Teer*, 7 (1931) 382.
47   J. KENDALL AND K. P. MONROE, *J. Am. Chem. Soc.*, 39 (1917) 1802.
48   H. SIEBENECK, *Viskosimetrische Tabellen und Tafeln*, Naturwissenschaftlicher Verlag Borntraeger, Berlin-Nikolassee, 1951.
49   S. A. TREMAZI, N. V. LOVEGREN AND R. O. FEUGE, *J. Am. Oil Chem. Soc.*, 42 (1965) 78.
50   M. G. R. CARTER AND T. MALKIN, *J. Chem. Soc.*, (1947), 554.
51   U. RIINER, *J. Am. Oil Chem. Soc.*, 47 (1970) 129.
52   U. RIINER AND K.-A. MELIN in A. RUTKOWSKI (Ed.), *International Symposium for the Chemistry and Technology of Rapeseed Oil and other Cruciferae Oils, Warsaw, 1970*, p.233.
53   H. J. LIPS, N. H. GRACE AND E. M. HAMILTON, *Can. J. Research*, 26F (1948) 360.
54   E. LARMOND, *Proc. Second Annual Meeting*, Rapeseed Assoc. in Canada, Saskatoon, 1969, p.63.
55   M. VAISEY AND K. SHAYKEWICH, *Rapeseed Oil Facts and Recipes*, Rapeseed Assoc. of Canada, Publication No. 6, 1969, p.8.
56   J. HRDLICKA AND J. POKORNY, *Prumysl Potravin*, 15 (1964) 400. In Polish; *C.A.* 64 (1966) 14871g.
57   R. K. DOWNEY, B. M. CRAIG AND C. G. YOUNGS, *J. Am. Oil Chem. Soc.*, 46 (1969) 121.
58   B. F. TEASDALE, Canada Packers Ltd, *Can. Patent* 726, 140 (1966).
59   Canada Packers Ltd., *Brit. Patent* 1,138,576 (1969).
60   The Procter & Gamble Co., *Brit. Patent* 1,138,092 (1968).
61   C. W. HOERR, *J. Am. Oil Chem. Soc.*, 37 (1960) 539.

62 *Fats and Oils in Canada, Semi Annual Review*, Department of Industry, Ottawa, 2 (No. 1) (1967) 84.

63 B. F. Teasdale in A. Rutkowski (Ed.), *International Symposium for the Chemistry and Technology of Rapeseed Oil and other Cruciferae Oils, Warsaw, 1970*, p.243.

64 S. Zalewski and F. A. Kummerow, *J. Am. Oil Chem. Soc.*, 45 (1968) 87.

65 D. R. Merker, L. C. Brown and L. H. Wiedermann, *J. Am. Oil Chem. Soc.*, 35 (1958) 130.

66 S. Rudischer, *Prumysl Potravin*, 15 (1964) 401. In Polish; *C.A.* 61 (1964) 15265g.

67 P. Seiden, The Procter & Gamble Co., *U.S. Patent* 3,353,964 (1967).

68 C. Kaczanowski and A. Grecki, *Tluszcze Jadalne*, 11 (1967) 209. In Polish; *C.A.* 68 (1968) 70408s.

69 C. Kaczanowski and A. Jakubowski, *Tluszcze i Srodki Piorace*, 6 (1962) 1. In Polish; *C.A.* 59 (1963) 14204h.

70 C. Kaczanowski and A. Jakubowski, *Tluszcze i Srodki Piorace*, 7 (1963) 65. In Polish; *C.A.* 61 (1964) 849c.

71 P. Seiden, The Procter & Gamble Co., *U.S. Patent* 3,298,837 (1967).

72 The Procter & Gamble Co., *Brit. Patent* 1,151,677 (1969).

73 *Rapeseed Dig.*, 3 (2) (1969) 4.

74 D. G. Caldwell, *Proc. Second Annual Meeting*, Rapeseed Assoc. in Canada, Saskatoon, 1969, p.82.

75 *Vegetable Oils and Oilseed, A Review*, Commonwealth Secretariat, London, No. 18, 1968.

76 H. J. Lips, N. H. Grace and S. Jegard, *Can. J. Research*, 27F (1949) 28.

77 N. H. Grace, H. J. Lips and A. Zuckerman, *Can. J. Research*, 28F (1950) 401.

78 The Procter & Gamble Co., *Neth. Pat.* 6,501,447 (1966); *C.A.* 65 (1966) 19232c.

79 C. O'Sullivan, *Proc. Symp. Fats and Oils Situation in Canada*, National Research Council of Canada, Ottawa, 1967, p.51.

80 L. L. Linteris and S. W. Thompson, *J. Am. Oil Chem. Soc.*, 35 (1958) 28.

81 N. Matsui, T. Tomita and T. Kawada, Kao Soap Co. Ltd., *Brit. Patent* 1,135,417 (1968).

82 I. Prokornyi, E. Maresh and A. Makhanichek, *Maslob.-Zhir. Prom.*, 24 (1958) 17. In Polish; *C.A.* 53 (1959) 3737a.

83 Unilever N.V., Neth. Pat. 6,810,982 (1968).

84 W. Bedenk, Procter & Gamble Co., *U.S. Patent* 3,037,864 (1962); *C.A.* 57 (1962) 7696a.

Non-Nutritional Uses of Rapeseed Oil and Rapeseed Fatty Acids

R. OHLSON

AB Karlshamns Oljefabriker, Karlshamn (Sweden)

CONTENTS

Rapeseed oil is primarily used as food, as shown in chapter 11, but it has many applications for other than nutritional purposes. Rapeseed oil, rapeseed fatty acids, and their derivatives are employed in these connections, as will be described in this chapter.

12.1 RAPESEED OIL

The specific uses of rapeseed oil are based on its content of long-chain molecules or molecules with double bonds. In the past rapeseed oil was used for illumination purposes, and mustard seed oil has been used as a fuel oil.

The rapeseed oil used for industrial products is usually refined. Certain modifications in properties can also be achieved, *e.g.* by blowing. This process has long been practised to increase viscosity and consists of passing a stream of air through the heated oil[1]. A blown rapeseed oil is appreciably soluble in paraffins, and the solubility varies inversely with the initial iodine value as well as with the extent of blowing. The alcohol solubility of a blown oil tends to be low and decreases with increased blowing time.

12.1.1 Rubber additives (factice)

This is probably the principal industrial use of rapeseed oil in the U.S.A. at present, and rapeseed oil is the classical oil for the production of factice in *e.g.* England[2], Germany[3], and Sweden.

Factice, originally introduced as a rubber substitute, is now frequently compounded with rubber. It is used in the rubber industry to increase stability towards ageing and changes in shape. Rapeseed oil and sulphur are reacted together to form a polymer in the same way as in the vulcanization process of rubber.

There are two varieties of factice: white and brown. White factice is prepared by reacting a relatively saturated oil, such as rapeseed oil, with liquid sulphur monochloride. The reaction takes place readily in the cold and is quite complex[2, 4-6]. There is no particular relation between the unsaturation of the oil and the amount of reagent absorbed. Substitution also appears to take place. White factice is a light-coloured, compressible, but more or less crumbly material, which has been used as an extender or modifier for rubber. An important use of this material is in the manufacture of erasers, where it confers the degree of friability essential in this product.

Brown factice is made[6, 7] by first blowing and then bodying a drying oil until thickened, followed by reaction with about 5–30 weight per cent sulphur in a closed vessel, with stirring at a temperature of about 120–175 °C for 1–2 hours.

Depending upon the percentage of sulphur, the temperature, and the reaction time, the product varies in consistency from a dark, viscous, and sticky semisolid to a hard and relatively brittle solid. The characteristics desired depend upon the specific use to which the factice is to be put. The chemical reactions that take place in the manufacture of brown factice are undoubtedly even more complex than those involved in making white factice, but again they may be compared to those which occur in the vulcanization of rubber.

Brown factice has been used not only as a rubber extender but also to modify the properties of drying oil products, such as varnishes and linoleum.

Factice-like products, with excellent flameproofing properties, were prepared by polymerizing rapeseed oil and treating the product with phosphorous halides[8]. The products could be used as additives for rubber, lacquers, varnishes, plastics, adhesives, and lubricants.

This is a traditional use of rapeseed oil, but although it has very favourable lubricating properties, current usage of rapeseed oil solely for this purpose seems to be rather limited. Changes in industrial techniques and the availability of synthetic derivatives have caused this decline. For instance, rapeseed oil was employed to lubricate steam locomotives, but other mixtures are now used[9].

An exclusive use of fatty oils is exceptional to-day, but compounded oils of fatty and mineral oils are commonly used. Actually the fatty oil orients itself on the metal surfaces and the mineral oil is sandwiched between the layers. The coefficient of friction of fatty oils is generally lower than that of mineral oils. Mixing of mineral oil, vegetable oil and fatty acids reduces also undesirable features, such as the tendency of vegetable oil to become gummy and to carbonize. In blends with mineral oils rapeseed oil is used as a cutting oil[14], because such blends have a higher dipole moment than either of the single components. There is probably a relation between this phenomenon and the prolonged tool life obtained with the use of rapeseed oil blends[14]. In lubricants for things that are shock loaded, *e.g.* locomotive axles, rapeseed oil also has special advantages as the viscosity of the oil can be increased to take care of the maximum load. Marine reciprocating engine bearings are usually lubricated with oils containing up to 20% blown rapeseed oil.

Working and forming of metals is generally divided into two main parts: the plastic deformation of the metal, and the frictional effects at the surface of the work as it passes the tool or die. Friction is undesirable not only because at the

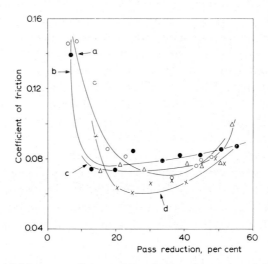

Fig.12:1. Variation of friction coefficient with reduction for various lubricants. a, Rapeseed oil; b, 70% rapeseed oil and 30% tallow; c, 60% rapeseed oil and 40% tallow; d, 50% rapeseed oil and 50% tallow.

extra energy that must be expended in overcoming it, but even more so because of the heavy working at the surface layers which in turn involves the danger of fatigue and surface fragmentation. Metal working compounds can be classified into three major types, according to the use to which they are put: cleaning and pickling, lubricating, and surface conversion treatment. In the metal working industry rapeseed oil-factice is used for slow processes and cutting oils of emulsion type with sulphated oil or soap for rapid processes.

Quenching of metals is the process of withdrawing heat from metal rapidly to obtain a required structure. Rapeseed oil is used as an additive in quenching oil[10], because it increases the cooling rate.

Rapeseed oil is used in mixtures with beef tallow to give an excellent cold rolling lubricant[11, 12]. The coefficient of friction varies with pass reduction (see Fig.12:1).

The high coefficient of friction at low reductions is attributed to a predominance of boundary lubrication, and the increasing efficiency of hydrodynamic lubrication with increasing pass reduction presumably accounts for the decrease in friction coefficient. As the load increases beyond the load carrying capacity of the hydrodynamic oil wedge, the efficiency of lubrication decreases and finally breaks down and, as a result, friction increases.

The steel corrosion rate in relation to the level of free fatty acids in refined rapeseed oil has been measured[13].

Rapeseed oil is as useful as ordinary commercial core oils for making castings[15]. Core oils are used as binding agents for the sand cores of hollow metal castings. The cores are prepared by mixing approximately 50 parts by volume of sand with 1 part of oil, molding the mixture in a wooden form, and baking at a temperature of about 200° to 230 °C until a hard, coherent mass is formed. The sand is usually dampened to make it more easily molded, and a water-soluble binder, such as casein or dextrin, is sometimes added to assist in maintaining the core in the proper form until it becomes hardened through heat polymerization of the oil[16].

The oils used in core making are usually bodied to different viscosities and are used without dryers.

The process of continuous casting is in a unique position in steel-making from the standpoint of lubricants. Without dependable and continuous lubrication, the process either slows down or comes to a complete halt. The prevention of ruptures or cracks in the solidified skin which forms in the mold is extremely important for the process, and proper lubrication is an important factor in the formation of the skin. The most widely used lubricant for this purpose is rapeseed oil[17, 78, 79]. About 5 oz. of oil are needed per ton of steel for adequate mould lubrication.

In the manufacture of additives for lubricants, rapeseed oil is preferably used in the form of sodium, lithium and calcium soaps[18], and patents are reported for nonfoaming lubrication grease[19], heat resistant bearing lubricants[20], antifriction bearing grease[21], heavy-duty gear lubricants[22], and for the improvement of the properties of these lubricants[23].

12.1.3 Fat-liquoring of leather in tanneries

Leather, the product of cattle hide, is processed either by vegetable tanning or by chrome tanning. Rapeseed oil finds use in the fat-liquoring step that always follows chrome tanning, primarily in the Eastern parts of Europe.

The purpose of the fat-liquoring of leather is to lubricate the fibers of the material to give it strength and flexibility. Actually this step is performed to restore oils that were removed during the tanning process. The fat-liquoring solution is composed of a mixture of the crude oil and a soap or sulphated oil[24-26]. The latter acts as the carrying agent for the crude oil, and the mixture is absorbed into the leather. The oil may also be oxidized before sulphonation and mixed with nonoxidized oils before or after[27]. The sulphonation is accomplished with 25 per cent sulphuric acid at 15–25 °C, and the resulting sulphoesters are washed with a 20 per cent solution of sodium sulphate[28].

12.1.4 Varnishes and lacquers

Rapeseed oil has been classed as semi-drying but it has too low an iodine value to be used directly. However, derivatives of rapeseed oil[29, 30] or its fatty acids[31] may be very suitable for this purpose.

Rapeseed oil has been used in specialized inks. An embossed effect with a transfer ink is obtained by the incorporation of 13 per cent blown rapeseed oil, which acts as a drying oil, and forms a film on the surface thereby preventing excessive penetration of the ink[32].

Blown rapeseed oil has also been recommended in processes for printing with a glossy finish on leather[33]. The blown oil is recommended as a wetting agent and as a vehicle for emulsion inks[33]. It is often used in the formulation of gold and silver sizes to impart the desired tack and adhesion[34]. In the preparation of metallic pigments, rapeseed oil behaves as a lubricant and prevents welding of the powdered metal during preparation[33]. Crude rapeseed oil has the peculiar property of bringing out the 'reflex' or bronzy cast of the iron blues in inks[34].

12.1.5 Textile chemicals

Chemicals based on rapeseed oil are employed for three general purposes in the textile industry: scouring, washing and dyeing.

Scouring. In the scouring of wool or worsteds, a solution of a rapeseed fatty acid amine condensate, mixed with various glycols and water, is employed to remove the oil and grease present in wool. A similar solution is used to remove the warp size gums and oils present in raw Greige goods, leaving a clean fabric for subsequent dyeing.

Washing. A solution containing a rapeseed oil fatty acid amine condensate is used to wash a fabric before and after printing. Prior to this operation, the fabric is washed to reduce cracking and bleeding of the print to be applied. After printing, the treated fabric is washed again to eliminate the gums and thickeners employed in connection with the printing, in order to obtain a soft, pliable fabric.

Dyeing. The amine condensate is also used as a dyeing assistant. It reduces, and most often completely eliminates, the formation of surface scum during the dyeing operation. The goods are left soft and conditioned for additional processing. Sulphonated erucic acid oils have been widely used in the dyeing of textiles[36].

Other uses in the textile industry for chemicals based on rapeseed oil are as finishing agents (soaps)[35] and for greasing wool (felting)[35].

12.1.6 Detergent additives

Sulphated rapeseed oil alcohols (18–22 carbon atoms) have shown very good foam[38] and detergent[39] properties, and are used for the preparation of washing compounds. The surface tension of a mixture of oleyl and erucyl alcohol sulphates in water is shown in Fig.12:2.

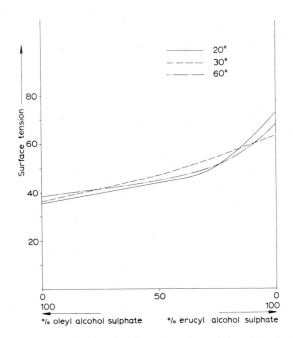

Fig.12:2. Surface tension of water solutions as a function of the ratio oleyl alcohol sulphate/ erucyl alcohol sulphate at 20°, 30° and 60°C.

Epoxidized fatty acid triglycerides are used as plasticizers and stabilizers in *e.g.* PVC and co-polymers of PVC and vinyl acetate, polyvinylidene chloride and chlorinated rubbers. They are also used for extension and modification of epoxy resins[40]. An epoxidized oil is used as plasticizer–stabilizer together with a primary plasticizer, *e.g.* dioctyl phthalate (DOP) and a metal stabilizer. Normally the amount of epoxy plasticizer used is 5–10 parts per hundred parts of resin for stabilizing and 15–60 parts as softener. The epoxy plasticizers have a stabilizing effect against heat and light and they also increase the effect of the primary stabilizer. The epoxy compound has a function in reacting with the hydrochloric acid from the decomposing PVC to give a chlorohydrin. The metal stabilizer then reacts with the chlorohydrin to give a metal chloride and a new epoxy ring[41].

A larger number of epoxy groups increases the solubility in PVC but cold resistance is decreased. A compound with an epoxy group in the center of the molecule has better properties than one having a terminal epoxy group.

A rapeseed oil with a high content of polyunsaturated fatty acids can easily be epoxidized and used as mentioned above.

12.1.8 Other reported applications

Rapeseed oil has been employed to protect neoprene against heat-aging[42] and as slip agent for paper tubes.

The addition of rapeseed oil factice instead of sulphur during calendering of elastic packing material produces a material with decreased tendency to stickiness[44].

Production of surface-active agents from rapeseed oil which have special applications in flotation have been described[37].

A plastic gasket compound with improved tensile strength and elasticity at low temperatures is prepared from polybutadiene, rapeseed oil, chalk, and asbestos or mineral or textile fibers[45].

Opaque iodinated oils have found use in röntgenographic exploration of the internal organs. Iodinated rapeseed oil has proved to be a highly satisfactory preparation which was tolerated by all types of tissues[46].

Rapeseed oil has been used to stabilize fungicidal and insecticidal suspensions in paraffin oils[43]. A fairly new use for rapeseed oil is as a spray for weed control instead of commonly used petroleum oils. One major advantage of using vegetable oils in postemergence weed sprays is that by themselves they offer little hazard of crop injury or harmful residues.

12.2 RAPESEED FATTY ACIDS

In some of the above mentioned technical uses it is possible to use the rapeseed oil fatty acids as raw materials instead of the original oil. No specific uses of the combined rapeseed oil fatty acids are reported, but industry requires large amounts of fatty acids derived from agricultural fats and oils in the preparation of the following products[36]:

soaps	surface active compounds
lubricating agents	polishes
alkyd resins	buffing compounds
candles	rubber goods

Sodium and aluminium salts of rapeseed oil fatty acids have been used as thickening agents for liquid fuels[49].

A coating for vinyl polymer films consisting of a sodium or potassium salt of a fatty acid with 12–22 carbon atoms per molecule improves the heat seal strength of the films[50].

Since the fatty acid chlorides are useful as intermediates in *e.g.* preparation of esters, a German patent[51] in this field should be referred to. Erucyl chloride can be condensed with aromatic hydrocarbons, cyclic terpenes, aniline, phenols, and nitrated aromatics. The products are used as pour point improvers for machine oils, gear oils and other lubricants.

The preparation of sulphated and hydroxy acids from rapeseed oil acids is described by Grace and Zuckermann[47]. A good description of the general possible reactions and uses of fatty acids is given by Markley[52] involving salt formation, esterification, dehydration, polymerization, halogenation (see also ref. 32), hydrogenation, and oxidation. The physical properties of fatty acid methyl esters are given by Gouw *et al.*[53]. Because of the high erucic acid content of rapeseed oil fatty acids, they offer special possibilities, as shown in the following.

12.3 ERUCIC ACID AND ITS CLEAVAGE PRODUCTS

On account of its special properties (long chain – high percentage of carbon), erucic acid (*cis*-13-docosenoic acid) is used *e.g.* as sodium salt in hot water detergents, and in water-repelling agents[54], and the corresponding amine as corrosion preventative. Erucic acid can easily react with *N*-bromosuccinimid to yield bromoacetoxy behenic acid[55]. The epoxide and other derivatives have also been prepared[56].

The production of hard waxes, suitable for the manufacture of shoe creams and floor polishes, by the esterification of 'behenic acid' with ethylene glycol, has been reported[57].

12.3.1 Erucyl amide

Certain long-chain fatty amides of erucic acid and rapeseed oil fatty acids have proved to be good plasticizers for vinyl chloride resins[58]. Physical characteristics of vinyl chloride–vinyl acetate copolymer plasticized with *N,N*-disubstituted amides of erucic acid are reported[59]. Erucyl amide is also used in polyethylenes and polypropylenes to solve intricate surface characteristic problems, such as antiblock slip, anti-static susceptibility to heat scaling and the ability to be formed or molded[80]. The plasticizer evaluation data show that erucic acid derivatives in general provide more desirable characteristics in the plastic stock than an equal concentration of DOP control plasticizer. This superiority is most apparent in low-temperature performance and in characteristics such as brittle point and volatility losses. Extractability of all the erucic acid derivatives in soapy water is rather high but decidedly inferior to that of DOP.

12.3.2 Products of ozonolysis

The use of ozone to effect oxidative cleavage of an unsaturated fatty acid is common. Ozonolysis of erucic (*cis*-13-docosenoic) acid yields pelargonic and brassylic acids:

$$CH_3(CH_2)_7CH{=}CH(CH_2)_{11}COOH + O_3 \rightarrow CH_3(CH_2)_7COOH +$$

$$\text{erucic acid} \qquad\qquad \text{ozone} \qquad \text{pelargonic acid}$$

$$+ HOOC(CH_2)_{11}COOH$$

$$\text{brassylic acid}$$

(see also ref. 60)

There is a great demand from industry for both of these products.

(a) Pelargonic acid

This acid is already industrially available as a cleavage product from the ozonolysis of oleic acid. Widespread use of this 9-carbon monobasic acid is reported in the following fields: plasticizers, alkyd resins, vinyl stabilizers, hydrotropic salts, pharmaceuticals, synthetic flavours and odours, flotation agents, insect repellants, and jet-engine lubricants[56, 61]. Pelargonic acid is applied primarily in the form of esters, amine condensates and metallic soaps.

Esters. The compatibility of certain pelargonic esters with a wide range of synthetic resins is much greater than that of esters of the higher molecular weight acids. Reaction with polyfunctional alcohols, branched chain 'oxo' and other alcohols give esters that possess properties required in superior vinyl and synthetic rubber plasticizers, and for fluids that are suitable for the rapidly expanding field of synthetic lubricants.

Amine condensates. Amine condensates can be made by reacting pelargonic acid with low molecular weight amines such as monoethanolamine or isopropanol-amine. Such condensates are characterized by good water solubility, which can be extended to blends made with other amine condensates. Thus, gelation tendencies of other low molecular weight amine condensates can be reduced by partial replacement with the pelargonic counterpart.

Metallic Soaps. The metallic soaps of pelargonic acid may be made by direct fusion or by decomposition. The latter method is generally preferred, giving lighter coloured end products.

The mixed barium–cadmium soaps of pelargonic acid have been found to be very effective as stabilizers for polyvinyl chloride resins, particularly when used in conjunction with a chelating type stabilizer. The use of this mixed soap has been found to be more effective than the customary barium–cadmium laurate, giving plasticized PVC resins with better resistance to ultra-violet and heat degradation.

Alkali metal salts of pelargonic acid are all very soluble in water, but are insoluble in most organic solvents, even with an excess of the fatty acid. Detergency of the short chain length salts is quite limited. The calcium salts and those of the heavy metals are insoluble in water.

Alkyd resins based on pelargonic acid have widespread use in the USA, especially in the surface coating of hardboard. Such coatings exhibit excellent colour retention, improved finish as well as improved resistance against stains.

(b) Brassylic acid

Brassylic acid (1,13-tridecanedioic acid) is a longer straight chain aliphatic dicarboxylic acid than those hitherto readily obtainable in quantity. This product is not yet available on an industrial scale, but Nieschlag et al.[62] have reported results on the continuous ozonolysis of erucic acid on a pilot-plant scale.

Related dicarboxylic acids, such as azelaic acid and sebacic acid, are used as plasticizers. (In 1959, 3650 tons of azelates and 4550 tons of sebacates were used for this purpose.) Nieschlag et al.[63] have shown that selected esters of brassylic acid are excellent low-temperature plasticizers for polyvinyl chloride with exceptional light stability. The reactivity ratio for copolymerization of vinyl chloride and 2-methylpentyl vinyl brassylate has been determined according to Chang et al.[64]. The availability of brassylic acid will expand the range of uses attainable with this important class of compounds[54].

Esters of brassylic acid are also reported to be suitable as lubricants over a wide range of temperature[65].

The macrocyclic ester ethylene brassylate has been prepared from dimethyl brassylate for use as a synthetic musk in the fixing of perfumes[66].

Another intriguing possibility is the production of new nylons from brassylic acid[67]. 13-Aminotridecanoic acid, m.p. 178 °C, the monomer for nylon 13, has been synthesized by several methods. Baruch, who first reported 13-aminotridecanoic acid, obtained it from behenolic acid[68]. Müller and Krauss[69] prepared this amino acid from 1,12-dibromododecane. Pied[70] synthesized a very pure sample of the amino acid from 11-bromoundecanoic acid. Neither of these synthetic methods utilized readily accessible starting materials. Green et al. reported[71] an extensive study of preparative routes to 13-aminotridecanoic acid

WATER ABSORPTION OF NYLONS-13[74]

Nylon	% of water absorbed at a relative humidity of	
	30%	65%
6,13-	0.66	1.41
7,13-	0.79	1.47
8,13-	0.71	1.33
9,13-	0.65	1.24
10,13-	0.63	1.40
11,13-	0.74	1.08
12,13-	0.53	0.74
13,13-	0.42	0.68

TABLE 12:2

MELTING POINTS FOR NYLONS FROM 1,13-TRIDECANEDIOIC ACID

Nylon	m.p., °C	Ref.
13	182–183	70
	180	72
5,13	176	73
6,13	199–208	74
	210	72
7,13	186–196	74
8,13	188–196	74
9,13	183	75
10,13	170	73
10,13	178–186	74
11,13	168–178	74
12,13	167–178	74
13,13	164–167	77
13,13	174	72

from erucic acid, methyl erucate or eruconitrile, all of which are plentiful chemical products.

The structures of nylon 13 and nylon 1313 indicate that they should be tough and resistant to abrasion. Due to their low water absorption (Table 12:1)[74] they should resist water better than do nylons 6 and 66, and they melt at lower temperatures (Table 12:2). The low melting points enable nylon 13 to be treated at lower temperatures than normal nylons, and this is a great advantage as nylons are very sensitive to oxidation in air. They should also be easier to mold or extrude to form rods, films and filaments.

In making nylon 13, erucic acid can be cleaved at its double bond by ozonolysis, followed by reductive amination to the *omega* acid, which is then polymerized by amide condensation. To make nylon 1313, erucic acid is cleaved by ozonolysis, followed by hydrolysis, to brassylic acid. Then part of the brassylic acid is converted to a diamine,

$$HOOC-(CH_2)_{11}-COOH \xrightarrow{NH_3} NH_4OCO-(CH_2)_{11}-COONH_4 \xrightarrow{4H_2O}$$

$$NC-(CH_2)_{11}-CN \xrightarrow{reduction} H_2N-(CH_2)_{13}-NH_2$$

and the acid and diamine are polymerized by condensation. Table 12:3 shows the physical properties of nylon 13 and 1313[72].

Other potential uses of brassylic acids include: polyesters, alkyd resins, lubricants, rubber additives, and surface active agents[61]. If brassylic acid with a purity of 99 per cent were available, more than 2,000 tons a year could be sold for testing in new nylon-type fibers with special properties not obtainable with the presently

TABLE 12:3

PHYSICAL PROPERTIES OF NYLON[72]

Property	Nylon-13	Nylon-13/13
Melting point, °C	180	174
Melt viscosity, 232°C, poises	550	500
Specific gravity	1.01	1.01
Tensile strength, p.s.i.	5,550	5,700
Tensile yield strength, p.s.i.	4,780	4,660
Elongation at break, %	130	130
Tensile modulus of elasticity, 10^3 p.s.i.	113	114
Flexural yield strength, p.s.i.	8,660	10,200
Tangent modulus of elasticity, 10^3 p.s.i.	151	176
Impact resistance, ft.-lb./in.	2.1	2.6
Hardness, Rockwell M	34	40
Hardness, Shore D	74	72
Torsional modulus, 10^3 p.s.i.	534	664
Compression stress at 1% deformation, p.s.i.	1,000	1,280
Compressive modulus of elasticity, 10^3 p.s.i.	100	128
Coefficient of expansion, 10^{-6} in./in./°C.	118	103
Deformation under load, %	1.6	1.0
Recovery, %	98.7	99.6
Heat deflection temperature, 264 p.s.i., °C	51	53
Water absorption, 23°C., %	1.04	0.75
Moisture regain, 50% relative humidity, 23°C., %	0.36	0.29
Abrasion resistance mg/1000 cycles	184	213
Coefficient of friction		
Static	0.40	0.39
Kinetic	0.30	0.25
Energy of rupture, ft.-lb./in.^{2}	5,280	6,060

available dibasic acids[61]. It should also be mentioned that in some cases mixtures of mono- and dibasic acids may be used directly in the production of polyester plasticizers without separation.

12.4 PITCH

The pitch from fatty acid distillation can be used in the lubrication industry[76].

REFERENCES

1   N. H. Grace and A. Zuckermann, *Can. J. Technol.*, 29 (1951) 71.
2   C. F. Flint, C. B. Featherstone and J. Donelly, *Trans. Inst. Rubber Ind.*, 33 (1957) 181.
3   A. H. Clark, C. F. Flint and D. L. McGee, *Trans. Inst. Rubber Ind.*, 37 (1961) 193.
4   F. Kirschof, *Kautschuk Gummi*, 16 (1963) 201, 431.
5   M. N. Baliga and T. P. Hilditch, *J. Soc. Chem. Ind.*, 67 (1948) 258.
6   F. Kirschof, *Kautschuk Gummi*, 6 (1953) 30.
7   H. J. Stern, *Rubber: Natural and Synthetic*, p.100.
8   W. Alexander, in Hauser: *Handbuch der gesamten Kautschuktechnologie*, Vol. I, 1934. Reprinted in Ann Arbor, 1948. p.389.
9   Henkel & Cie G.m.b.h., *Brit. Patent* 1,132,792 (1967).
10   G. H. Hurst, *Lubricating Oils, Fats and Greases*, Scott Greenwood and Son, London 1911.
11   T. Watanabe, *Japanese Patent* 15,454 (1962).
12   J. Billigman and W. Fichtl, *Stahl Eisen*, 78 (1958) 344.
13   I. leMay, K. D. Nair and M. L. J. Hoffman, *Lubrication Eng.*, 23 (1967) 415.
14   A. Rutkowski, K. Babuchowski and J. Batura, *Tluszcze, Srodki Piorace, Kosmet*, 12 (1968) 123.
15   E. H. Kadmer, *Oel u. Kohle*, 39 (1940) 836.
16   Bureau of Mines, Mineral Dressing and Met. Lab. Ottawa Invest., No. 2477 (1948).
17   J. J. Obrzut, *Iron Age*, (1968) July 25, p.105.
18   I. S. Evans, J. D. Pardo and M. P. Nicholaichuk, *Natl. Lubricating Grease Inst. Spokesman*, (1962) 146.
19   A. J. Morway, *U.S. Patent*, 2,825,692 (1958).
20   A. J. Morway, *U.S. Patent*, 2,825,694 (1958).
21   A. J. Morway, *U.S. Patent*, 2,872,416 (1959).
22   A. J. Morway and E. W. Ball, *U.S. Patent*, 2,937,144 (1960).
23   P. R. McCarthy and T. R. Orem, *U.S. Patent*, 2,844,537 (1958).
24   F. Kolos and L. Wiedner, *Bör-es Cipotech.*, 5 (1955) 81.
25   V. Pilc, *Veda Vyzkum Prumyslu Kozedelnem*, 2 (1957) 59.
26   B. Rewald, *J. Soc. Chem. Ind.*, 56 (1937) 403.
27   V. Pilc, M. Skrabal and J. Ondracek, *Czech. Patent* 94,914 (1960).
28   V. Pilc, *Kozarstvi*, 11 (1961) 322.
29   H. Sallinger, *German Patent*, 1,010,678 (1957).
30   A. Uzzan, Y. Laushart and C. Triziz, *Rev. Franc. Corps. Gras*, 6 (1959) 416.
31   J. D. von Mikusch-Buckberg, M. R. Mills and K. H. Mebes (Unilever N.V.), *Swedish Patent*, 190,652.
32   C. Ellis, *Printing Inks*, Reinhold Publ. Co., New York, 1940.
33   B. L. T. Farkeas, *U.S. Patent* 1,439,623 (1922).
34   C. H. J. Wolfe, *Manufacture of Printing and Lithograph Ink*, McNair-Dorland, New York (1933).
35   J. Boer, C. P. Jansen and A. Kentie, *J. Nutr.*, 33 (1947) 339.

36 A. ZUCKERMAN AND N. H. GRACE, *Can. Chem. Process.*, 33 (1949) 588.
37 W. FINDLEY AND B. H. ROBIN (Swift & Co.), *U.S. Patent* 2,993,919 (1961).
38 E. KÖNIG, P. HAHN, K. STICKDORN AND H. G. BRAUN (VEB Deutsches Hydrierwerk Rodleben), *German Patent* 1,274,118 (1968).
39 B. E. SZMIDTGAL AND H. PASTERNAK, *Tluszcze, Srodki Piorace, Kosmet.*, 12 (1968) 202.
40 R. B. GRAVER, *J. Paint Technol.*, 39 (1967) 71.
41 E. W. LINES, *Plastics Technol.*, March 1961, p.51.
42 K. L. SELIGMAN AND P. A. ROUSSEL, *Rubber Chem. Technol.*, 34 (1961) 869.
43 L. K. O. COX AND N. W. CRITCHLOW, (Esso Research & Engineering Comp.), *Brit. Patent* 860,942 (1961).
44 W. KIRCH AND W. GLANDER (Chemieprodukte G.m.b.H.), *German Patent* 1,007,056 (1957).
45 E. GRÖBEL, H. RITTER AND K. H. KAUERT, *German (East) Patent*, 60,106 (1968).
46 M. A. GILASER AND G. W. RAIZISS, *Radiology*, 20 (1933) 471.
47 N. H. GRACE AND A. ZUCKERMANN, *Can. J. Technol.*, 29 (1951) 276.
48 I. GREEN, S. MARKINKIEWICZ AND P. R. WATT, *J. Sci. Food Agr.*, 6 (1955) 274.
49 A. BEREZNIAK, *Biul. Wojskowej Akad. Tech.*, 6 (1957) 32, p.21; *C.A.* 53 (1959) 4711.
50 F. E. EASTES (Grace & Co) *U.S. Patent* 3,419,421 (1968).
51 B. BLASER AND H. WEDELL (Henkel & Cie G.m.b.H.) *German Patent* 961,531.
52 K. S. MARKLEY (Ed.), *Fatty Acids-Their Chemistry, Properties, Production and Uses*, 2nd Ed., Interscience Publ., New York, 1960–67.
53 T. H. GOUW AND J. C. VLUGTER, *J. Am. Oil Chem. Soc.*, 41 (1964) 142.
54 U.S. Dept. Agr. Crops Res. Div., ARS-34, Sep. 1962.
55 A. JOVTSCHEFF, *Chem. Ber.*, 95 (1962) 2629.
56 I. L. KURANOVA, Y. D. SHENIN AND G. V. PIGULEVSKII, *Ztr. Obshch. Khim.*, 33 (1963) 2988.
57 K. TUMA AND A. KORBELYI, *Czech. Patent* 89,576 (1959).
58 R. R. MOD, F. C. MAGNE, E. L. SKAU, *J. Am. Oil Chem. Soc.*, 45 (1968) 385.
59 R. R. MOD, F. C. MAGNE, E. L. SKAU, H. J. NIESCHLAG, W. H. TALLENT AND I. A. WOLFF, *Ind. Eng. Chem. Prod. Res. Develop.*, 8 (1969) 176.
60 S. OHARA AND Y. SHINOZAKI, *Yukagaku*, 5 (1956) 222.
61 J. H. BRUUN AND J. R. MATCHETT, *J. Am. Oil Chem. Soc.*, 40 (1963) 1.
62 H. J. NIESCHLAG, I. A. WOLFF, T. C. MANLEY AND R. J. HOLLAND, *Ind. Eng. Chem. Prod. Res. Develop.*, 6 (1967) 120.
63 H. J. NIESCHLAG, J. W. HAGEMANN, I. A. WOLFF, W. E. PALM AND L. P. WITNAUER, *Ind. Eng. Chem. Prod. Res. Develop.*, 3 (1964) 146.
64 S. CHANG, T. K. MIWA AND W. H. TALLENT, *J. Polymer Sci.*, A1, 7 (1969) 471.
65 S. W. CRITCHLEY (Geigy Co.), *Brit. Patent* 896,436 (1962).
66 F. VONASEK, E. TREPKOVA, *Czech. Patent* 108,762 (1963).
67 *Chem. Eng. News*, 42 (Nov. 30, 1964) 31.
68 J. BARUCH, *Ber.*, 26 (1893) 1870.
69 A. MÜLLER AND P. KRAUSS, *Ber.*, 65 (1932) 1354.
70 J. F. PIED, *Ann. Chim. (Paris)*, 5 (1960) 469.
71 J. L. GREENE JR., R. E. BURKS JR. AND I. A. WOLFF, *Ind. Eng. Chem. Prod. Res. Develop.*, 8 (1969) 171.
72 R. B. PERKINS, J. J. RODEN, A. C. TARNQUARY AND I. WOLFF, *Modern Plastics*, May 1969, 136.
73 D. D. COFFMAN, C. J. BERCHET, W. R. PETERSSON AND E. W. SPANAGEL, *J. Polymer Sci.*, 2 (1947) 306.
74 B. RÅNBY AND T. RINGBERGER, to be published.
75 W. P. SLICHTER, *J. Polymer Sci.*, 35 (1959) 77.
76 W. F. MAESS, Z. E. BUCHSPIESS PAULENT AND F. STINSKY, *Seifen-Öle-Fette-Wachse*, 90 (1964) 18.
77 J. L. GREENE JR., E. L. HUFFMAN, R. E. BURKS JR., W. C. SHEENAN AND I. A. WOLFF, *J. Polymer Sci. A*1., 5 (1967) 391.
78 K. SAIGO, *Iron Steel Engr.*, 40 (1964) 73.
79 R. EASTON, *Iron Steel Engr.*, 40 (1964) 121.
80 Humko Products Chemical Division, *Fatty Amides, Their Properties and Applications*, 1964.

287

CHAPTER 13

Rapeseed Lecithin

U. PERSMARK

AB Karlshamns Oljefabriker, Karlshamn (Sweden)

CONTENTS

13.1 INTRODUCTION

Rapeseed 'lecithin' is a current expression for a chemically complex mixture of substances, which originates from the polar lipid matrix of membraneous structures. These substances occur in various subcellular organelles in the seeds, in conjunction with *e.g.* triacylglycerols, carbohydrates, proteins, glucosinolates etc.[1-3]. Rapeseed oil manufactured by processing of rapeseed, *i.e.* by pressing and solvent extraction, contains, along with neutral oil (triacylglycerols), 'lecithin' components, and the amounts present will to a certain extent depend both qualitatively and quantitatively on the processing parameters. Thus what is commonly known as 'lecithin' includes besides phospholipids in the proper sense also other phosphorus- as well as nonphosphorus-containing compounds.

The lack of uniform terminology in the literature may also cause confusion about the meaning of the term *lecithin*. Besides in the general sense of 'gums' obtained during the processing of vegetable oils (see below), 'lecithin' is also used in the following ways:

(1) To designate the acetone precipitable lipids from crude lecithin, which are soluble in ethanol.

(2) As a synonym for phosphatidyl choline.

(3) In commercial nomenclature to also designate the 'lecithin' from other oil-seed crops, in particular that from soyabeans (ref. 1, p.601). Other designations for lecithin, *e.g.*, polar lipids, phospholipids, and phosphatides also occur. Therefore it is advisable to consult the original papers in order to trace the actual meaning of the terms used. Uniform and pure compounds and substances are classed according to their known or supposed chemical structures using the nomenclature as discussed and presented in ref. 2.

13.2 INDUSTRIAL PRODUCTION OF RAPESEED LECITHIN

Technical rapeseed lecithin is obtained as a by-product in the production and processing of rapeseed oil. Crude oil from solvent extraction or from pressing and solvent extraction contains a certain amount of mucilaginous materials or so-called gums, which are removed in the degumming procedure, generally consisting of a treatment of the crude oil with 2–4 per cent water at moderate temperatures. This procedure leads to a precipitation of the gum, from which crude lecithin is subsequently obtained after separation from the oil. (For details, see Chapter 9.)

Crude rapeseed lecithin is a brown or dark brown highly viscous paste (Table 13:1). The amount of crude lecithin produced by extraction plants at least partly

TABLE 13:1

CHARACTERISTICS OF TECHNICAL RAPESEED LECITHIN*

Water	1 (% wt.)
Acid value	20–25 (mg KOH/g)
Acetone insoluble material	65–70 (% wt.)
Colour (Gardner)**	10–11
Viscosity (η)*** 25 °C	200–300 P
50 °C	30–40 P

* Approximate figures.
** 5 per cent solution in mineral oil.
*** 'Falling ball' method.

depends on the processing parameters, and probably on differences in the source. Normally the yield is likely to be 0.2–0.5 per cent*. An adequate figure in Sweden at the present time is 0.3–0.4 per cent*[4]. Goss[5] has reported a 0.2–0.4 per cent

---

* Calculated on total crude oil yield, *i.e.* prepressed and solvent extracted oil.

yield of crude rapeseed lecithin as calculated on seeds. A considerably lower figure frequently cited in the literature (see *e.g.* ref. 6, p.30) is about 0.1 per cent phosphatides in 'crude' rapeseed oil[7].

The precipitation of phosphatides from the crude oil with water is not complete, as judged from the phosphorus content of the degummed oil†. The remaining phosphorus-containing compounds in the oil are commonly designated non-hydratable phosphatides, and they are structurally different from those which constitute the main part of the precipitated lecithin sludge. These non-hydratable phosphorus compounds are still less known than the hydratable as far as their chemical structures are concerned, and have not yet been systematically investigated in the case of rapeseed oil. In soyabean oil for instance these non-hydratable phosphatides have been found to consist mainly of phosphatidic acids, glycero-phosphates and inorganic phosphates, primarily as Ca–Mg salts, as reported by Nielsen[8]. Most likely the same types of compounds are present in the non-hydratable portion of the phosphorus compounds in rapeseed oil.

The phosphorus content of degummed rapeseed oil manufactured in Sweden is of the order of 0.015 per cent (equivalent to 0.4 per cent lecithin). On the basis of the phosphorus content before degumming, approximately 0.03 per cent[4], the

TABLE 13:2

COMPOSITION OF RAPESEED 'LECITHIN'

| | Per cent (by weight) | | | |
Ref.	(22)	(23)	(24)	(26)
Compound		*	**	***
Phosphatidyl choline	29	22	16	37
Phosphatidyl ethanolamine	15	15	18	13
Phosphatidyl inositol	4	18	8	39
Phosphatidyl serine	1	—	—	—
Lyso-phosphatidyl ethanolamine	—	—	2	—
Sterol glycosides	—	—	8	—
Acidic compounds (unknown)	—	16	—	—
Non-polar lipids	—	—	38	—

* Purified rapeseed gum. Original composition (per cent): phospholipids 51, triacylglycerols 16, non-lipids 9 and water 24.
** Technical Swedish 'lecithin', mainly from *B. napus*.
*** Acetone extraction residue of technical rapeseed lecithin containing (per cent): neutral oil (26.2), phosphatides (71), water (2.5) and ethyl ether insoluble material (0.3).

---

† A frequent and generally accepted method for calculating the content of phosphatides is the measurement of the content of phosphorus, which is subsequently multiplied by a factor of about 25. This way of calculation must however generally be considered to give very approximate figures (*cf.* following sections).

over-all precipitation efficiency is about 50 per cent. The corresponding figure for soyabean oil given by Nielsen[8] is about the same.

The figures given represent an average calculated on the basis of the total oil yield, *i.e.* prepressed and solvent extracted oil. The phosphorus content in pressed oil is about 0.02 per cent (0.5 per cent 'lecithin') and about 0.05 per cent (1.2 per cent lecithin) in the solvent extracted oil with the former representing 2/3 and the latter 1/3 of the total oil yield. Taking into account the fact that only the extracted part is normally degummed, efficiency increases to about 70 per cent, *i.e.* from about 0.05 per cent P to 0.015 per cent P[4]. (See also ref. 9.)

Industrially produced crude lecithin contains about 30–40 per cent triacylglycerols (refs. 5 and 10 and Tables 13:1 and 13:2). The reason for this inclusion of oil in the precipitated 'lecithin' sludge has been explained by Desnuelle and co-workers[11] by the assumption that so-called mixed double-layers are formed on the surface of the water droplets in contact with the crude oil. These double-layers consist of neutral acylglycerols and phosphatides, with a maximal stability of the mixed double-layers at a 1:2 molar ratio between the neutral acylglycerols and phosphatides, which may thus explain the inclusion of about one third neutral oil into the lecithin sludge.

13.3 LABORATORY PREPARATION

13.3.1 *Extraction of lipid material from the seeds*

The total amount of phosphorus-containing compounds in rapeseed oils obtained from seeds extracted on a large scale, as well as the yield of crude lecithin, has to some extent been referred to previously. The figures given concerned hydrocarbon solvent extraction. If other extraction media are taken into consideration, which can be used primarily in laboratory scale preparations, completely different results can be obtained. Shinozaki *et al.*[12] thus report a content of 2 per cent phosphatides insoluble in acetone and soluble in chloroform/methanol 7:1 from rapeseed *(Brassica napus)* extracted with $CHCl_3/CH_3OH$ 2:1

The influence of the properties of the solvent on the extractability of phosphorus compounds can be very strikingly demonstrated[13]: Rapeseed without any pretreatment extracted (on a laboratory scale) with n-hexane yielded about 9 parts of oil, and a subsequent extraction of the residue with chloroform/methanol (2:1) yielded about one part of oil. The phosphorus contents of the two extracts, however, showed that the chloroform/methanol extract contained ten times the phosphorus of n-hexane extracted oil, calculated on the total yield of oil[13a]. The same observations have also been reported for soyabeans[14]. Compared to an oil from industrial pressing and extraction, the phosphorus content of the n-hexane extract (laboratory scale) was considerably lower (0.05 per cent and 0.015 per cent respectively). Since the properties of the solvents should be approximately identical, other parameters than solvent must influence the extractability of phosphorus. These must primarily arise from the processing pretreatments of rapeseed in extraction plants, humidity conditioning, heat treatments, etc. Nielsen[8] for instance has shown that the water content in soyabeans influences the phosphorus content in the extracted oil in such a way that a higher humidity in the seeds produces a higher phosphorus ratio in the oil. Since technical extraction procedures are more efficient than laboratory extraction with n-hexane with respect to the yield of oil, this seems to indicate that the less available fatty material contains larger amounts of phosphorus then the readily extractable oil.

13.3.2 Isolation of the polar lipids from crude extracted oil

The further isolation of phosphatides is performed on a large scale by precipitation with water, as mentioned above. The conditions of smaller scale preparations are however much more variable. Besides water treatment, which of course is generally applicable, numerous other modifications of precipitation and isolation procedures have been described[2, 3]. Precipitation with acetone is commonly used with or without the addition of water. Other precipitation agents are for example acids and certain salts, and occasionally dialysis has been used. Numerous extraction procedures with suitable solvents or solvent mixtures are also frequently applicable with or without a precipitation treatment. Modern chromatographic techniques such as column and thin layer chromatography are nowadays frequently used for the isolation of the polar lipids (see reviews in refs. 2, 15 and 16).

13.4 ANALYSIS OF RAPESEED LECITHIN

13.4.1 Introduction

Before proceeding with more detailed discussion of the analysis and composition of rapeseed lecithin, it is advisable to start with some general remarks about the limitations of the data presented in the following sections.

(1) Polar lipids, which constitute the major part of the 'lecithin', are on the whole a rather unknown and undefined material, as far as their origin in the seed is concerned.

(2) The constitutional characteristics of these components depend as pointed out above on several processing parameters and methods during the manufacturing of oil from the seeds.

(3) Many different varieties and strains often not specified in the literature are usually included under the name 'rape'. Constitutional differences between species must also be considered, together with variations arising from environment and cultivation.

13.4.2 Separation and analysis of rapeseed oil phospholipids

The isolated crude lecithin is eventually further purified. Commercial (or technical) lecithin contains, as mentioned above, about one-third neutral oil which can be removed by *e.g.* solvent fractionation.

The classical procedure still in use for the separation of vegetable phosphatides is based upon a solvent fractionation in ethanol. The two fractions thus obtained, *i.e.* the soluble and the insoluble parts, were designated lecithin and cephalin respectively, which previously were considered to be two different, uniform substances. Rather soon, however, it became clear that the fractionation did not give such well defined separation, but at the most two groups of substances which in addition were rather ill-defined. Furthermore the two fractions were sometimes more or less contaminated with each other[17]. Folch[18], for instance, isolated pure phosphatidyl ethanolamine (cephalin), which proved to be readily soluble in ethanol, while Bear[19] showed that synthetic phosphatidyl ethanolamine containing saturated fatty acids was on the contrary only slightly soluble in ethanol. There are probably many factors governing the solubilities, which are more or less marked for each sample, and in natural mixtures certain mutual interactions are of importance.

Thus, Rewald[10] investigated rapeseed phosphatides by fractionation in ethanol and found that the main portion of the phosphatides was insoluble in ethanol. He further characterized a third intermediate fraction, insoluble in cold but soluble in warm ethanol, and established that the phosphatides were associated with polysaccharides. Heiduschka and Neumann[20], using the same fractionation method, likewise reported that the insoluble phosphatides were the dominant substances in rapeseed 'lecithin'. They did not however, contrary to Rewald, observe any association with polysaccharides. Rapeseed lecithin was also investigated at approximately the same time by Hilditch and Pedelty[21]. They reported 'cephalins' as the major constituent of rapeseed lecithin in agreement with Rewald and with Heiduschka and Neumann.

Sulser[22] analyzed a commercial rapeseed lecithin sample. After hydrolysis of the phosphatides, the water soluble compounds which primarily characterize the various phosphatides were analysed by paper chromatography. The amounts of phosphatides thus found are shown in Table 13:2.

The major phospholipids of Canadian rapeseed gum have been investigated by Weenink and Tulloch[23]. The crude gum was first purified by solvent fractionation and acetone precipitation and subsequently separated on a DEAE-cellulose column followed by a silicic acid column separation. The fractions thus obtained were further analyzed by *e.g.* thin layer, paper and gas chromatography, and by IR spectroscopy (Table 13:2). Besides the components in Table 13:2 minor unidentified constituents were also reported to be present. One of these was reported to be a phytoglycolipid fraction amounting to about 9 per cent and containing two major components qualitatively identified as C_{22} and C_{24} hydroxy acids. A qualitative test also indicated sterol glycosides in the phosphatidyl choline fraction[23]. The presence of sterol glycosides in rapeseed phosphatides has also been reported by Aylward and Nichols[25].

Commercial Swedish rapeseed lecithin (mainly from *B. napus*) was investigated by Persmark[24]. After preliminary fractionation and precipitation with acetone, the components were repeatedly separated on silicic acid columns by elution with

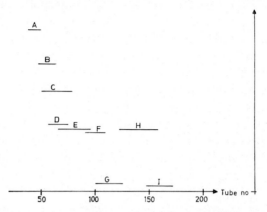

Fig.13:1. Elution pattern of rapeseed lecithin components: A, neutral oil; B, sterol glycosides; C, phosphatidyl ethanolamine; D, phosphatidyl inositol; E and H, phosphatidyl choline; F, lysophosphatidyl ethanolamine; G and I unidentified.
Chromatographic system: silicic acid column, eluent: gradient $CHCl_3$–CH_3OH.[24]
The ordinate represents relative migration distances from TLC according to Fig.13:2.

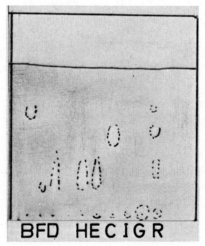

BFD HE C I G R

Fig.13:2. TLC of rapeseed components. Designations according to Fig.13:1.
R = original rapeseed lecithin.
System: Silica gel G (Merck), eluent: $CHCl_3$–CH_3OH–H_2O (65:25:4).
Visualization: 2′,7′-dichlorofluorescein[24].

a gradient mixture mixture of CHCl₃/CH₃OH (Fig. 13:1). Further identification of the fractions was carried out by TLC (Fig.13:2), paper chromatography after hydrolytic splitting, IR, NMR, GLC and elemental analysis. Besides neutral oil (CHCl₃ fraction on the silicic acid column) four major components were found (Table 13:2). Among the minor constituents isolated, all of them not fully identified, was lysophosphatidyl ethanolamine, accounting for about 2 per cent.

The polar lipids of various Cruciferae seeds have been investigated by Shinozaki et al.[12]. The seeds were extracted with chloroform/methanol 2:1. The crude lipid extracts thus obtained were treated with acetone, and the acetone insoluble part further fractionated by CHCl₃/CH₃OH followed by separation on a silicic acid column. The phosphatides of *Brassica napus* were thus reported to contain 29 per

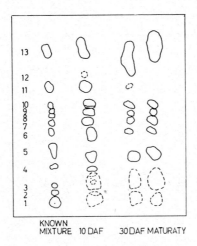

Fig.13:3. TLC of rapeseed polar lipids from plants at various stages of maturation[27].
System: silica gel G – diisobutyl ketone: acetic acid: water (40:25:5).
Identifications:

1. Unresolved mixture
2. Lysophosphatidyl choline
3. Lysophosphatidyl ethanolamine
4. Phosphatidyl inositol
5. Phosphatidyl choline
6. Digalactosyl glyceride
7. Phosphatidyl ethanolamine
8. Sterol glycoside
9. Unknown phospholipid
10. Monogalactosyl glyceride
11. Esterified sterol glycoside
12. Phosphatidic acid (?)
13. Non-polar lipids

DAF = days after flowering

cent phosphatidyl ethanolamine and 22 per cent phosphatidyl choline. The presence of various carbohydrates, *viz*. glucose, inositol and even saccharose was demonstrated in the acetone insoluble fraction. The existence of sterols, especially in mustard *(B. nigra)*, was also reported.

Linow and Mieth[26] investigated and compared various analytical data for rapeseed and soyabean phosphatides. The technical phosphatides were fractionated in acetone, ethanol and ethanol/chloroform, and the fractions obtained compared

TABLE 13:3

FATTY ACID COMPOSITION OF RAPESEED LECITHIN COMPONENTS (GAS CHROMATOGRAPHIC DATA)

Fraction/component	Fatty acid (per cent)										Ref.
	16:0	16:1	18:0	18:1	18:2	18:3	20:0	20:1	22:0	22:1	
Glycerol phosphatides*	23.9	7.5	—	40.0	17.7	5.4	5.5**				10
Phosphatides (total)*	9.9	—	—	27.4	29.4	10.5	7.1**			15.7	10
Phosphatides (total)***	20.6	2.0	1.0	25.2	37.9	8.8	1.5			3.0	26
Phosphatidyl choline	11.1	1.7	1.6	33.1	36.4	7.8	4.2			4.1	26
Phosphatidyl ethanolamine	1.4	1.3	21.6	43.4	11.3	2.8			2.9		26
Phosphatides (total)	16.2	1.0	0.9	24.8	45.2	9.3	—	1.9		—	23
Phosphatidyl choline	8.8	1.3	.8	34.5	44.9	6.2	—	2.4		—	23
Phosphatidyl ethanolamine	1.2	1.0	23.1	51.2	8.2	—	1.7				23
Phosphatidyl inositol	43.4	2.8	1.5	15.0	36.4	8.5	—	—			23
Unidentified fraction ('acidic' compounds)	16.2	1.0	.9	24.8	45.2	9.3		1.9		—	23
Phosphatidyl choline	12	2	1	30	40	9	—	1		5	24
Phosphatidyl ethanolamine	1	1	28	38	11	—	1	—	4		24
Phosphatidyl inositol	34	1	1	18	33	10	—	—		3	24

* B. napus
** Probably 20:1 (author's note)
*** Acetone extraction residue

with regard to phosphatides, fatty acids, carbohydrates, tocopherols, copper and oxidative substances. They concluded that only small differences existed in the composition of the phosphatides of soyabean and rapeseed (Table 13:2). Differences in the fatty acid pattern, taking into account the normal differences between the oils, were found in particular with respect to linolenic acid (lower in soyabean phosphatides). Further differences were reported for minor substances such as carbohydrates, copper and oxidative material. Composition data for rapeseed phosphatides are shown in Table 13:2.

13.4.3 Fatty acid composition

The fatty acid patterns of rapeseed phosphatides are in most respects different from those of the triglycerides. Most data given in the literature are in agreement concerning the absence or very low content of the most typical fatty acids of rapeseed oil, *i.e.* the long chain monoenoic fatty acids, especially erucic acid. Corresponding relations concerning the erucic acid content in neutral *vs.* polar lipids have also been reported for crambe seed[28] and with respect to ricinoleic acid in castor seed[29].

Early data on fatty acids in rapeseed phosphatides are derived from work by Heiduschka and Neumann[20] and Hilditch and Pedelty[21]. The major fatty acids reported were palmitic, oleic and linoleic acids. With the exception of an erucic acid content of about 20 per cent in the 'cephalin' fraction reported by the latter authors 'no' erucic acid was found. Later data based on gas chromatographic determinations reported by Shinozaki *et al.*[12], Weenink and Tulloch[23], Linow and Mieth[26], and Persmark[24] are compiled in Table 13:3. The fatty acid composition of Polish rapeseed oil phosphatides fractionated on a silicic acid-celite column has been investigated by Niewiadomski and Stolyhwo[30]. The erucic acid content of the phosphatides was found to amount to only about one-tenth of that of the neutral oil. McKillican[27] reported however about 10 per cent erucic acid in the fatty acids of the glyco- and phospholipid fraction, whereas Appelqvist[31] has reported about 4 per cent in the total 'polar' lipids of *B. napus*.

13.5 UTILIZATION OF RAPESEED LECITHIN

The gross composition of rapeseed lecithin (Table 13:2) does not to any substantial degree differ from that of soyabean lecithin[25, 32], and it thus seems probable that even the physical properties are approximately equivalent. Jakubowski and Bialecka[32a] have reported that the surface activity of rapeseed lecithin was less than that of soyabean lecithin only at relatively low levels (0.01 per cent). The difference could be explained, according to the authors, by a lower activity of 'cephalin' in rapeseed lecithin.

In spite of their close resemblance, however, rapeseed lecithin has not been given the same widespread use in the food industry as soyabean lecithin. (For a review see ref. 1, p.609.) The main reason for this is that rapeseed lecithin has been found inferior to soyabean lecithin in colour as well as in flavour, taste and general appearance. Nevertheless it has been reported by Stanley (ref. 1, p.639), though not in detail, that rapeseed lecithin has been successfully utilized by margarine and chocolate manufacturers in England and Germany. Presumably it is technically possible to produce acceptable preparations of rapeseed lecithin which are comparable to soyabean lecithin by the use of suitable refining treatments. Efforts in this direction will be briefly described below.

A method for purification of vegetable lecithins based on acetone washing described by Rewald[33] has not been quite successful. Thus Pardun[34] reports that although somewhat better flavour was obtained compared to the untreated rapeseed lecithin, no essential improvement in colour was achieved. As an alternative Pardun[34] proposes a purification procedure of lecithin in a binary solvent mixture. The method was originally worked out for soyabean lecithin but later patented for other vegetable lecithins, e.g. that from rapeseed[35]. Good recoveries and products were reported[34]. Preparations of modified vegetable phosphatides, including rapeseed phosphatides, by treatment of the hydratable lecithins with dilute acids at pH 2–5, yielded a product which was reported to have improved emulsifying properties[36]. In Poland a new degumming process for producing high-grade lecithin has been developed[37]. No subsequent refining is required, the moisture content is less than 2 per cent, the acid number 4–8 and the flavour is described as 'perfect'. A process by Bolleman[38] including steam deodorization considerably improved the flavour quality, but colouring matter was not eliminated and thus an additional bleaching was necessary in order to improve colour characteristics. For the latter treatment hydrogen peroxide has been frequently used[39], but this method of bleaching has been reported not to have any substantial effect on rapeseed lecithin[40]. Furthermore, a considerable increase in the peroxide value is obtained. In order to obtain a lighter lecithin, bleaching of the micella by active earth has been proposed[40]. Large scale applications of this method have however not been reported. Hydrogenation may also be a possible method for improving qualitative characteristics and properties. Although used to some extent[3], no specific application to rapeseed lecithin has been found.

In spite of the fact that several possible methods for the improvement of rapeseed lecithin quality have been described and investigated, at the present time the main question concerns economy. Since there is a good supply of soyabean lecithin on the market the process must be cheap, and preferably also provide a lecithin with better properties than products which are now commercially available.

REFERENCES

1 J. STANLEY, *Soyabeans and Soyabeans Products*, Vol. II, K. S. MARKLEY (Ed.), Interscience Publ., New York, 1951.
2 G. B. ANSELL AND J. N. HAWTHORNE, *Phospholipids*, Elsevier Publishing Company, Amsterdam/London/New York, 1964.
3 H. WITTCOFF, *The Phosphatides*, Reinhold Publ. Corp., New York, 1951.
4 J. DAHLÉN, personal communication.
5 W. H. GOSS, *The German Oilseed Industry*, Hobert Publ. Co., Washington D.C., 1947.
6 D. SWERN (Ed.), *Bailey's Industrial Oil and Fat Products*, 3rd edn., Interscience Publ., New York/London/Sydney 1964.
7 H. P. KAUFMANN, *Fette und Seifen*, 48 (1941) 53.
8 K. NIELSEN, *Studies on the Nonhydratable Soyabean Phosphatides*, Nytt Nordiskt Förlag, Arnold Busck, Copenhagen, 1956.
9 H. P. KAUFMANN, J. BALTES, J. HEINZ AND P. ROEVER, *Fette, Seifen, Anstrichmittel*, 52 (1959) 35.
10 B. REWALD, *J. Soc. Chem. Ind. (London)*, 56 (1937) 403T.
11 P. DESNUELLE, *Progress in the Chemistry of Fats and Other Lipids*, Vol. 1, R. T. HOLMAN, W. O. LUNDBERG AND T. MALKIN (Eds.), Pergamon Press, London 1952.
12 Y. SHINOZAKI, S. OHARA AND H. KONDO, *Nippon Nogeikagaku Kaishi*, 37 (1963) 553; *C.A.* 63 (1965) 7348 g.
13 U. PERSMARK, unpublished data.
13a M. E. MCKILLICAN AND J. A. G. LAROSE, *J. Am. Oil Chem. Soc.*, 47 (1970) 256.
14 W. C. BULL AND T. H. HOPPER, *Oil and Soap*, 18 (1941) 219.
15 D. J. HANAHAN, *Lipid Chemistry*, John Wiley & Sons Inc., New York/London, 1960.
16 G. V. MARINETTI (Ed.), *Lipid Chromatographic Analysis*, Vol. 1, Marcel Dekker, New York, 1967.
17 H. MACLEAN, *Biochem. J.*, 9 (1915) 351.
18 J. FOLCH, *J. Biol. Chem.*, 146 (1942) 35.
19 E. BAER, J. MAURUKAS AND M. RUSSEL, *J. Am. Chem. Soc.*, 74 (1952) 152.
20 A. HEIDUSCHKA AND W. NEUMANN, *J. Prakt. Chem.*, 151 (1938) 1.
21 T. P. HILDITCH AND W. H. PEDELTY, *Biochem. J.*, 31 (1937) 1964.
22 H. SULSER, *Mitt. Lebensm. Hyg.*, 45 (1954) 251.
23 R. O. WEENINK AND A. P. TULLOCH, *J. Am. Oil Chem. Soc.*, 43 (1966) 327.
24 U. PERSMARK, *J. Am. Oil Chem. Soc.*, 45 (1968) 742.
25 F. AYLWARD AND B. W. NICHOLS, *Nature*, 184 (1959) 1319.
26 F. LINOW AND G. MIETH, *Nahrung*, 11 (1967) 663.
27 M. E. MCKILLICAN, *J. Am. Oil Chem. Soc.*, 43 (1966) 461.
28 Y. SHINOZAKI AND S. ISHII, *Nippon Nogeikagaku Kaishi*, 42 (1968) 207.
29 S. VENKOB RAO, M. M. PAULOSE AND B. VIJAYALAKSHMI, *Lipids*, 2 (1967) 88.
30 H. NIEWIADONSKI AND A. STOLYHWO, *Roczniki Technol. Chem. Zywnisci*, 12 (1966) 189; *C.A.* 65 (1966) 10816.
31 L.-Å. APPELQVIST, *Hereditas*, 61 (1969) 9.
32 J. A. BERNOTAVICS, *Biscuit Maker and Plant Baker*, (1963) 596.
32a A. JAKUBOWSKI AND A. BIALECKA, *Przemysl Spozywczy* 14 (1960) 474; *C.A.*, 55 (1961) 1824f.
33 Hanseatische Mühlenwerke A.G., *Brit. Patent* 412,224 (1932).
34 H. PARDUN, *Fette, Seifen, Anstrichmittel*, 64 (1962) 536.
35 Unilever Ltd. *Brit. Patent* 933,814 (1959).
    Lever Bros. Co., *U.S. Patent* 3,047,597 (1962).
    Unilever N.V., *Belg. Patent* 590,731 (1959).
36 Unilever N.V., *Brit. Patent* 1,118,373.
37 *Confectionary Production*, (1967) 788.
38 H. BOLLEMAN, Hanseatische Mühlenwerke A.G., *Brit. Patent* 259,166 (1925).
39 *German Patent* 511,851, *C.A.*, 25 (1931) 1336.

Brit. Patent 356,384, *C.A.*, 26 (1932) 4418.
US Patent 1,893,393, *C.A.*, 27 (1933) 2225.

40   P. MÖLLER AND M. GABRIELSSON, unpublished data.

Nutritional Value of Rapeseed Oil

E. AAES-JØRGENSEN

Department of Biochemistry Royal Danish School of Pharmacy, Copenhagen (Denmark)

CONTENTS

A discussion of the biological value of a nutrient for human beings often culminates in a number of general statements because the final conclusions in most cases are based on a limited selection of all the information necessary to give a specific answer. Thus, the basic experiments normally must be carried out with laboratory animals. This involves two difficulties from the beginning, namely, what is the normal optimum condition of the animals, and can the results from animal experiments be converted safely to the situation in human beings? It is worthwhile to recall that experience has shown many examples of difficulties and disappointments due to the lack of correlation between man and his experimental model, *cf.* Barnes[12].

Further, for practical reasons semisynthetic diets are commonly used in animal experiments. Such diets may be adequate and realistic as far as the amounts of calories, proteins, lipids, salts and vitamins are concerned, but certainly differ greatly regarding the selection of ingredients composing a diet typical for animals or humans. Another difficulty is related to the choice of criteria used for evaluation of the nutrient under study. Two commonly used parameters are growth rate of young animals, and lack of development of gross symptoms indicating dietary deficiencies. Both of these conditions are rather non-specific and in many cases far from sufficient for a deeper understanding of nutritional problems.

Regarding dietary lipids especially, it has until recently been a serious handicap for research in this field that the analytical techniques available have not permitted a thorough characterization of their detailed composition. However, based on the almost explosive development which has taken place during the last 10–20 years in this area of research, much more precise information is now available. Thus, the importance of lipid components in living organisms at all levels of development is being unraveled at a rapid rate. With these general remarks in mind, studies on the nutritive value of rapeseed oil will be discussed in some detail.

14.2 RAPESEED OIL/ERUCIC ACID IN EXPERIMENTAL DIETS

The suitability of rapeseed oil as a dietary fat has been discussed for many years. One of the main subjects in most of these discussions has been the fact that rapeseed oil has a remarkably high content of erucic acid, *i.e.* Δ^{13}-*cis*-docosenoic acid. The nutritional value of rapeseed oil in diets for experimental animals has been reviewed previously, *e.g.* by Beare[13], Potteau[115], Thomasson, Gottenbos, Van Pijpen and Vles[140], and Jacquot, Rocquelin and Potteau[89].

The nutritive value of rapeseed oil has primarily been studied in experiments with rats. Thus, it appears reasonable to review first the studies carried out with this species.

Discussion on the dietary value of rapeseed oil reached one of its peaks in the 1940's in connection with a debate on the superiority of butter over margarine in nutritive value, and vice versa. Much work has been concentrated on a possible relationship between the presence of erucic acid and the nutritive effects of rapeseed oil. However, these discussions were hampered by the fact that available analytical techniques were rather primitive, especially regarding the unsaturated fatty acids. In 1941, Boer[30] found young rats growing at a slower rate on a diet containing rapeseed oil than on a similar diet with butter. Von Beznak, von Beznak and Hajdu[27] suggested that the weight inhibiting effect of dietary rapeseed oil was not related to its content of erucic acid but to the presence of a toxic compound. However, this postulate was invalidated in a subsequent experiment by the same authors[28]. In studies on the growth promoting effect of summer butter, Boer et al.[31, 32] used rapeseed oil for comparison in some of the experiments. Again rapeseed oil promoted a poor growth rate in young rats. These experiments were primarily concerned with a suggested specific growth effect of vaccenic acid, which Boer's group claimed to be responsible for the special effect of summer butter.

These experiments prompted Deuel et al.[66] to study the dietary effects of vaccenic acid and of rapeseed oil. Vaccenic acid was shown to be without specific effects on the growth of young rats. The gain in weight of animals on diets containing 10% rapeseed oil was in most cases significantly lower than the increase in weight of the animals receiving butterfat or cottonseed oil over a 6-week period. The efficiency of utilization of the rapeseed oil was poorer than that of the other two dietary fats. It was suggested that this may be attributable to a poor digestibility of rapeseed oil, which again may be due to its high content of erucic acid. They also mention that crude rapeseed oil has an especially unappetizing taste. However, it is very difficult to judge whether such an argument has any importance in experiments with rats.

Von Euler et al.[70] found the growth rate of young rats fed on rapeseed oil inferior to that of animals reared on butterfat. In the beginning of the 1950's, a growth depressing effect of rapeseed oil was also shown by Carroll[38, 39] and by Carroll and Noble[48]. As these studies primarily centered on the effect of rapeseed oil on the adrenals they will be discussed later (see p.324).

The growth of young animals may be influenced not only by the type of dietary fat but also by the amount of fat in the diet (Barki et al.[11]). Thomasson[135] considered this fact and carried out a large experiment with 20 different types of fats varying in amounts from 10 to 73 calorie per cent (*i.e.*, 8–48 weight% fat in the diet) over a period of 6 weeks. Compared to butterfat, a significantly poorer growth was obtained on diets containing 20 to 73 calorie per cent of rapeseed oil. Of the rats receiving the diet with 73 calorie per cent of rapeseed oil all were dead after 37 days of experiment (mean value: 17 days). The author suggests that the rapeseed oil effect on growth may be ascribed to the presence of fatty acids with 20 or more carbon atoms, *i.e.* erucic acid in particular. It is further suggested that these fatty acids do not differ from other fatty acids in caloric value, but that their metabolic rate is different. This latter statement is based on the fact that the calculated food efficiency constants were about the same for all dietary lipids tested. Finally, it is pointed out that the biological value of oils and fats cannot be judged only on such criteria as growth and food intake, because the animals on 50 calorie per cent rapeseed oil lived significantly longer than those on 50 calorie per cent butterfat (see p.344).

The suggestion that the growth-retarding effect of rapeseed oil may be related to its high content of erucic acid was further studied by Thomasson and Boldingh[138]. Diets were prepared with 20 to 73 calorie per cent rapeseed oil (about 50% erucic acid), 30 calorie per cent Nasturtium seed oil (86% erucic acid), 30 per cent of the fatty acids of the two oils, and a mixture of groundnut oil and trierucin (50:50), or of groundnut oil and erucic acid (50:50). With 73 calorie per cent of rapeseed oil in the diet, all experimental animals died within a mean life span of 16 days. No death occurred within any of the other groups. By interpolation it appeared that 30 calorie per cent of Nasturtium seed oil had the same effect on growth as 57 calorie per cent of rapeseed oil. All of the experiments supported the conclusion that the poor growth obtained with rapeseed oil is actually and exclusively due to the presence of erucic acid. The unsaponifiable fractions of rapeseed and groundnut oils did not affect the growth of the animals.

Carroll and Noble[49] verified the growth- and appetite-inhibiting properties of erucic acid and its methyl and ethyl esters. They also found similar effects of eicosenoic acid, whereas oleic acid permitted relatively good growth. In another experiment, Carroll and Noble[50] used a stock diet (Masters Meal), the composition of which is not given. To this diet was added 25% rapeseed oil, 10, 15, or 30% crude erucic acid, or 15% oleic acid. Normal growth curves were obtained in all cases, whether the diets were given to young or adult male rats. However, the animals suffered a progressive reduction in spermatogenesis. Female rats grew similarly well, and their estrous cycle remained undisturbed, but there was some interference with parturition. Beare, Murray and Campbell[20] studied the effect of a Canadian rapeseed oil on young Wistar rats, which were fed for 9 weeks on diets with 10 and 20% fat. The dietary fat consisted of corn oil, rapeseed oil or mixtures thereof containing 10, 20, 40, or 80% rapeseed oil. When the ordinary oils were fed as 10% by weight of the diet, there were no significant differences in weight gain. At the 20% level there was a significant reduction in weight in

304

both sexes with the 80% rapeseed oil mixture, equivalent to 16% of the total diet, and still further reduction for rapeseed oil as 20% of the diet.

Murray et al.[104] studied the growth effect of various long-chain fatty acids in young male and female rats. They used diets with 5% of pure methyl esters of oleic, eicosenoic and erucic acids compared to diets with 5% corn oil or no fat. After 10 weeks on the respective diets, no significant growth depression was found amongst the males, whereas the growth of the females was depressed significantly by each of the pure esters. However, the authors mention unpublished data which indicate that a higher dietary level of erucate results in a significant growth depression in males. Symptoms of EFA-deficiency occurred only among rats fed the fatty acid esters, indicating an acceleration of the depletion of EFA in these animals.

Roine and Uksila[120] found that young rats, given a free choice, eat a little more of a diet with rapeseed oil than with soybean oil at a 10% dietary fat level. However, at a 20% level the animals showed a distinct preference for the soybean oil diet. Erucic acid at a 20% level was disliked as much as was rapeseed oil. Diets with 10 and 20% butter were preferred to similar diets with rapeseed oil. The growth rate for the first 30 days on diets with 10% fat was about the same with rapeseed oil, soybean oil and butter. During the following 2 weeks growth was slower on rapeseed oil than on the other two diets. At the 20% level, growth rate was distinctly slower on rapeseed oil than on the other two types of fat. Food intake was also lowest in the rapeseed oil group. With 20% erucic acid in the diet, growth was as poor as with rapeseed oil in spite of a definitely higher food intake. Experiments with rapeseed and soybean oils emulsified in skim milk indicated that the poor growth on rapeseed oil was not due to low emulsifiability of this oil in the digestive tract. The authors conclude that rapeseed oil in relatively small amounts in the diet of young rats has no deleterious effect on growth rate. When the amount of rapeseed oil or erucic acid becomes higher, the food is less palatable and growth rate is retarded. At very high rapeseed oil levels the animals lose weight and even die. Growth retardation also occurred on a feed restriction regimen, which implies that reduced food intake is not the only reason for poor growth on rapeseed oil.

Beare et al.[22] compared the utilization of rapeseed oil (Golden strain) and corn oil in young and adult Wistar rats. The oils were given as 20% of the diet over a period of 5 weeks. Further, paired feeding of the two diets was carried out with young rats. In all three experiments, animals fed rapeseed oil showed smaller weight gains and less food consumption than those on corn oil, but when body weight gains were adjusted for food consumption by covariance analyses, the differences largely disappeared. They concluded that there was no practical difference in the utilization of the two oils in rats. The chief effect of rapeseed oil was an apparently depressed appetite early in the feeding period followed by adaption to the oil during the experiment.

So far very little attention had been paid to possible differences in the fatty acid composition of different varieties of rapeseed oil used in nutritional experiments. Beare, Gregory and Campbell[18] studied the effect of the Polish, Golden or Swedish varieties of rapeseed oil on growth of young male and female Wistar rats during 9 weeks compared to the effect of corn oil or mixtures of ethyl erucate and corn oil (51:49 and 26:74, respectively). The oils were added at a level of 20% by weight to a purified basal diet. The weight gain of the rats was found to be universally related to the dietary content of C22 acids, essentially erucic acid. Thus, the results confirmed the suggestion that the growth retarding effect of rapeseed oil is due to its content of erucic acid. In general, the differences in weight gains could be explained by the effect of rapeseed oil on food consumption.

The fatty acid spectrum of rapeseed oil is characterized by a higher content of erucic acid and a lower content of saturated fatty acids (especially palmitic acid) than most other vegetable oils. Beare et al.[16] studied the effects of saturated fat in rats fed on rapeseed oil diets. Young male Wistar rats were fed each test oil as 20% by weight in a semisynthetic diet for 4 weeks. The dietary fats comprised two Canadian-grown varieties of rapeseed oil (Polish and Swedish) supplemented with palm oil to reduce the concentration of erucic acid and increase that of palmitic acid, or a mixture of lard and olive oil (3:2). The animals fed Polish rapeseed oil ate less and grew significantly poorer than those on the other two diets. Analyses of covariance indicated that the variation in weight gain could be accounted for by food consumption. It was evident that the diet containing Swedish rapeseed oil and palm oil was more acceptable to the rats than that containing Polish rapeseed oil.

The saturated fatty acids comprised approximately 30% of the total fatty acids in both the mixture of Swedish rapeseed oil and palm oil and the mixture of lard and olive oil. This proportion (1:2) of saturated to unsaturated fatty acids in diets containing 10% of the fat as linoleic acid was shown by Hopkins, Murray and Campbell[86] and by Murray et al.[103] to give the greatest weight gain in rats. The proportion 1:2 of saturated to unsaturated fatty acids is approximately equal to that found in fat depots of rats.

In the same series of experiments Beare et al.[16] added methyl laurate, myristate, palmitate or oleate to Swedish rapeseed oil to dilute its content of erucic acid to the same level as that of Polish rapeseed oil. These dietary fat mixtures were compared with Polish rapeseed oil. None of the added methyl esters increased the body weight gain beyond that obtained with Polish rapeseed oil. A mixture of Swedish rapeseed oil and palm oil promoted greater weight gains than those obtained with Polish rapeseed oil of a similar erucic acid content (20%). The relative effects of Polish and Swedish rapeseed oils alone in depressing appetite and weight gain were directly related to the content of erucic acid. Animals fed on a diet containing an oil low in palmitate (3% palmitic acid) with no erucic acid grew approximately as well as those fed on Polish rapeseed oil (2.7% palmitic and 19.5% erucic acid). Feeding of rapeseed oil (Swedish), corn oil or the methyl or ethyl esters of the fatty acids of these two oils showed that the methyl esters produced a significantly lower food intake and weight gain than the respective glycerides or ethyl esters. The ethyl esters of Swedish rapeseed oil were more acceptable to the animals than the original oils but were not so efficiently utilized.

Acceptability of the diet was improved by increasing the saturation of rapeseed oil by the addition of palm oil, and lowered by decreasing the saturation of an oil containing fatty acids of no more than 18 carbon atoms. Nutritional properties of such vegetable oils are therefore influenced by the degree of saturation as well as by the chain length of their fatty acids. It is concluded that the effect of rapeseed oil in experimental diets can be ascribed both to its low content of saturated fatty acids and its high content of erucic acid.

All the experiments described so far have shown that rapeseed oil has a poorer nutritive value than most other oils and fats. Because of its high content of erucic acid, it has been natural to suggest that this particular acid may be responsible, at least to some extent, for the low dietary value of rapeseed oil. For this reason it was very interesting when Stefanson et al.[132] and Downey and Craig[69] succeeded in breeding *Brassica napus L.* with no erucic acid in the seed oil. Enough of this oil, called canbra oil, was available in 1967 to be tested in animal experiments.

Rocquelin and Cluzan[119] compared the effect of two types of rapeseed oil, one with 45% erucic acid, and one with no erucic acid (1.9% docosaenoic acid), to that of arachis oil. The oils were given as 15% of the diet to young male and female rats for 6 months. Arachis oil and the rapeseed oil free from erucic acid gave the same growths in both sexes. The ordinary rapeseed oil caused growth retardation in both sexes compared to the two other oils.

Another study with canbra oil was carried out by Craig and Beare[57]. Male weanling rats of the Wistar strain were fed *ad libitum* for 24 days on diets with 20% rapeseed oil, olive oil, olive oil with lard (9:16, w/w), canbra oil, canbra oil with tripalmitin or tristearin (89:11, w/w), canbra oil interesterified with either of these triglycerides, canbra oil with palm oil (37:13, w/w) or canbra oil with lard (3:7). Canbra oil contained only 2.7% more saturated fatty acids than rapeseed oil and no docosaenoic acid. Food intake and gain in body weight were low on rapeseed oil, whereas both of these criteria were significantly higher on canbra oil and similar to those obtained with olive oil or a mixture of lard and olive oil. No further enhancement of weight gains was attained by the addition of tripalmitin, tristearin, palm oil or lard to canbra oil. Interesterification of the simple glycerides with the canbra oil also caused no significant change. Thus, in this rather short-term experiment, canbra oil appeared to be as nutritionally satisfactory as any other oil tested.

14.2.2 Experiments with partially hydrogenated rapeseed oil

Partially hydrogenated oils are widely used, especially in human diets. It is well known that many such treated oils are very low in essential fatty acids, and they have therefore become of special interest in relation to studies on essential fatty acid deficiency[1, 5, 82]. However, rapeseed oil has rather reasonable amounts of linoleic and linolenic acids. Thus, a partial hydrogenation should leave fairly high amounts of essential fatty acids in the oil. Partially hydrogenated rapeseed oil has been studied to some extent in relation to the general effect of rapeseed oil in experimental animals.

Von Euler et al.[71] studied the effect of margarine on the growth, reproduction and longevity of rats. The experiments lasted up to three years and thus the composition of the margarine changed during the experimental period. The content of, e.g., hydrogenated rapeseed oil varied from zero to 47.5%. They found no specific effects which could be related to the hydrogenated rapeseed oil in the margarine. Carroll[39] used hydrogenated rapeseed oil in some of his studies on the effect of rapeseed oil on adrenal cholesterol. Young male rats of the Sprague-Dawley strain were fed for 4 weeks on Master's fox chow alone, or mixed with 25% rapeseed oil or hydrogenated rapeseed oil. At the end of this short-term experiment the increase in body weight was similar for the rats on Master's fox chow alone or mixed with hydrogenated rapeseed oil, whereas a marked reduction in weight gain was noticed in the animals given rapeseed oil in the diet.

In 1961, Beare, Craig and Campbell[17] studied the nutritive effect of partially hydrogenated rapeseed oil on rats. They used partially hydrogenated rapeseed oils (m.p. 42 °C, 20 °C, 39 °C, and 41.5 °C), partially hydrogenated soybean oil, and soybean oil. Semi-synthetic diets containing 20% of the oils were fed to young Wistar rats for 6 to 8 weeks. Hydrogenation of rapeseed oil (42 °C) increased its growth-promoting action, whereas hydrogenation of the soybean oil decreased it. In a second experiment with partially hydrogenated rapeseed oil (m.p. 40 °C) in the diet for 6 weeks no differences were observed in weight gain in either sex. Feeding of rapeseed oil partially hydrogenated under conditions used in shortening manufacture and randomly rearranged (m.p. 39 °C), or under conditions used in margarine manufacture (m.p. 41.5 °C) had depressing effects on growth and food consumption compared to the effects of unhydrogenated rapeseed and corn oils. Of the latter two oils the rats grew far better on corn oil than on rapeseed oil.

Gas chromatography of the methyl esters of the partially hydrogenated rapeseed oils revealed that the partially hydrogenated rapeseed oil, which increased weight gain, was still relatively rich in linoleic acid. The other two samples of partially hydrogenated rapeseed oil contained lesser amounts of linoleic acid. However, all three samples contained appreciable amounts of $C_{18:2}$ positional isomers. Partial hydrogenation also decreased the content of $C_{22:1}$ acid and increased the amount of stearic acid. The authors suggest that the alterations in the C_{18} fatty acids resulting from hydrogenation of rapeseed oil are responsible for the differing responses in weight gain.

Determination of the gross calorific value of rapeseed oil as heat combustion in a bomb calorimeter was found to be practically the same as that of lard, (Kabelitz[91]). Determination of the available caloric contents of rapeseed oil, partially hydrogenated rapeseed oil and corn oil by rat growth were carried out by Middleton and Campbell[100]. Growth responses to the three oils and to lard were similar, and the calculated caloric content of the oils showed no significant difference from that of lard under the experimental conditions employed.

Roine and Uksila[120] compared the effect of rapeseed oil, soybean oil and partially hydrogenated rapeseed oil on the growth rate of rats. On a restricted feeding regimen, *i.e.*, the amount of food was restricted to the amount in grams eaten by the group which ate the least, no significant differences were found at a 15% dietary fat level between rapeseed oil, soybean oil or hydrogenated rapeseed oil. At the 20% dietary fat level, rapeseed and hydrogenated rapeseed oils resulted in significantly lower growth rates than soybean oil, but the lowest growth was obtained with erucic acid.

Craig *et al.*[59] fed two types of rapeseed oil with erucic acid contents of 23 (Polish) and 42% (Swedish), their hydrogenated products produced under selective and non-selective conditions, corn oil, and a lard-olive oil mixture (3:2) to rats at a 20% level in the diet. The experimental period was 4 weeks. Selective hydrogenation caused a decrease in the content of linoleic and linolenic acid in the rapeseed oils without the production of saturated acids, whereas non-selective hydrogenation increased the content of saturated and monoenoic acids. The two types of rapeseed oil adversely affected the food intake and body weight gain in proportion to the content of erucic acid. The hydrogenated and non-hydrogenated rapeseed oils (42% erucic acid) produced similar results, but the non-selectively hydrogenated (Polish) rapeseed oil promoted weight gains which were greater than those produced by the same oil selectively hydrogenated. Growth and diet consumption were lower, in all cases, on the rapeseed oils than on the control diets. When weight gains were adjusted for food consumption by covariance analyses, there were no significant differences between the dietary fats. It is suggested that the reduced weight gain of rats fed rapeseed oil may be the result of two factors: the high content of erucic acid which is rather sparingly deposited, and the low level of saturated acids, particularly palmitic acid.

Cheniti *et al.*[53] studied the nutritive value of sunflower oil, rapeseed oil, partially hydrogenated rapeseed oil with and without a supplement of sunflower oil as a source of essential fatty acids (7 g sunflower oil/kg food). The oils were given at a level of 20% of the diet to young rats for 7 months and 4.5 months. The growth rate on partially hydrogenated rapeseed oil was significantly poorer than on sunflower oil, but was brought to about the same level as obtained with rapeseed oil by supplementation of the diet with essential acids, *i.e.* 7 g of sunflower oil per kilogram of diet. Both of these latter two groups grew somewhat less than the controls. Calculations of the dietary efficiency (grams of weight gain/grams of diet ingested × 100) as a function of time showed a levelling off on all four diets. However, the efficiency was highest for the sunflower diet throughout the experiments. The authors conclude that the decreased efficiency of hydrogenated rapeseed oil may not only be a matter of an increased requirement for essential fatty acids, but also related to the presence of *trans* isomers and of behenic acid.

In a third experiment, lasting only 39 days, young rats were fed on a fat-free diet supplemented with about 8% of ethyl oleate, erucate or behenate and 0.4% ethyl linoleate. Erucate gave the lowest weight gain and efficiency. The biological value of rapeseed oil was inferior to that of other vegetable oils. The authors suggest that this inferiority is located at the metabolic level, in part as a direct effect of specific components in the rapeseed oil, *e.g.*, erucic acid, and in part indirectly as an interference with other dietary elements.

14.2.3 Experiments with heat-treated rapeseed oil

Polymerized or heat-treated rapeseed oil has been studied in only a few experiments. Beare et al.[20] used 10 and 20% corn oil or rapeseed oil heated to 200 °C for 120 hours in diets for young rats. The heat-treated corn oil and rapeseed oil produced no changes in weight gain when fed at the 10% level, but exhibited some growth-retarding effects at the 20% level. In this connection it should be mentioned that Crampton, Farmer and Berryhill[60] found that rapeseed and corn oils heat-polymerized at 276 °C for 30 hours in a stream of carbon dioxide had a growth-depressing effect when fed to rats as 10% of the diet.

Japanese workers have also reported the effect of dietary polymerized rapeseed oil. Matsuo[99] compared the growth rates of rats fed on diets with 20% rapeseed oil or 20% rapeseed oil polymerized at 250 °C for 50 hours. Some of the rats on rapeseed oil showed a slow growth rate which was attributed to incomplete absorption of the oil. With the polymerized rapeseed oil the rats did not grow properly, and the cyclic ethyl esters separated from the polymerized rapeseed oil were toxic to rats, although the toxicity was lower than that of a similar fraction from cuttlefish oil. Akiya et al.[3] heated soybean oil, rice bran oil and rapeseed oil at 200 °C for 12 hours. Administration of 1.5 g of these oils per day to rats caused growth retardation in the case of the heat-treated rapeseed oil, but not with the two other oils.

Diets containing 10% heated rapeseed oil or hydrogenated vegetable fat had a depressive effect on the growth of rats during a 6 months' test period (Nikonorow and Homrowski[107]). These studies were followed up recently by Homrowski[84] in studies on heat-treated rapeseed oil, margarine, and hydrogenated shortening. Both of the latter two products contained rapeseed oil. Rapeseed oil and hydrogenated shortening were heated to 170 °C and 200 °C respectively, for 30 minutes, or to 170 °C for 12 hours. Margarine fat was heated to 130 °C or 170 °C for 30 minutes, or to 130 °C for 3 hours. The fresh and heated fats were fed as 10% of the diet to rats for 12 months. All rats gained weight, but statistically significant differences were found in the growth of animals fed on heated fats compared to those on the non-heated products or normal laboratory diet. Fresh rapeseed oil alone also retarded growth. The author found no significant weight deficit in rats receiving 18.5% erucic acid from unheated margarine or 40.5% of this acid from hydrogenated shortening, the latter amounting to approximately the same content as in fresh rapeseed oil. He suggests that the effect of rapeseed oil on growth may be connected with a component other than erucic acid, or the variations in chemical composition of margarine and hydrogenated vegetable shortening may influence the reduction of the growth depressive effect of erucic acid when contained in these products.

14.2.4 Effect of dietary rapeseed oil on other species than the rat

In general, the rat has been the experimental animal par excellence in studies on the nutritional effects of rapeseed oil. However, in the literature on rapeseed oil some examples where other species have been used are also found. The widest spectrum of experimental animals used for rapeseed oil studies in one laboratory is that of Unilevers' Research Laboratory in Holland, as reported by Thomasson et al.[140]. They have tested rapeseed diets on rats, mice, guinea pigs, hamsters, and ducklings. In the following a brief account of the results obtained in various species of experimental animals is given.

Pigs have been used as experimental animals in some studies involving dietary rapeseed oil. Paloheimo et al.[111] carried out two experiments with young pigs. They added 100 g/kg diet of rapeseed or soybean oil and divided the experiments into a period of restricted feeding (about 4–5 weeks) and a period of *ad libitum* feeding (about 3–5 weeks). During the period of restricted feeding, no marked differences were seen between the two groups, but during the *ad libitum* feeding in the second experiment the animals on soybean oil showed a better appetite and grew more than those fed on rapeseed oil. Further, the latter animals had a larger water consumption than those on soybean oil. Crampton et al.[61] found that swine were not able to utilize dietary rapeseed oil efficiently. Thus, a significant growth depression occurred in swine fed on a diet containing 20% rapeseed oil when compared to those on a similar diet containing butterfat.

Jacquot et al.[88] kept *young boars* on a semi-synthetic diet containing 15% rapeseed oil (high in erucic acid (45%) and linolenic acid (20%)), or 15% of a mixture of olive oil and palm oil (no erucic acid, low in linolenic acid and high in oleic acid). Both diets contained 15% linoleic acid. The animals in the two experiments were fed *ad libitum* for 119 and 161 days, respectively. The gain in weight was grossly the same on the two diets, but food efficiency was lowest on rapeseed oil. These experiments are also described by Thoron[141].

Dogs were used as experimental animals by Lindlar[95] in a study on the presence of erucic acid in organs and depots after rapeseed oil feeding. He found large quantities of erucic acid in the depot fat, lesser amounts in various organs, and none in brain, bone marrow and erythrocytes. He did not find erucic acid in the glycerophospholipids. Crampton *et al.*[61] found puppies to thrive well on a diet with 20% rapeseed oil. Bernhard *et al.*[24] gave 3 dogs a load of rapeseed oil (60–100 g/day) during 50 to 76 days. They found erucic acid accumulated in the depot fat, whereas it could for the most part not be detected in the phospholipids. Carroll[40] found a normal growth rate in dogs on a 25% rapeseed oil diet for 24 days.

Chickens reared on a diet containing rapeseed oil had depot fat which differed sensibly from ordinary chicken fat (Chudy[55]). Normal growth of chickens on a diet containing 25% rapeseed oil for 4 weeks was observed by Carroll[40]. Sell and Hodgson[127] kept chickens on rations with 4 or 8% rapeseed oil, tallow, soybean oil or sunflower seed oil for 8 weeks. Rapeseed oil supported weight gain to the same extent as the other vegetable oils. Equivalent levels of tallow, rapeseed oil and sunflower oil resulted in ration metabolizable energy values which were approximately equal but significantly lower than with soybean oil in the diet.

Ducklings grew well on diets with 20 cal.% rapeseed oil, but at 40 cal.% their growth rate was significantly poorer than that of ducklings on 40 cal.% butterfat. With 60 cal.% rapeseed oil a very pronounced growth retardation occurred, indicating a progression of this effect with increasing content of rapeseed oil in the diet (Thomasson *et al.*[140]).

Guinea pigs as test animals for dietary rapeseed oil were used by Crampton *et al.*[61]. A diet with 20% rapeseed oil reduced growth in guinea pigs compared to those on a similar diet containing 20% butterfat. Guinea pigs given a diet containing 25% rapeseed oil did not gain weight during an experimental period of 4 weeks (Carroll[40]). Thomasson *et al.*[140] found a significant growth retardation in guinea pigs fed on a 50 cal.% rapeseed oil diet compared to those on a similar diet with butterfat.

Mice fed on a diet containing 25% rapeseed oil showed no gain in weight after 4 weeks on the diet, Carroll[40]. Thomasson *et al.*[140] found that mice grew significantly more poorly on diets with 40 cal.%, or more, of rapeseed oil than on butterfat.

Hamsters showed diminished growth rate with increasing rapeseed oil content in the diet. The difference between butterfat and rapeseed oil groups was significant at 40 cal.% and higher (Thomasson *et al.*[140]).

Rabbits have generally not been used in rapeseed oil experiments. However, Wigand[145] gave diets containing 8% of rapeseed oil, butterfat or corn oil to rabbits. He found no significant differences in growth rate among the 3 groups of animals during experimental periods of 7.5 and 15 weeks. Carroll[40] found a normal growth rate in rabbits fed on a diet with 25% rapeseed oil for 4 weeks.

Turkeys showed a poor growth on a 10% rapeseed oil diet (Joshi and Sell[90]).

Geese were given barley for 12 days, then an addition of 40 ml rapeseed oil daily for another 16 days, and finally barley only for another 12 days (Chomyszyn[54]). The subcutaneous and internal fat assumed the characteristics (iodine number, melting point, smell) of rapeseed oil after only 8 days on the diet. These characteristics disappeared again after a week on barley alone.

14.3 DIGESTIBILITY AND ABSORPTION

In order to evaluate the nutritive value of dietary fats it is necessary to consider their digestibility and absorbability. As stated by Deuel[65], these two terms are interpreted by many people as synonymous, connoting the same physiological phenomenon. The overall utilization of dietary fat is often defined as the coefficient of digestibility calculated from the ingested amount of fat and the fecal excretion of fat: [Fat ingested − Fat excreted (corrected for metabolic fat)] × 100 /Fat ingested. Fat absorption commonly describes the removal of fat from the small intestine as a function of time. Thus, the absorption coefficient usually describes the degree of absorption as measured by the amount of the ingested fat present in the gut several hours after ingestion of a single dose. Therefore, the absorption rate can be determined only in experiments of short duration.

That rapeseed oil can be digested and absorbed, at least to some extent, was shown as early as 1868 by Radziejewski[116]. By feeding rapeseed oil to rats, erucic acid was found in the depot fat. Erucic acid was discovered in 1849 by Darby[63]. The first studies of digestibility of rapeseed oil in humans were carried out in 1918, by Holmes[83]. The average coefficient of digestibility in four persons was found to be 98.8. More rational and elaborate experiments on the utilization of dietary rapeseed oil in experimental animals were started in 1940 by Deuel *et al.*[67]. They compared the rate of absorption of various natural fats, including rapeseed oil, in rats. In order to obtain the most uniform and comparable results, the absorption rates were calculated on the basis of body surface area. Of the oils tested, *i.e.* cotton seed oil, butterfat, coconut oil and rapeseed oil, the latter was found to have the slowest absorption of all.

During the following years a number of reports indicated poor growth of rats on rapeseed oil. This prompted Deuel *et al.*[64] to study the digestibility of crude and refined rapeseed oil in rats. The coefficient of digestibility was found to be 77 and 82 for crude and refined rapeseed oil, respectively. The calculation of the coefficient included correction for metabolic fat, *i.e.* excreted fat in feces from animals on a fat-free diet. This endogenous fat, 'metabolic fat', comprises not only triglycerides and fatty acids, but also soaps which are not extractable with ether. This had been ignored in previous studies, but was shown by Augur *et al.*[8] and Crockett and Deuel[62] to be a factor of importance.

Further, the amounts of fecal soaps are related to the composition of the salt mixture. Thus, calcium–magnesium rich salt mixtures increase the amount of excreted soaps (Cheng et al.[52].) A similar effect is seen with poorly digested fats. The presence of fecal soaps implies that acidification before extraction is necessary. It has been later shown by Hopkins et al.[86] that acidification with 5% acetic acid in petroleum ether is preferable to acidification by means of hydrochloric acid which results in the extraction of considerable amounts of non-lipid material also. It should be pointed out that in order to calculate the actual food intake in experiments with *ad libitum* feeding, it is very important that possible feed spillings are carefully controlled.

Deuel et al.[66] studied the possibility that a relationship between the poor growth of rats on rapeseed oil could be correlated with a decreased food intake compared to the intake on diets containing other types of fat. In an experiment over 6 weeks, they calculated diet efficiency as animal weight gain in grams × 100/ caloric consumption. The rapeseed oil diets had a lower efficiency of utilization than other vegetable oils tested. The authors suggest that this finding is probably related to an unsatisfactory digestibility of the rapeseed oil. When corrections were made in calculating the caloric values of the diets for the digestibility of the fats, the differences in the efficiency of utilization between the rapeseed oil diet and those containing butterfat or cottonseed oil largely disappeared. The less efficient utilization of rapeseed oil by rats was attributed to the poor digestibility of this fat which may be due to its characteristically high content of erucic acid.

As mentioned above, earlier studies in man showed a high coefficient of digestibility for rapeseed oils (Holmes[83]). However, in these studies the possible presence of fecal soaps was not considered. Therefore, Deuel et al.[68] repeated the experiments in man with rapeseed oil and cottonseed oil diets, and found coefficients of digestibility of 99 and 96.5, respectively. They concluded that a species difference exists between man and the rat with respect to the utilization of rapeseed oil.

The fatty acid composition of dietary and chyle fat was studied in a child with a chylothorax condition (Fernandes et al.[72]). The assumption was made that similar absorption coefficients of fatty acids indicated the absence of splitting of the glyceride molecules. On this background it was suggested that certain fats, e.g. rapeseed oil, olive oil and sunflower seed oil, are absorbed almost without hydrolysis.

Thomasson[135] studied the biological value of 22 different types of dietary fats in rats in feeding experiments over 6 weeks. The various oils and fats showed differences in growth promoting effect. Average body weight gain decreased as the amount of rapeseed oil in the diet increased. However, calculation of food efficiency based on weight gain and food calorie intake, without correction for food wastage, showed approximately the same efficiency constant for all fats and oils. This indicated that the increase in weight per calorie of food intake is independent of the type of fat and the level at which it is administered. The author

314

therefore suggested that the growth retardation effect of rapeseed oil can possibly be related to absorption and/or metabolism of its long-chain fatty acids. This was investigated by Thomasson and Boldingh[138], who concluded that the unfavorable growth obtained with rapeseed oil must be ascribed to the erucic acid content of this oil.

Further studies by Thomasson[137] on the rate of intestinal absorption of different fats, including rapeseed oil, showed a significant correlation between the rate of absorption of the dietary fats and their growth-promoting action. Rapeseed oil was the slowest absorbed oil after 3 hours and the fourth slowest after 6 hours. After 6 hours half of the administered fat was removed from the gastro-intestinal tract when common dietary fats and oils were tested, but after 8–9 hours when rapeseed oil was used. A dose of 400 mg butterfat/100 cm² body surface given by stomach tube to rats had disappeared from the gut to the extent of 50% after 5 hours whereas this time was almost doubled for a similar dose of rapeseed oil. The author suggests that the decreased rate of absorption of rapeseed oil may be due to a feed-back mechanism which originates from other phases of the metabolic pathway. However, this is not supported by preliminary experiments by Gottenbos and Hornstra[75]. They found that the absorbed fat appears in the chyle of cannulated thoracic duct rats more slowly on tubing rapeseed oil than when corn oil is administered. If the rates of absorption in cannulated and non-cannulated rats are comparable, the decreased rate of intestinal absorption may be partly or entirely a local process.

In an experiment with food restriction, 7.5 g food containing 20 cal.%, i.e. 2.2 g rapeseed oil or butterfat, given to rats weighing about 250 g for 4 weeks led to no significant difference in mean body weights between the two groups (Thomasson and Gottenbos[139]). The authors also carried out paired feeding trials with young rats fed on diets containing 72 cal.% butterfat or rapeseed oil. The rapeseed oil diet was available ad libitum. After 4 weeks, the weight gain was 18 g and 32 g per animal on the rapeseed oil and butterfat diets, respectively. In a control group receiving the butterfat or rapeseed oil diets ad libitum the absorption values were 98% and 94%, whereas in the restricted feeding experiment the percentages were 98% and 95% for butterfat and rapeseed oil, respectively. When the rats were trained to consume 1, 2 or 3 g of fat per day during ½–1 hour the absorption values were 98% and 96% (1 g fat), 96% and 89% (2 g fat), and 95% and 87.5% (3 g fat) for butterfat and rapeseed oil, respectively. The authors suggest that this difference of 3–7% may explain the growth differences observed in the paired feeding experiment, but probably not the considerable difference in growth and food consumption observed on ad libitum feeding.

In a study on the effect of monoenoic acids on the cholesterol content of the adrenals, Carroll[39] noticed that at a 10% dietary level the absorption of fatty acids from the gut decreased with increasing chain length. Thus, 85% of oleic acid, 55% of erucic acid, and only 25% of nervonic acid were absorbed. This topic was also touched upon in a following experiment (Carroll and Noble[49]). Sprague-Dawley rats were fed on a semisynthetic diet containing 0–15% by weight of different fatty acids or their methyl or ethyl esters. The coefficient of digestibility, representing essentially the percentage of ingested fatty acids which is absorbed or otherwise metabolized by the animal, ranged from 52 to 59 for erucic acid and ethyl erucate, respectively. These coefficients were corroborated in a following study on the digestibility of individual fatty acids in Sprague-Dawley rats (Carroll[41]). He found that the digestibilities of mono-unsaturated acids were approximately the same as those of saturated fatty acids with six less carbon atoms. The average coefficients of digestibility were 53, 62 and 59 for erucic acid, methyl and ethyl erucate, respectively.

However, fatty acids are normally present in food as triglycerides. Thus, Carroll and Richard[51] studied possible differences in digestibility between triglycerides and non-esterified fatty acids. Triglycerides of unsaturated fatty acids were more completely digested than the corresponding non-esterified fatty acids. Thus, the coefficients for trierucin and erucic acid were 63 and 48, respectively. Palmitic and stearic acids were poorly digested either as triglycerides or as free acids. Further, the calcium content of the diet was found to play an important role in determining the digestibility of fatty acids. The poor digestibility of the free fatty acids may be caused by the formation of calcium soaps and calcium-phosphate containing complexes which are not readily absorbed. This suggestion may be of special importance for erucic acid, which is slowly absorbed (Thomasson[137]). On the other hand, the digestibility of erucic acid appeared to improve as the protein level in the diet was increased. Finally, the digestibility of erucic acid seemed to be lower in old than in young rats (Carroll and Richards[51]).

Murray et al.[104] fed young Wistar rats on a semisynthetic diet supplemented with 5% of, e.g. methyl erucate and methyl eicosenate. They found the coefficients of digestibility to be about 90 and 95, respectively, and suggested that the discrepancies between these high figures and the much lower figures found by some other workers may in part be explained by the fact that their calculations[104] were made on the excreted fatty acids and not on the total lipid excreted. It should also be mentioned that Murray et al. used Wistar rats, while e.g. Carroll used Sprague-Dawley rats. That species difference may be of some importance not only when comparing experimental animals with man, but also within different strains of the same species has been shown on several occasions, see p.338.

Carroll[40] compared the digestibility of rapeseed oil in rats (Sprague-Dawley), guinea pigs and rabbits. He used a fox chow diet supplemented with 25% rapeseed oil by weight and found coefficients of digestibility to be 58, 61 and 58, respectively. Beare et al.[22] fed Wistar rats on a diet with 20% rapeseed oil and found an apparent digestibility (no correction for endogenous fat) of 93. Beare et al.[21] investigated the response of two strains of rats, inbred Wistar and Sprague-Dawley, to dietary rapeseed oil and corn oil. The digestibility of corn oil was not significantly different in the two strains of rats, but the digestibility of rapeseed oil was significantly lower in the Sprague-Dawley than in the Wistar (coefficients 65 and 83, respectively). Roine and Uksila[120] mention in their discussion that they have unpublished data indicating that the digestibility of rapeseed oil is dependent on the age of the rats, being lowest in young animals and increasing gradually with age.

Alexander and Mattson[4] found almost the same degree of absorbability of rapeseed oil and soybean oil in Holtzman rats reared on diets containing 20 or 73 cal.% of the two oils. On the other hand, both oils were less well utilized when they were gelatinized with ethyl cellulose. The absorbed caloric efficiency of the two diets was calculated as the grams gain in body weight per 1000 calories eaten minus the calories lost in the feces as fat. The values for absorbed caloric efficiency were about the same for all types of diets. However, for the diets high in fat, those containing soybean oil were slightly superior to those containing rapeseed oil. The authors feel that it is impossible to determine from their data whether this is a real difference or whether it is the result of a combination of factors such as food consumption, palatability of the ration or a slight decrease in absorbability.

Savary and Constantin[123] studied the absorption of erucic acid in Wistar rats, and its incorporation into lymphatic chylomicrons. Their data showed that neither free erucic acid nor trierucin is quantitatively absorbed from the gut. Further, very little erucic acid was found in the phospholipids and cholesterol esters.

Paloheimo and Jahkola[110] studied the digestibility of rapeseed oil and soybean oil in two young pigs. The digestibility of the oils was investigated at three different levels: 100, 200, and 300 g of oil/animal/day. The animals received the two oils in alternate periods. The coefficient of digestibility was about 100 for both oils.

Wigand[145] found a coefficient of digestibility of 94 in rabbits fed on a semi-synthetic diet with 8% rapeseed oil, compared to a coefficient of only 58 in Carroll's experiments with 25% rapeseed oil[40].

A further underlining of differences among species regarding the digestibility of rapeseed oil is also illustrated by experiments on very young guinea pigs, puppies and pigs by Crampton et al.[61]. With 20% rapeseed oil in the diet the coefficients of digestibility (uncorrected for endogenous fat) were 72, 94, and 78, respectively. The coefficient for pigs is interesting compared to the results obtained by Paloheimo and Jahkola[110]. The discrepancy may be the result of differences in age and strain. Sell and McKirdy[128] studied the digestibility of rapeseed oil, 10% by weight in the diet, in young chickens. The coefficient of digestibility was found to be 90.

14.3.1 Effect of dietary components other than fat on the digestibility of rapeseed oil

In the preceding discussion it has occasionally been indicated that the digestion and absorption of rapeseed oil and erucic acid may be influenced by other dietary components. Deuel et al.[64] found that dietary rapeseed oil increased the fecal content of soaps. This indicates that it is not the hydrolysis of the oil which is slow but primarily an effect of a low absorption rate for erucic acid. Studies by Cheng et al.[52] with an ordinary salt mixture (Osborne-Mendel) and the mixture without calcium and phosphate in diets with 15% rapeseed oil to adult female rats showed a considerable decrease in the amount of feces, and in the content of fecal soaps when calcium and phosphate were excluded from the diet. Further, the total amount of 'metabolic fat' was about equal on diets with the two types of salt mixtures, but relative to the amounts of feces it was about twice as high on the low calcium-phosphate salt mixture.

Carroll and Richards[51] found that the calcium content of the diet plays an important role in determining the digestibility of fatty acids. On a calcium-free diet the coefficient of digestibility of erucic acid was 92, whereas with a calcium–phosphate ratio of 1.4 : 1 in the salt mixture (5% of the diet) it was 21. With only 2% salt mixture high in calcium the coefficient was 48. An increase of the dietary content of calcium primarily resulted in a large increase in the fecal soaps, and of the calcium phosphate containing fatty acid complexes. Further, high dietary protein (65%) improved the digestibility of erucic acid compared to a low protein level (5.5%). Young animals appeared to digest erucic acid better than older animals. These studies were followed up by Richards and Carroll[118], who found that the ether-soluble calcium–phosphate complex is formed with a number of saturated fatty acids, but not with e.g. palmitic and stearic acids. These experiments confirmed the structure of such complexes proposed by Swell et al.[133, 134].

The poor digestibility of rapeseed oil may also be related to its emulsifiability in the digestive tract. Augur et al.[8] found that 3% lecithin lowered the susceptibility to diarrhea in rats fed on diets with cottonseed oil or hydrogenated cottonseed oil. Simultaneously, lecithin increased the digestibility of the hydrogenated oil, probably because of an increased rate of absorption. Roine and Uksila[120] emulsified rapeseed oil and soybean oil with skim milk powder in skimmed milk. The milk preparation contained 3.1% fat, equivalent to 38% of the dietary calories. This preparation fed to rats caused diarrhea, but the milk preparation given together with a low-fat basal ration was tolerated. After 30 days of experiments the rats on the rapeseed oil milk preparation grew significantly more poorly than those on the soybean oil preparation. The authors conclude that retarded growth on the rapeseed oil diet is not due to poor emulsifiability in the digestive tract. Thomasson and Gottenbos[139, 140] found poor growth of rats on diets high in rapeseed oil (72%) to be due to the inhomogeneity of the diet, which was a result of the oil being separated off. This could be prevented by the addition of 1% sawdust (saturated with water) which resulted in improvement of growth and decreased mortality. Several years later Alexander and Mattson[4] found that gelatinization of the liquid oil with ethyl cellulose overcame this difficulty.

318

14.3.2 Effect of rapeseed oil/erucic acid on 'metabolic fat'

In the discussion on digestibility and absorption of dietary fat the fact that not only is dietary fat digested in and absorbed from the digestive tract, but a secretion of fat out into the gut is also taking place has been referred to. Some of this fat is reabsorbed, and some is found in the feces. Thus, this endogenous or 'metabolic fat' poses a problem in itself.

In 1884, Müller[105] studied feces from fasting dogs and from a fasting human (German: Hungerkünstler) and found a considerable amount of lipids to be present. He suggested that the fecal lipids under these circumstances were due to a fat secretion. In the 1920's, Hill and Bloor[77], Sperry and Bloor[131], and Sperry[129] showed that a fat secretion occurs into the gut in dogs kept on a fat-free diet. That this fat secretion does not occur by the way of the bile was shown in experiments with dogs on a fat-free diet (Sperry[130]). Schönheimer and Hilgetag[126] found cetyl alcohol, which is endogenous, to be secreted through the mucosa into the gut.

Norcia and Lundberg[109] studied the effect of dietary fat (tripalmitin, triolein or a tripalmitin–triolein mixture) on the composition of fecal fat. They concluded that endogenous fat was unaffected both in quantity and composition by the dietary fats, although the depot fat was markedly different depending on whether the rats had been preconditioned on tripalmitin or olive oil. The composition of the depot fat and changes in composition induced by dietary changes exerted no influence on the composition of the endogenous fecal fat. From these data they concluded that endogenous fat does not result from secretion of fat, from desquamation of epithelial cells of the intestinal mucosa, but is synthesized by intestinal bacteria. A similar view was advanced by Holasek[79, 80, 81]. He found that rats on a low-fat diet with cellulose and terramycin produced feces in which longer chain fatty acids were essentially absent. Similar results were obtained in rats with a biliary fistula, and with humans treated with antibiotics. He concluded that fecal fat was not secreted into the gut but was of bacterial origin.

However, this conclusion is invalid since fat secretion does take place. Thus, Bernhard et al.[25] fed rats for prolonged periods with daily doses of 2–3 ml of rapeseed oil followed by a similar period of feeding a normal fat-free diet. They found in these rats, as well as in rats with a biliary fistula and the same dietary regimen, that erucic acid occurred in the feces long after the rapeseed oil feeding was stopped. They concluded that erucic acid is absorbed, some is deposited and can be subsequently secreted into the gut together with other endogenous fatty acids. A similar secretion was observed in rats with biliary fistula. The collected bile contained no erucic acid, indicating that the secreted fat is not transported out into the gut by means of the bile. The fat secretion is more marked in the absence of bile as the fat absorption is then hindered. In this case the fatty acids from the fecal lipids contained much more stearic acid and erucic acid than the fat of the wall of the intestine or of the depots. In the case of a fat-deficient diet the fecal lipids consist of modified body fat together with dietary fat which has not been absorbed. When the supply of bile to the intestine is inadequate body fat appears in increased amounts in the feces.

Carroll and Noble[49] found erucic acid to increase the excretion of endogenously produced cholesterol in Sprague-Dawley rats with little change in the cholesterol concentration in the carcass except for increased concentrations in the adrenals and liver. Other homologues of oleic acid, namely eicosenoic and nervonic acids, produced similar changes in fecal cholesterol excretion. Ester cholesterol accounted for much of the observed increases, when most fatty acids were fed, whereas a fat-free diet, or diets with short-chain fatty acids or oleic acid resulted in excretion of the sterols in the free form. Apparently the fecal cholesterol did not originate in the liver since no increase was observed in the biliary excretion of cholesterol. The fecal content of other sterols was not increased.

In a following study, Carroll[43] examined the effect of erucic acid on incorporation of acetate-1-^{14}C into cholesterol and fatty acids. Erucic acid consistently stimulated the incorporation of labeled acetate into cholesterol by rat liver slices. The incorporation of acetate into cholesterol by liver slices was much greater on erucic acid than on a fat-free diet or diets with palmitic, stearic, oleic, or linoleic acids. Erucic acid did not stimulate acetate incorporation into cholesterol by the mucosa. On this basis Carroll modified the idea expressed above to suggest that it is conceivable that the excess fecal cholesterol excreted by rats on erucic acid diets may originate in the liver, but the mechanism behind this effect is unknown. It might be that the primary effect of erucic acid is exerted on the liver to stimulate cholesterol synthesis, and the increased fecal excretion may be merely a means of removing excess sterol.

Murray, Campbell, Hopkins and Chisholm[104] found increased excretion of unsaponifiables in feces of rats on diets with erucic acid. Studies on the amount of fecal fat were also made by Buensod and Favanger[34]. They used deuterated dietary fatty acids and found a clear-cut increase of endogenous fat excretion with palmitic and stearic acids, but not with tripalmitate or tristearate.

It appears reasonable to suggest that the excretion of endogenous lipids into the gut is influenced to some extent by the dietary fat but it is difficult to estimate this effect. In studies on the digestibility of dietary fats, corrections should be made for 'metabolic fat'. Further, the correction factor should be kept as low as possible. Thus, short-chain fatty acids cause less endogenous fat excretion than a fat-free diet, whereas long-chain fatty acids increase the amount of fecal lipids, and certain monoenoic acids increase the content of fecal sterols. It should be pointed out that coprophagy apparently has not been considered in studies concerning erucic acid and rapeseed oil.

14.3.3 Appetite and palatability related to rapeseed oil diets

Is there a relation between appetite and/or palatability and the consumption of diets with rapeseed oil? This problem has been mentioned briefly in preceding sections of this chapter, *e.g.*, pages 303, 304, 305, 306 and 311. In the present section a short survey is made of some of the experiments in which this problem has been considered more specifically.

Beare *et al.*[16] concluded from studies of rapeseed oils and their ethyl esters that the effect of dietary rapeseed oil on appetite is not solely related to its content of erucic acid. Acceptability of the diet was improved by increasing the saturation of rapeseed oil by the addition of palm oil, and lowered by decreasing the saturation of an oil containing fatty acids of not more than eighteen carbon atoms.

Beare *et al.*[22] compared diets with 20% rapeseed oil or corn oil during 5 weeks by *ad libitum* and paired feeding. The reduced rates of weight gain and food consumption on the rapeseed oil diet were ascribed to an apparent depressed appetite early in the feeding period followed by an adaption to the oil during the experiment.

Thomasson and Gottenbos[139] found the poor growth of rats fed on a diet with 72 cal.% rapeseed oil to be due to the inhomogeneity of the diet, which was a result of the oil being separated off. Addition of 1% water-saturated sawdust to the diet improved growth and decreased mortality. A similar but less dramatic effect was observed on a similar diet with sunflower seed oil. However, it is the experience of many laboratories, that rats fed *ad libitum* eat less food spontaneously when the diet contains appreciable quantities of rapeseed oil. Thomasson and Gottenbos[139, 140] added butter flavour substances to rapeseed oil diets without noting any improvement of food consumption in rats. They concluded that the difference in food intake is not caused by an appetite-stimulating factor in butter. By administration of the dietary fats per stomach tube a possible influence of a taste factor was eliminated[139, 140]. These studies argue against the existence of an organoleptic rapeseed oil or butter factor. Further, the tubed rapeseed oil appeared to result in a reduced rapeseed oil intake about 15–20 minutes after incubation of the fat indicating an 'induction period'.

In another experiment, eating habits were controlled. With rapeseed oil and butterfat diets the rats of both groups had an average of 10 meals per day, each lasting for about 5 minutes. However, the animals consumed 13 g and 9 g of the butter and rapeseed oil, respectively. Thus, the difference in food intake arises during the meals. The authors conclude that the physiological mechanism underlying the food-intake-reducing action of rapeseed oil is unknown, but is apparently not localized in the taste organs. They suggest that the reduced consumption of rapeseed oil diets is related to a low rate of transport and/or metabolism of rapeseed oil fatty acids, especially erucic acid. In order to follow up this idea they studied the intestinal absorption of rapeseed oil and butterfat. The absorption rate was about twice as high for the former of the two fats as that of the latter. The decreased rate of absorption for rapeseed oil was suggested to be the result of a feed-back mechanism originating from other phases of the metabolic pathway. However, this view was not supported by studies of chyle from cannulated thoracic duct rats given rapeseed oil or corn oil, which indicated that rapeseed oil enters the blood stream more slowly than corn oil. Thus, these results indicate that the rate of intestinal absorption is partly or entirely a local process. The authors point out that this does not rule out the possibility of rapeseed oil also reducing appetite *via* other mechanisms.

At high levels of oils in semi-synthetic diets there is a tendency for the oil to separate off from the rest of the diet. This problem was remedied by Thomasson and Gottenbos[139] by the addition of 1% water-saturated sawdust. Alexander and Mattson[4] found that in diets with 48% oil the solid ingredients in the diet settled to the bottom of the feeding cup, and the animals were therefore presented with a ration covered with a layer of liquid oil. Thus, the poor growth and early death of the rats on diets high in rapeseed oil were attributed to the unpalatable form in which the diet was fed. This separation phenomenon was overcome by gelatinizing the liquid oil by means of ethyl cellulose. Under these circumstances the authors found the nutritive value of rapeseed oil to be about equal to that of soybean oil. However, rapeseed oil at high dietary levels was still associated with low food consumption and poor weight gain compared to soybean oil at similar levels.

Beare and Beaton[15] fed weanling female Wistar rats on a fat-free diet which was consumed largely during the night. The intake of this diet was recorded for two 24 h periods, 4 days apart. During each day the rats were dosed orally twice with 2 ml of corn oil or rapeseed oil. The rats force-fed rapeseed oil did not lose their appetite for other food consumed later in the same 24 h period. In another experiment, male Wistar rats had the ventromedial area of the hypothalamus electrolytically lesioned. These lesions are known to induce hyperphagia. The animals were fed a diet containing 20% corn oil for 5 days and then a diet containing 20% rapeseed oil for 6 days. The change of diet from corn oil to rapeseed oil caused a significant reduction in food intake but not in the hypothalamic-hyperphagic group. The authors conclude that the decrease in food consumption usually observed with rapeseed oil is fairly rapidly and temporarily mediated and probably involves hypothalamic control. However, the appetite-reducing effect of rapeseed oil and the mechanism involved is still an unsolved problem.

14.4 PHYSIOLOGICAL CHANGES INDUCED BY RAPESEED OIL

14.4.1 Effect of rapeseed oil on depot fat

It is well-known that dietary fat has some influence on the composition of depot fat. Studies on the relationship between rapeseed oil and depot fat have of course been mainly concerned with the presence of erucic acid, which normally occurs in very low amounts, if at all, in depot fat.

Radziejewski[116] found erucic acid in depot fat from rats fed on a rapeseed oil diet. However, more thorough studies could first be performed in conjunction with recent developments in analytical techniques.

Bernhard et al.[25] fed rats on an ordinary diet and dosed the animals with 2–3 ml rapeseed oil/animal/day for two weeks. During the following period olive oil and then coconut oil were given instead of rapeseed oil. Shortly after the beginning of the olive oil period, some of the rats had a bile fistula. The mixed fatty acids from the body fat of these latter animals contained on the average 5% erucic acid. In a later experiment by Wagner et al.[143], rats were fed either a choline-sufficient or a choline-deficient diet containing 10% rapeseed oil for 18 days. The depot fat from the eviscerated carcasses contained an average of 7.7 and 7.6% of erucic acid, respectively. The liver fatty acids of the controls amounted to 78 mg with 3.8% erucic acid, whereas the liver of the choline-deficient rats had on the average 348 mg fatty acids with 7.8% as erucic acid.

Hopkins et al.[85] studied the deposition of monoenoic fatty acids in male and female rats. They added 5% methyl oleate, methyl eicosenoate, methyl erucate or corn oil to an essentially fat-free diet which was fed for 12 and 20 weeks. The bodies were then separated into three portions, viz. pelt, viscera and carcass. Eicosenoic and erucic acids were deposited to a considerable extent when fed as the respective methyl esters. However, erucic acid was deposited in much lesser amounts than was eicosenoic acid. The quantities deposited were nearly the same in males and females, although the latter animals were maintained on the diet for an additional period of eight weeks. The authors suggest that the mono-unsaturated acids, oleic, eicosenoic and erucic, were deposited without alteration in chain length or degree of saturation.

Wagner et al.[144] kept adult male rats on a rapeseed oil diet for 6 months. The contents of erucic acid and several other fatty acids were determined in neutral fats and phospholipids from numerous organs and depot fat. A true accumulation of erucic acid could not be determined. A study of the biological half-life of erucic acid in lipids from various organs and depots showed it to be about 18–31 days. The authors conclude that erucic acid behaves like a fatty acid normally utilized in metabolism.

Bernhard et al.[24] studied the distribution of erucic acid in organ- and depot-fat in adult dogs fed 60–100 g rapeseed oil per day for 50–76 days. Erucic acid occurred primarily in the triglycerides of depot fats, whereas it could mostly not be detected in the phospholipids from various organs. These observations agree with the results of Beare[14], who found that dietary fatty acids have more influence on the carcass fat than on the liver fat. Erucic acid represented 5.6% of the liver fatty acids which are partially located in the phospholipids, while the carcass fat contained twice this amount. Beare *et al.*[23] noticed that a pyridoxine deficient rapeseed oil diet reduced the amount of fat in the carcass compared to a similar corn oil diet. Erucic acid was somewhat more concentrated in the carcass fat of rats not supplied with pyridoxine than in those that were.

Sell and Hodgson[127] found substantial portions of eicosenoic and erucic acids in the adipose tissue of chicks receiving 4 or 8% dietary rapeseed oil for 8 weeks.

Craig *et al.*[58] fed Wistar rats on diets with 20% rapeseed or corn oil for 21 weeks. The content of eicosenoic acid in the cutaneous, abdominal and carcass fat resembled that of the diet, *i.e.* 9–10% and 11.6%, respectively. The amount of erucic acid was much lower in the depots (6–7%) than in the dietary rapeseed oil (37%). The authors suggest that erucic acid is converted *in vivo* to oleic acid. The same authors[29] studied the carcass fat from rats kept for 4 weeks on diets with rapeseed oil (Polish or Swedish, 23% and 42% erucic acid, respectively). Only small amounts of erucic acid were incorporated into the body fats, whereas the relative amounts of eicosenoic acid were of the same order in the body fats as in the dietary oils. Beare *et al.*[16] studied the effect of palmitic acid on the intake of rapeseed oil and thus of erucic acid. They used Polish rapeseed oil (2.8% palmitic, and 24% erucic acid), a mixture of Swedish rapeseed oil and palm oil (26.6% palmitic, and 19.3% erucic acid), Swedish rapeseed oil plus methyl palmitate (44.3% palmitic, and 20.3% erucic acid), and Swedish rapeseed oil (3.7% palmitic, and 39.8% erucic acid). The variation in palmitic acid had no effect on the deposition of erucic acid.

Craig and Beare[56] found erucic acid in lipids from skin, head, carcass, and 'organs' (liver, heart, lungs, spleen, intestine) when rapeseed oil was fed to rats. The concentration of erucic acid in the fatty acids from the head and organs was lower than that from the skin and carcass, which is in agreement with previous data showing that erucic acid is less readily incorporated into phospholipids than into glycerides. Eicosenoic acid occurred in each body fat in somewhat higher concentrations than did erucic acid, although the dietary fat contained only half as much eicosenoic as erucic acid. In this experiment, 11-docosenoic acid was also studied and the incorporation of this acid was similar to that of erucic acid. Thus, apparently the rat did not distinguish between the two isomers.

Thoron[141] fed boars on a diet with 15% rapeseed oil for 7 months. Although the dietary level of erucic acid was high and the experiment lasted for 7 months, erucic acid occurred in the depot fat, liver and testes only in very small amounts: 2, 1, and 0%, respectively.

14.4.2 *Effect of rapeseed oil on the suprarenal glands*

It has long been known that rapeseed and rapeseed oil in the diet of experimental animals result in changes in their adrenal glands. Kennedy and Purvis[92] noticed hypertrophy of the adrenal cortex with a large increase in the lipid content of the adrenal cortical cells in rats fed on rapeseed. Abelin[2] found an increase in the concentration of adrenal cholesterol to 6% in rats on a diet containing rapeseed oil, whereas other dietary fats failed to produce a similar effect. Carroll[38] fed young Sprague-Dawley rats rapeseed as 45% by weight of the diet and found a slight increase in adrenal size and very marked increase (3–4 fold) in both the percentage and the absolute amount of adrenal cholesterol. Nearly the same effect was obtained by feeding rapeseed oil, turnip seed oil, or Nasturtium seed oil as 25% of the diet. Various other fats and oils fed at the same level resulted in a very small increase of the cholesterol content. The increased cholesterol is mostly in the esterified form. Rapeseed oil had very little effect on the concentration of ascorbic acid in the adrenals. Feeding rapeseed oil to hypophysectomized rats did not cause a significant increase in their adrenal cholesterol, and the adrenals atrophied as a result of the hypophysectomy.

Carroll[39] suggested in view of the high content of erucic acid in the above seed oils, that this acid may be responsible for the effect on the adrenals. Thus, rapeseed oil and the fatty acids from rapeseed oil were fractionated and tested during 3 weeks in male rats fed on diets containing 10–25% of the various fractions. The results indicated that methyl erucate, and to a lesser degree also nervonic acid (a higher homologue of erucic acid), cause a considerable increase in the weight of the adrenals and in their content of cholesterol. A number of individual fatty acids were also tested, *e.g.* oleic, eicosenoic, ximenic, lumequic, and *trans*-nervonic acids, none of which showed an effect on the adrenals comparable to that of erucic and nervonic acids.

Carroll and Noble[48] studied whether rats fed on rapeseed oil differed from control rats in any way that could be correlated with altered adrenal cortical function. However, the accumulation of adrenal cholesterol resulting from feeding rapeseed oil appeared not to be associated with marked changes in adrenal function, and the adrenals responded normally to stimulation. When rapeseed oil feeding was combined with other regimens, *e.g.* thyroid, stilbestrol and cortisone, known to affect adrenal size and cholesterol content, the effects of the combined treatments seemed to be additive. In a subsequent study, Carroll and Noble[49] extended their investigations. Erucic acid was found to increase the cholesterol concentration in the adrenals and liver, and it also increased the excretion of endogenously produced cholesterol in the feces, whereas it caused only small changes in the cholesterol concentration in the remainder of the carcass, which was also noted for eicosenoic and nervonic acids.

Carroll[40] also studied the relation between dietary rapeseed oil and adrenal cholesterol in different species. He found the cholesterol content of the adrenals significantly increased in rats (Sprague-Dawley) and mice, whereas only a small change or no such change was observed in rabbits, guinea pigs, chickens and dogs. The differences in response to rapeseed oil do not appear to be due to species differences in digestibility of rapeseed oil. The author points out that dietary cholesterol tends to be deposited more readily in the latter four species than in the rats and mice.

Beare et al.[20] found no weight increase of the adrenals in Wistar rats after feeding diets with 10 or 20% rapeseed oil or thermally oxidized rapeseed oil for 10 weeks. Similar results were obtained in a subsequent experiment with young and older rats fed on rapeseed oil diets (Beare et al.[22]). However, rapeseed oil produced a significantly greater adrenal cholesterol concentration than did corn oil in adult rats fed 20% of the oils in the diet. Alexander and Mattson[4] found that diets high in rapeseed oil (73 cal.%) produced higher adrenal cholesterol values in Holtzman rats than did similar diets containing soybean oil. Beare et al.[21] looked into the possibility of a discrepancy between two strains of rats, Wistar and Sprague-Dawley, regarding the effect of dietary rapeseed oil on their adrenals. After 6 weeks on diets with 20% rapeseed oil or corn oil the adrenals of Sprague-Dawley rats were larger in the animals fed rapeseed oil than in those fed corn oil, while the adrenals of Wistar rats were not influenced by the dietary oil. Significant differences in total adrenal cholesterol paralleled those of adrenal weight, and in the Sprague-Dawley rats corresponded to those previously reported by Carroll[38]. Feeding of 20% partially hydrogenated rapeseed oil to Wistar rats significantly reduced the adrenal weight compared to rats on diets with 20% rapeseed or corn oil.

Bernhard et al.[24] found only small amounts of erucic acid in the organs of dogs fed 60–100 g rapeseed oil per day during 2–3 months. Of the many organs studied the highest concentration of erucic acid occurred in the adrenals. Carroll[44] compared the distribution of erucic acid in different lipid classes of the adrenals of Sprague-Dawley rats fed on diets containing 15% erucic acid for 3 weeks or 6 months. Erucic acid was found in large amounts in adrenal cholesterol esters and in moderate amounts in adrenal triglycerides, whereas only small amounts were found in adrenal phospholipids. The increase in adrenal cholesterol caused by feeding erucic acid appears to be primarily due to the accumulation of cholesterol erucate. Erucic acid accounted for 35% of the total fatty acids in the triglyceride fraction, whereas the phospholipids contained only 1.2% erucic acid in the experiments over 3 weeks. After 6 months the adrenal cholesterol esters contained 45% of the total fatty acids as erucic acid, while in the triglycerides it had increased to 25%. In the other lipid fractions the amount of erucic acid had not increased further during the long feeding period.

14.4.3 Effect of rapeseed oil on ovaries, testes, reproduction and milk secretion

That rapeseed oil in the diet may be of some importance in relation to reproduction was indicated by studies of Kennedy and Purves[92], who found that female rats fed on rapeseed oil matured sexually much later than normal females, as they would not become pregnant until 5–7 months of age. Further, the ovary and the adrenals are the only organs that contain most of their cholesterol in the esterified form. Preliminary studies by Carroll and Noble[18] indicated that the ovaries of rats fed on rapeseed oil contained more cholesterol than those of controls. Therefore, immature female rats were fed on a diet containing 25% rapeseed oil for 60 days, during which period the rats reached maturity. Compared to the controls the rats on rapeseed oil grew less well, their vaginas opened one to two weeks later, but at the same approximate body weight. After opening of the vagina, daily vaginal smears were taken and no irregularities in the cycles were observed. At the end of the feeding period the cholesterol content of the ovaries from the rats on rapeseed oil was significantly higher than that of the controls. A similar feeding experiment with 25% rapeseed oil for 28 days was carried out on adult female rats. Again the sex cycles were normal, but the ovarian cholesterol concentration was significantly increased by the rapeseed oil diet. In both of these experiments the ovaries were smaller in the treated than in the control animals.

A very comprehensive study on the influence of dietary rapeseed oil, erucic acid and other fatty acids on fertility in rats has been carried out by Carroll and Noble[50], who studied reproduction in rats through several generations. Experiments with young female rats indicated that erucic acid or oleic acid in the diet resulted in decreased fertility in several of the animals. Some of the animals did not become pregnant, while in other cases fetal resorption occurred, or there was interference with parturition resulting in death of the mothers in labor. However, in most cases the animals became pregnant and the young were born alive, but there was a high mortality amongst the offspring. Many litters died shortly after birth. The death rate was also high in the following weeks, and many of those surviving until after weaning were stunted in growth. It became apparent that the failure to survive was attributable to deficient mammary gland development and lactation in the mother, which was confirmed by histological examinations. Mating of adult female rats fed 15% erucic acid for more than 8 weeks resulted, with a few exceptions, in pregnancy, but poor survival of the young. Feeding of otherwise complete diets with 10% or more, by weight, of erucic acid to male rats from weaning resulted in degeneration of tubules of the testes after approximately 3 months, and sterility was total after 5 months. Diets with 15% oleic acid (containing small amounts of linoleic acid) or 25% rapeseed oil did not cause deleterious changes of the spermatogenic tissue. Similar experiments with adult rats indicated that sterility also develops, but the impairment possibly occurs at a slower rate.

Efforts to restore normal function to the testes by again placing the animals on a stock diet produces a degree of improvement which suggests that recovery may occur if the degeneration has not reached an advanced stage. The authors point out that the defects resemble those which occur in animals deficient in essential fatty acids or vitamin E. However, they are aware that the erucic acid diet did not produce all the effects of essential fatty acid deficiency[1, 5, 82], and they offer the suggestion that erucic acid may interfere with reproduction by interfering with the metabolism of the essential fatty acids. It should be mentioned that the weight of the testes was markedly affected in growing animals that were fed the erucic acid supplemented diet. In contrast, the weight of the seminal vesicles and prostate appeared to be relatively normal.

Beare et al.[22] noticed that the testes in young Wistar rats fed on 20% rapeseed oil for 5 weeks were significantly smaller than the testes of rats on a similar diet with corn oil. Histological studies of the testes revealed no alteration in cellular characteristics, but a reduction in tubular size. Similar results were obtained in experiments with older rats. In a subsequent study, Beare et al.[18] compared the effect of diets with rapeseed oils, corn oil, or a stock diet (Fox chow) on reproduction in weanling Wistar rats. The females showed no irregularities in their estrous cycles that could be attributed to diet. During the production of three litters it appeared that the rats fed a commercial diet had a superior reproductive performance, particularly in the first litter. Statistical analyses revealed insignificant differences between the effects of the diets containing corn oil and rapeseed oil, both with respect to the number of young born and the number weaned. With rapeseed oil in the parental diet, the weight of the weanling rats was considerably lower. The authors suggest that this is related to decreased milk production resulting from the reduced food intake of animals receiving rapeseed oil and to the presence of some erucic acid in the milk. In the male rats of these latter experiments, no disturbances were observed in spermatogenesis as they sired three litters. However, it should be recalled that rapeseed oil contains a substantial fraction of essential fatty acids.

Beare et al.[21] compared the weight of the testes from young Wistar and Sprague-Dawley rats reared on diets containing 20% rapeseed or corn oil for 6 weeks. They found the same weight of the testes from the Sprague-Dawley rats on the two experimental diets, but smaller testes in the Wistar rats fed on rapeseed oil compared to those on the corn oil diet. In an experiment with 20% partially hydrogenated rapeseed oil, rapeseed oil or corn oil in the diet of young Wistar rats for 6 weeks, Beare et al.[17] found a much lower weight of the testes from rats on the partially hydrogenated oil than on the other two dietary oils.

Noble and Carroll[108] surveyed the literature on dietary erucic acid, vitamin E and essential fatty acids in relation to reproduction in rats. They kept male and female rats on diets containing 10 or 15% erucic acid for rather long periods of time. In male rats testicular changes were induced resulting in patchy to total degeneration of the seminiferous tubular epithelium. The females had several sterile matings, and most of the young born rarely survived until the time of weaning and at that time they were greatly underweight. Normal development of mammary gland tissue occurred during pregnancy, but there was a rapid involution after parturition. The time on the erucic acid diet, as well as the concentration, appeared to be important for the development of the symptoms. Preliminary experiments on recovery were carried out using corn oil, *dl*-*a*-tocopherol acetate, vitamin A acetate and the oxytocic hormone of the posterior pituitary gland, but they were not conclusive.

Paluszak[112] found atrophy of the testes, epididymes and the vesicular gland in rats fed on a diet containing 50 cal.% of fully hydrogenated rapeseed oil. However, this effect is ascribed to a deficiency of essential fatty acids.

Recently, Bourdel and Jacquot[33] studied the effect of rapeseed oil on the testes of rats. They conclude that rapeseed oil does not have any harmful effect on the genital tract of male rats. Thoron[141] found that the relative weight of testes and epididymes was much smaller in pigs fed on rapeseed oil than in those fed on a mixture of olive and palm oil.

Hilditch and Thompson[76] studied the influence of different dietary oils on the fatty acid composition of cow's milk. Feeding of 4 oz. of rapeseed oil twice daily to cows for a week or more resulted in an increase in the proportion of oleic acid and the appearance of erucic acid to the extent of 3.7 weight% of the total milk fatty acids.

Beare *et al.*[19] carried out an experiment on reproduction in Wistar rats over four generations. The rats were fed on 'Fox Chow' alone, with 20% rapeseed oil, corn oil, or a 3:2 mixture of lard and olive oil. The addition of 20% corn oil or the mixture of lard and olive oil to ground fox cubes had no effect on the number of offspring, but the weanling weights of the offspring were greatest on the fox cube diet. Fewer young were born of rats fed on rapeseed oil, and their average weight at weanling was the lowest in the four groups of animals.

In order to obtain an impression of the importance of the mothers' milk for the development of the suckling young the fatty acid composition of milk samples was analyzed. The milk samples were collected from the stomachs of nursing offspring of four females on each diet from the last litter. The fatty acid analyses showed that the fatty acid content in the milk varied with the dietary fat of the mother. Milk from the mothers on fox cubes (low fat content) contained primarily saturated fatty acids, whereas dietary corn oil increased the content of linoleic and oleic acids to such an extent that they became the dominating fatty acids. With the lard-olive oil mixture in the diet the fatty acid par excellence in the milk was oleic acid. Dietary rapeseed oil resulted in milk containing eicosenoic as well as erucic acid, although in smaller amounts than in the original oil. Erucic acid accounted for circa 6% of the total calories of this latter milk. The authors suggest that although the effect of this amount of erucic acid in the milk on the appetite of the young is not known it is conceivable that it might have a significant effect on food intake by being slowly absorbed.

In a following study, Beare *et al.*[14] followed the influence of dietary fat on the fatty acid composition of the milk from nursing rat mothers. A diet with 20% rapeseed oil was fed to mothers for 14 to 15 days following parturition. The authors point out that the rats which had been fed erucic acid for several generations (see above, ref. 19) secreted milk containing somewhat less than half of the amount of that fatty acid produced by the rats of the present study, which had received rapeseed oil only while nursing the young. They suggest that some adaption is possible in the utilization of erucic acid. Further, the fatty acids of milk probably more closely reflect the dietary pattern than do those of the tissues.

14.4.4 Effect of rapeseed oil on the thyroid gland

Anti-thyroid compounds are known to occur in a variety of seeds from plants of the Cruciferae family. The active compound was originally thought to be one of the mustard oils (Gadamer[74], Schneider[125] and Schmid and Karrer[124]). Amongst the compounds with goitrogenic effect is *l*-5-vinyl-2-thiooxazolidone (Astwood et al.[6, 7], Carroll[37]). In studies of the toxic reactions produced in rats by phenyl-thiourea and α-naphthylthiourea, Carroll and Noble[47] found that rats which had been pretreated with anti-thyroid compounds or fed certain goitrogenic diets developed a marked resistance to these toxic compounds. The goitrogenous substances are apparently found in much higher concentration in rapeseed meal than in rapeseed oil (Turner[142], Lips[97]). Crambe meal given to chicks results in enlarged thyroids, poor growth and high mortality. Ammoniation of the meal caused improved growth and all body organs including thyroids appeared normal on autopsy (Kirk et al.[93]) (Cf. Chapter 14).

It is of course of interest to know whether this goitrogenic effect is present also in rapeseed oil. Carroll and Noble[50] noticed that in some of the male rats fed on diets with 15% erucic acid the thyroid gland was enlarged and weighed up to 25 mg; a weight of 13 mg is about normal for rats of comparable body weight. The gland appeared brownish in color, and on section showed a slightly increased colloid content, but no evidence of hyperactivity. A similar weight increase of the thyroid was also seen in some of the female rats on the same diet. Niemi and Roine[106] studied the effect of rapeseed oil on the thyroid function of rats. They used a diet with 50 cal.% rapeseed oil, or a mixture of saponified rapeseed oil fatty acids (91.3%) plus glycerol (8.7%), or the unsaponifiable fraction of rapeseed oil (0.8%) plus soybean oil (99.2%). The authors conclude that rapeseed oil, and specifically its unsaponifiable fraction, contains a factor which interferes with the ability of the thyroid gland to accumulate iodine from the plasma, but which does not seem to prevent the binding of the iodine in organic form. Thus, the thyroid becomes deficient in iodine, the synthesis of thyroxine is reduced and secretion of the thyrotrophic hormone increased. The result is hypothyroidism, despite the presence of an adequate amount of iodine in the diet.

Roine et al.[121] fed pigs on a diet with 28 cal.% rapeseed oil or soybean oil. The pigs appeared in general more sensitive than rats to a high fat diet, as histological examinations of the thyroid gland revealed definite changes indicative of hyperfunction in both series of experiments. In the same experiment, rats were fed on diets providing 0, 15, 30, 50, or 70% of the calories as rapeseed oil. The control rats received 30 or 70 cal.% of soybean oil. No changes were observed in the thyroids of any of the rats.

Sell and Hodgson[127] found a small, if any, effect on the weight of the thyroid gland of feeding cockerels on a diet containing 4% or 8% rapeseed oil for 8 weeks. The rapeseed oil supplied about 2.5% of erucic acid to the total ration when the oil comprised 8% of the diet. They concluded that the goitrogenic material normally found in whole rapeseed and rapeseed oil meals probably is not present in the rapeseed oil tested.

14.4.5 Effect of rapeseed oil on the liver

The weight of the liver, the fatty acid composition of the liver fat, and in some cases the content of liver cholesterol have been studied in relation to rapeseed oil diets.

Carroll[38] found some increase in the weight of the liver and a marked increase in its content of esterified cholesterol when Sprague-Dawley rats were fed on a diet with 25% rapeseed oil for 4 weeks. In long-term experiments with Wistar rats fed 50 cal.% of rapeseed oil, the liver in some of the animals showed a slight indication of fat deposition and degeneration in the central part of the lobes (Thomasson[136]). The effect of feeding erucic, eicosenoic, oleic, or stearic acid to Sprague-Dawley rats for 3 weeks was studied by Carroll and Noble[49]. Erucic acid increased the level of cholesterol in the liver and adrenals but not in the remainder of the carcass. The other fatty acids tested had little or no effect on tissue cholesterol levels. By feeding rapeseed oil to several different species, Carroll[40] revealed a significant increase in the liver cholesterol of rats and some effect in guinea pigs after 4 weeks on the diet, whereas no such effect was seen in mice, rabbits, chicks or dogs.

Beare et al.[20] found that the liver weight of male rats was increased by feeding of rapeseed or corn oil heated to 200 °C for 120 hours. No such effect was observed in female rats. Wagner et al.[143] studied the content of erucic acid in the liver of rats fed on choline-free and choline-containing diets containing 10% rapeseed oil for 18 days. Choline deficiency results in fatty liver. They found 7.8 and 3.8% of the liver fatty acids to be erucic acid in the choline-deficient and control animals, respectively. The same authors[144] examined the content of erucic acid in neutral and phospholipids of livers from rats fed 10% rapeseed oil containing 42% erucic acid for 6 months. Erucic acid occurred in the neutral lipids of all organs tested, but no special accumulation of this fatty acid was seen, although the dietary content was high. The percentage of erucic acid appeared lower in the liver than in any of the other organs tested. Erucic acid apparently was not present in the phospholipids. Similar results were obtained in dogs (Bernhard et al.[24]).

Absolute liver weights were significantly greater for corn oil (20%) fed animals at 3, 4, and 5 weeks than for those on a diet with a similar amount of rapeseed oil (Beare *et al*.[22]). However, statistical adjustment of both groups to the same body weight indicated that the rapeseed oil group had significantly heavier livers.

No differences in liver cholesterol levels were observed in Holtzman rats fed on diets containing rapeseed oil or soybean oil for 8 weeks (Alexander and Mattson[4]).

Carroll[43] studied the effect of erucic acid on the incorporation of acetate-1-C[14] into cholesterol by liver slices. The liver preparations were from young male Sprague-Dawley rats fed semi-synthetic diets containing either no fat or various individual fatty acids for 3 to 4 weeks. Acetate incorporation into cholesterol by liver slices was much greater in animals fed erucic acid either as the free acid or as trierucate, than in those fed no fat, palmitic, stearic, oleic, or linoleic acid. There was, however, in most cases no corresponding effect on the incorporation of acetate into fatty acids. The concentration of cholesterol in the liver was not increased appreciably by erucic acid in the present experiment. Wood and Migicovsky[146] found increased incorporation of acetate into cholesterol by liver homogenates and in intact rats after feeding either erucic acid, rapeseed oil, corn oil or oleic acid. Saturated material such as coconut oil and lauric acid had the opposite effect with respect to the amount of liver cholesterol and *in vivo* incorporation. The saturated lipids had no significant effect on synthesis in homogenates. The effect of the oils in the diet was rapid, the stimulating effect of rapeseed oil being observed after 3 days of feeding. In the literature several studies on the stimulation of various oils and fatty acids on the synthesis of cholesterol are reported, *e.g.* Mukherjee and Alfin-Slater[101], Linazarro *et al*.[94], and Hill *et al*.[78]. The change in cholesterogenesis produced by dietary fat apparently occurs within a short period of time, but it appears not to be a specific effect of erucic acid.

Wistar and Sprague-Dawley rats were compared in their response to 20% rapeseed or corn oil in the basal diet or ground fox cubes for 6 weeks (Beare *et al*.[21]). Sprague-Dawley rats had larger livers than Wistar rats, but in neither strain was the liver weight significantly affected by the dietary oil. However, adjustment of the liver weights for body weights by a covariance analysis showed that those of rats fed rapeseed oil were significantly heavier, and the strain effect was negligible. The total lipid concentration of the livers was unaffected by the dietary oils in the two strains.

Partially hydrogenated rapeseed oil as 20% of the diet of Wistar rats caused a slightly lower liver weight, but a significantly lower concentration of liver lipids than rapeseed or corn oil diets (Beare *et al*.[17]). In a subsequent study, Beare[14] again found that dietary rapeseed oil does not influence the weight of the liver compared to other vegetable oils. Further, dietary fatty acids had more influence on the carcass fat than on the liver fat, where a large proportion of the fatty acids are constituents of the phospholipids which have a relatively constant composition. Erucic acid amounted to 5.6% of the liver fatty acids compared to 11.3% of the carcass fatty acids in rats fed rapeseed oil. Craig *et al.*[58] made a detailed examination of the tissue fatty acids of Wistar rats fed a diet containing 20% rapeseed oil or corn oil for 21 weeks. The content of oleic acid was 48% higher in the dietary corn oil than in the rapeseed oil, but in the liver the oleic acid concentration was highest on the rapeseed oil. Linoleic and linolenic acids showed a more direct relationship to the proportions of these acids in the dietary oils. Of the mono-unsaturated C_{20} and C_{22} acids present in rapeseed oil, and essentially absent in corn oil, eicosenoic acid more closely reflected the content in the dietary fat than did erucic acid. Eicosenoic acid occurred in the greatest concentrations in the phospholipids, whereas erucic acid did not appear in this lipid fraction. From the fact that erucic acid was not found in tissue fat to the extent that it existed in the dietary fat, and that an unusually high proportion of oleic acid was found in rats fed rapeseed oil, the authors suggested a conversion of erucic to oleic acid.

Beare *et al.*[16] added palm oil or methyl esters of fatty acids to rapeseed oil to study the effect of altering the intake of saturated fat on *e.g.* the fatty acid composition of rat organs. Liver weights did not differ significantly irrespective of the type of dietary fat. When rapeseed oil was present in the diet, the level of arachidonic acid in the liver was less than when the diet contained no erucic acid. Thus, it appeared that in the presence of erucic acid the *in vivo* conversion of linoleic acid to arachidonic acid was decreased. Beare *et al.*[23] studied the relative effects of rapeseed oil and corn oil in rats subjected to adrenalectomy, cold, or pyridoxine deprivation. The weights of the livers of rats maintained in the cold (4 °C) were lower than those of unoperated or sham-operated animals at room temperature, irrespective of the dietary fat. Adrenalectomy did not influence the liver weight, whereas vitamin B_6-deficiency resulted in very small livers, the weight of which was not further affected by the type of dietary oil.

Craig and Beare[56] gave rapeseed oil containing 27.8 mole% erucic acid or a synthetic oil containing 30.9 mole% 11-docosenoic acid to young male Wistar rats for 4 weeks. Both of these isomeric acids occurred in all tissues, but in lower concentrations in the liver lipids than in the other organs studied. The same authors[57] compared the effect of rapeseed oil and canbra oil (no erucic acid) on the fatty acid composition of the livers from rats. Canbra oil which contained 2.7% more saturated fatty acids than rapeseed oil, was associated with 7% more saturated acids in the total liver fatty acids. The level of monounsaturated fatty acids in the liver was highest with rapeseed oil.

14.4.6 Effect of rapeseed oil on blood cholesterol

A possible relationship between dietary rapeseed oil or erucic acid and blood cholesterol has been studied showing very little effect of erucic acid on blood cholesterol, and thus, only a few of the more pertinent data on this subject are mentioned below.

Carroll[38] found no effect on the contents of free and esterified blood cholesterol of 25% rapeseed oil in the diet of young rats compared to controls on a commercial diet. Similar results were obtained by Carroll and Noble[49] by adding 0.15% oleic or erucic acid to a fat-free diet fed to male Sprague-Dawley rats for 3 weeks. Comparison of the effect of rapeseed oil on cholesterol metabolism in different species (Carroll[40]) indicated no change in the blood cholesterol of rats, mice, rabbits and dogs, whereas a tendency to hypercholesterolemia was observed in guinea pigs and chickens.

Linko[96] found a hypocholesterolemic action of rapeseed oil in man on a diet containing butter as the only lipid.

Beare et al.[22] compared the utilization of rapeseed oil and corn oil in young and adult rats. They found no specific influence of either of the two oils on plasma cholesterol in young or adult Wistar rats. Similar results were obtained by comparing Wistar rats with Sprague-Dawley rats (Beare et al.[21]). On the other hand, partially hydrogenated rapeseed oil fed to Wistar rats resulted in significantly lower serum cholesterol levels than those obtained with rats fed on either rapeseed oil or corn oil. The highest serum cholesterol value were observed on the corn oil diet (Beare et al.[14, 17]).

Horst et al.[87] found no influence on plasma lipids and cholesterol of diets containing cholesterol alone or with rapeseed oil although the liver cholesterol was increased. Similar results were obtained by Rozynkowa et al.[122], but an almost fat-free diet or one containing hydrogenated rapeseed oil both caused a marked increase in serum cholesterol.

In experiments with Holtzman rats blood plasma cholesterol levels in general were similar whether the diets contained soybean oil or rapeseed oil, and whether the diets supplied 20 or 73 Cal.%. However, on a purified diet high in soybean oil and gelatinized with ethyl cellulose, a very high cholesterol value was observed. The authors offer no explanation for this result, whereas a low value for the comparable semipurified diet without ethyl cellulose reflects the depressed condition of the rats. This has to do with the separation of the oil from the rest of the dietary ingredients.

In an experiment on the effect of erucic acid on cholesterol metabolism Sprague-Dawley rats were fed on a commercial diet with and without a supplement of 15% erucic acid for 3 weeks or 6 months (Carroll[44]). The feeding of erucic acid did not significantly affect the weight of plasma lipid fractions. Erucic acid was found in fairly large amounts in the plasma triglycerides (17.8%), only in small amounts in phospholipids (5.7%), and least in the plasma cholesterol (3.5%). It is interesting that some erucic acid was found in the phospholipids. The small content of erucic acid in cholesterol esters is in agreement with the general impression that this acid has very little effect on the plasma cholesterol.

334

This topic has been discussed to some extent for specific organs in preceding sections of this review. It is the aim of this section to present observations made on a variety of organs in relation to dietary rapeseed oil or erucic acid.

In his studies on the longevity of rats fed on diets containing rapeseed oil or butterfat, Thomasson[136] noticed slight indications of fat deposition and degeneration in the central parts of the liver lobules of some of the animals fed on rapeseed oil. The livers from rats on butterfat were normal. The increase of liver fat did not prevent the rapeseed oil group from attaining a longer life span. No important abnormalities were noticed in the other organs. However, due to the floods in Holland in 1953, most of the autopsy material from this experiment was lost. The author points out that this may be the reason why no other ageing diseases were observed.

Histological studies of mammary gland tissue from female rats fed on rapeseed oil, erucic acid or oleic acid diets revealed deficient development (Carroll and Noble[50]). Diets containing 10% or more erucic acid resulted in degeneration of the tubules of the testes after approximately 3 months and total sterility after 5 months. However, this is probably related to a deficiency of essential fatty acids.

Murray et al.[104] found no histopathological symptoms in the lungs, heart, arteries, liver, kidneys and bladder of rats fed on a diet containing 5% corn oil or the methyl esters of oleic, eicosenoic, or erucic acids for 12 weeks.

Roine et al.[121] made histological studies on the thyroid, heart, liver, spleen, kidneys, adrenals, stomach, small and large intestine, aorta and striated muscle from rats and pigs. Both animal species were kept on diets containing rapeseed oil. The rats fed on diets containing 50 Cal%, and especially 70 Cal%, rapeseed oil for 6 weeks showed interstitial inflammatory changes in the myocardium. Similar changes were not seen when 30 Cal% or less rapeseed oil or even 70 Cal% soybean oil diets were used. With the exception of the heart no other examined organ revealed definite pathological changes in any of the rats. In the pigs histological examinations of the thyroid of all the animals fed on either rapeseed oil or soybean oil revealed definite changes indicative of hyperfunction. Similarly, all pigs showed evidence of interstitial myocarditis and inflammatory reactions in the gastric mucosa. The authors conclude that the pigs appeared in general to be more sensitive than rats to diets high in fat.

Young Wistar rats fed on 20% rapeseed oil for 5 weeks had testes which were significantly smaller than the testes from rats on a similar diet with corn oil (Beare et al.[22]). Histological examinations of the testes revealed no changes in cellular characteristics but a reduction in tubular size. Similar results were obtained in older rats.

Raulin et al.[117] described a myocarditis of nutritional origin in rats. However, the authors conclude that erucic acid is not responsible and suggest that it may be an effect related to the unsaponifiable fraction of rapeseed oil or the imbalance between saturated and monoenoic acids in this oil. Rocquelin and Cluzan[119] made histological studies on the heart and aorta, liver, kidneys, spleen, pancreas, adrenals, lungs, testes, fat depots, and gastro-intestinal tract from rats fed on diets containing rapeseed oil either with 44% or 0% erucic acid. Severe myocarditis was observed in 7 months old male or female rats on either of the two types of rapeseed oil. The frequency of myocarditis was higher in the males than in the females. The authors conclude that their results indicate that myocarditis in long-term experiments with rats is not a specific effect of erucic acid but may be related to a too low content of saturated fatty acids, an imbalanced ratio between saturated and monoenoic acids, or to the unsaponifiables in the rapeseed oils. It should be also mentioned that both types of rapeseed oils after 3 months induce a weight increase of the heart, liver, kidneys and spleen of the male rats, whereas similar changes were much less prominent or not observed in females.

In one case caries has been studied in relation to dietary rapeseed oil (Wysocki and Ziomski[147]). With 3% rapeseed oil for 13 months all rats had caries. Similarly, caries was also observed with lard, butter or margarine as dietary fat but not with soybean oil.

Alexander and Mattson[4] found no abnormalities on gross autopsy of the heart, lungs, liver, spleen, kidneys, thymus, and thyroid from Holtzman rats fed on a diet containing 20 or 73 Cal% gelatinized rapeseed oil. However, a slight hyperplasia was noticed in the adrenals of the animals that received the high level of dietary fat.

Homrowski[84] found no histopathological changes in the liver, spleen, kidneys and adrenals of rats fed for 12 months on diets containing 10% rapeseed oil or various types of heated fats containing rapeseed oil.

Histopathological surveys in relation to rapeseed oil diets have also been carried out in other species than the rat. Thomasson et al.[140] have repeated their original long-term experiment[136] but in the reference at hand[140] the histopathological data of these animals are not yet available. However, in the same study mice which were fed on diets containing 40 Cal% or more rapeseed oil showed lower mortality over a period of 89 weeks than the corresponding animals fed butterfat. The latter group displayed significantly more cases of chronic kidney disease than the rapeseed oil group. The anatomical investigations demonstrated that the frequency of tumors was significantly higher in the butterfat group than in the rapeseed oil group. This may be the result of increased food intake in the mice reared on butterfat compared to those on rapeseed oil. These differences in pathology between the rapeseed oil and the butterfat animals may be the cause of the shorter life-span of the latter animals.

A short-term experiment with guinea pigs fed on 50 Cal% oil for 6 weeks revealed hypertrophy of the liver and spleen, increased erythropoiesis and a slight fatty infiltration of the heart. None of these symptoms was observed in the guinea pigs fed a similar butterfat diet. In hamsters 60 Cal% rapeseed oil induced a pale heart muscle, and lipid accumulation occurred in the heart and the liver. Many ducklings fed on diets containing 40 Cal% or more of rapeseed oil developed hydropericardium, and fat infiltration was found in heart and skeletal muscles. Analysis of this fat demonstrated the presence of large quantities of erucic and eicosenoic acid. Hydropericardium also developed in ducklings fed trierucate. Further, the ducklings had mottled livers, in a few cases swollen kidneys were observed, and the average weights of the liver, heart and spleen were increased.

14.6 EFFECT OF RAPESEED OIL ON VITAMIN REQUIREMENT

The liver storage of vitamin A was determined in young Wistar rats fed on a diet containing either 20% rapeseed oil or corn oil. Vitamin A in doses of approximately 200 I.U. in 0.1 ml of rapeseed oil or corn oil was administered by mouth twice weekly (Beare et al.[22]). Analyses of the livers after 5 weeks showed that the storage of administered vitamin A was similar with both oils. Thus, there appeared to be no adverse effect of rapeseed oil on the absorption of vitamin A, and consequently no indirect effect of the vitamin on weight gain.

Carroll[42] found that young male Sprague-Dawley rats fed a semi-synthetic diet containing 15% erucic acid but lacking fat-soluble vitamins level off at a lower weight than those fed similar diets containing no fat or other individual fatty acids. Further, erucic acid hastened the onset of a deficiency of essential fatty acids. Growth was resumed in the erucic acid fed animals when vitamin A acetate or corn oil was administered, but neither was effective alone. The corn oil could be replaced by methyl linoleate. Rats fed an erucic acid diet containing fat-soluble vitamins still level off at a lower weight than rats fed a similar diet containing oleic acid. Vitamin A acetate alone caused some additional growth in rats which had levelled off on diets containing no fat or palmitic acid, but it was more effective when combined with corn oil. The author suggests that the requirement of Sprague-Dawley rats for vitamin A is increased by dietary erucic acid. Beare et al.[21] compared the effect of erucic acid on the liver storage of vitamin A in young Sprague-Dawley and Wistar rats. The animals were deprived of vitamin A for 18 days. Then some of the animals were killed and the amount of vitamin A in the liver was determined. The remaining rats were fed purified vitamin A-free diets containing 20% rapeseed or corn oil, and were dosed orally with either 75 or 150 I.U. of vitamin A in 0.1 ml of rapeseed oil or corn oil for 14 consecutive days. Two days later the animals were killed. No significant effect of the strain of rat or the type of dietary oil on the liver storage of vitamin A was observed. The rate of depletion of vitamin A from the liver following a single dose of 1200 I.U. of vitamin A was also studied indicating no significant differences attributable to the dietary oils.

Murray *et al.*[102] studied the effect of corn oil, olive oil, and rapeseed oil on the rate of vitamin A depletion of liver and kidneys. Initial stores of vitamin A in the liver and kidneys were 92 ± 4 and 1.5 ± 0.1 I.U., respectively. During the course of vitamin A depletion during two months following a single oral dose of 1000 I.U. vitamin A in 0.1 ml of cottonseed oil, the total content of vitamin A in the liver was similar for each dietary oil. As the liver stores of vitamin A declined, the relatively small amount in the kidneys increased throughout the depletion period but less markedly in the group fed rapeseed oil. The authors conclude that rapeseed oil has only a slight influence on vitamin A metabolism.

Beare *et al.*[23] found that pyridoxine deprivation of rats had a more adverse effect on weight gain and food intake when the animals were fed on a diet containing 20% rapeseed oil than when corn oil was used as dietary fat, although the total vitamin B_6 in the liver was independent of the type of dietary oil. Further, rats deprived of vitamin B_6 and fed rapeseed oil deposited a smaller proportion of fat than those fed corn oil. Erucic acid was somewhat more concentrated in the carcass fat of rats not supplied with pyridoxine than in those that were.

Carroll[38] found the concentration of ascorbic acid in the adrenals to be somewhat lower in rats fed rapeseed oil than in the controls (fox chow), but the absolute amount was nearly the same in the two groups of animals.

Removal of one adrenal from normal rats produced a marked depletion in the concentration of ascorbic acid and cholesterol of the second adrenal. Carroll and Noble[48] compared this response in rats fed on rapeseed oil to that of normal rats. They found that the lowering of ascorbic acid and cholesterol is much the same in both groups in spite of a marked difference in cholesterol content between the treated and control groups.

14.7 IMPORTANCE OF STRAIN, SEX AND AGE OF EXPERIMENTAL ANIMALS

In preceding sections of this chapter it has been mentioned that strain, sex and age of the experimental animals may be of some significance for the evaluation of the effect of rapeseed oil and erucic acid in nutrition. Some of the more important experimental data on this topic are briefly reviewed.

Wistar and Sprague-Dawley rats showed a poorer growth rate on a diet containing rapeseed oil than on corn oil (Beare *et al.*[21]), but throughout the experimental period of 6 weeks the Wistar rats grew much less than the Sprague-Dawley rats on rapeseed oil. However, after 4 weeks the differences in weight gain were dependent on earlier growth rate and not on the gains obtained from 4 to 6 weeks.

The coefficient of digestibility of rapeseed oil, erucic acid and erucates is low in Sprague-Dawley rats. This has been shown in several experiments by Carroll and coworkers[39, 41, 51, 40]. A much higher coefficient was observed in Wistar rats (Beare *et al.*[22] and Murray *et al.*[104]). A similar difference in the digestibility of corn oil was not found in the two strains of rats. Alexander and Mattson[4] used Holtzman rats in their experiments. They found almost the same degree of absorbability of rapeseed oil and soybean oil.

The adrenals of Sprague-Dawley rats were larger on diets containing rapeseed oil than on a similar corn oil diet, whereas no influence of the two oils was observed on adrenal weight in Wistar rats. Further, significant differences in total cholesterol paralleled those of adrenal weight (Beare *et al.*[21]). Dietary rapeseed oil produced higher adrenal cholesterol values in Holtzman rats than did soybean oil (Alexander and Mattson[4]).

Rapeseed oil or corn oil did not interfere with the weight of the testes in Sprague-Dawley rats, but in Wistar rats the testes were smaller on the rapeseed oil diet (Beare *et al.*[21]). Severely impaired spermatogenesis was observed after 5 months in Sprague-Dawley rats fed on a commercial diet to which was added 10% or more of erucic acid. No such effect was noticed in Wistar rats fed on a diet containing rapeseed oil equivalent to 8.5% erucic acid for about 11 months (Beare *et al.*[18]). However, the latter two experiments involve not only a strain difference but also a very important difference in dietary linoleic acid.

The sex of the experimental animals appears to be of little or no importance for the evaluation of the nutritive value of rapeseed oil and erucic acid. Generally growth depression and deposition of erucic acid and eicosenoic acid are similar in either sex. It may be mentioned that, *e.g.*, Beare *et al.*[20] found increased liver weight in male Wistar rats fed on a diet containing rapeseed oil or corn oil heated to 200 °C for 120 hours, whereas no such effect was observed in the females. Rocquelin and Cluzan[119] found severe myocarditis in male and female rats after 7 months on a diet containing rapeseed oil (40% erucic acid) or canbra oil (no erucic acid). However, the frequency of myocarditis was higher in the males than in the females.

The age of the experimental animals appears to be of minor importance in nutrition studies concerning rapeseed oil and erucic acid. However, Roine and Uksila[120] claim that the digestibility of rapeseed oil is lower in young than in adult rats. On the other hand, Carroll and Richard[51] found the digestibility of erucic acid to be better in young than in older rats.

14.8 METABOLISM OF ERUCIC ACID

The preceding sections of this chapter have illustrated that feeding of rapeseed oil to experimental animals produces a variety of effects, *e.g.* growth retardation, occurrence of small amounts of erucic acid in body lipids, changes in cholesterol metabolism, and reproduction disturbancies. Erucic acid is a very prominent fatty acid in rapeseed oil but is not a common acid in animal lipids except when it has been included in the diet. For these and several other reasons a number of experiments have centered on the metabolism of erucic acid.

Bernhard and Vischer[26] isolated erucic acid from rapeseed oil and saturated it with deuterium, thus forming deuteriated behenic acid. The ethyl ester of this saturated fatty acid was fed to rats either as such or mixed with olive oil (1:1) for 7 and 8 days, respectively. Approximately 40% of the dietary behenic acid was absorbed. Isolation of the fatty acids from liver, gut and body fat showed that stearic, palmitic, myristic and oleic acids were labeled with deuterium. The highest concentration of deuterium in the fatty acids isolated occurred in stearic acid. Behenic acid was found only in trace amounts in the body fat. The authors interpreted their results as an indication that behenic acid is metabolized by β-oxidation.

Thomasson[135] suggested that the long-chain fatty acids in rapeseed oil are metabolized to the same extent and in the same way as e.g. the C_{18}-acids, but that the rate at which metabolism takes place is different for the various fatty acids. Hopkins et al.[85] studied the deposition of eicosenoic and erucic acids in comparison to oleic acid when fed as the methyl esters to male and female Wistar rats. Eicosenoic and erucic acid were found to be deposited in the body fat. However, erucic acid was deposited in considerably smaller amounts (circa 12% of the fatty acids) than was eicosenoic acid (circa 28%). Neither of the two acids appeared to be as readily deposited in body fat as the linoleic acid of corn oil. There was no evidence of conversion of eicosenoic acid to arachidic or stearic acids, or of erucate to behenic, arachidic or stearic acids in any appreciable amounts. Similarly, there appeared to be little or no conversion of eicosenoic acid to oleic acid or of erucic acid to eicosenoic acid. Wagner et al.[144] kept adult rats on a rapeseed oil diet for 6 months. A true accumulation of erucic acid could not be determined. The decrease of the biological half-life of erucic acid, after termination of the rapeseed oil diet, was calculated to be about 18–31 days in various organs and depots. They concluded that erucic acid behaves like fatty acids normally utilized in metabolism.

Carroll[44] found that feeding of erucic acid to Sprague-Dawley rats resulted in large amounts of erucic acid in the cholesterol esters of the adrenals and moderate amounts in the triglycerides of adrenals and plasma. Further, eicosenoic acid was observed in some cases in adrenal cholesterol esters and in triglycerides of both adrenals and plasma. The level of oleic acid relative to other fatty acids was increased in most lipid classes of adrenals and plasma. The author suggests that erucic acid has undergone stepwise β-degradation with partial stabilization at the C20- and C18-acids. It was noticed also that arachidonic acid was more concentrated in the control rats than in those which had received 15% erucic acid in a commercial diet. Carroll[45] then studied the metabolism and fate of erucic acid and nervonic acid in rats. Palmitic acid-1-^{14}C, erucic acid-2-^{14}C, and nervonic acid-2-^{14}C were administered to young adult rats perorally or by tail-vein injection. The experiments showed that nervonic acid is absorbed at a slower rate than erucic acid and palmitic acid. Tissue lipids of rats given palmitate generally contained a higher percentage of the radioactivity administered than lipids from corresponding tissues of rats given erucic or nervonic acids. When administered by mouth palmitate contributed more radioactivity to expired carbon dioxide during the first 6 hours than erucic acid or nervonic acid. When the acids were given by tail vein carbon 2 of erucic acid and nervonic acid contributed more radioactivity to respiratory carbon dioxide than carbon 1 from palmitic acid. The earliest samples collected from the rats given palmitate had the highest activity observed for that fatty acid, but there was a definite time lag before peak activity was reached in rats given erucic acid or nervonic acid. The author suggests that the rate of oxidation of the second carbon atom in erucic acid to carbon dioxide may be different from that of carbon atoms further down the carbon chain.

Craig et al.[58] made a detailed examination of the tissue fatty acids of rats fed on either rapeseed or corn oils for comparably long periods of time. The fact that erucic acid was not found in the tissue fat to the extent that it existed in the dietary fat and that an unusually high proportion of oleic acid was found in rats fed rapeseed oil suggested a conversion of erucic acid to oleic acid. Beare et al.[16] found the level of arachidonic acid in the liver to be less when rapeseed oil was present in the diet than when the diet contained no erucic acid. The authors felt that in the presence of erucic acid the conversion of linoleic acid to arachidonic acid was depressed. Beare et al.[23] concluded from experiments in rats deprived of pyridoxin that the lack of this vitamin implies a decreased rate of metabolism of erucic acid. However, they also point out that the greater accumulation of this acid may be related to the presence of small proportions of C_{16}-acids in the fat. Carroll[46] studied the distribution of radioactivity among lipid classes of liver at different time intervals after administration of labeled oleic acid, erucic acid and nervonic acid in rats. 1-^{14}C-oleic acid, 2-^{14}C-erucic acid, and 2-^{14}C-nervonic acid were administered by tail vein. The distribution of radioactivity in liver lipids was determined at intervals from 15 minutes to 6 hours after injection. Nervonic acid, and to a lesser extent erucic acid, were deposited in the liver as free acids and were metabolized more slowly than oleic acid. Further, the results indicate that as chain-length increases fatty acids are metabolized more slowly and accumulate to a greater extent as free acids. This tendency appears to be more marked with saturated than with monoenoic acids.

In 1967 Craig and Beare[56] undertook a study in vivo of positional isomers of docosenoic acids. Two groups of Wistar rats were fed on a diet containing rapeseed oil (27.8% Δ^{13}-docosenoic acid, erucic acid, and 13.6% Δ^{11}-eicosenoic acid) or a 'synthetic' oil (30.9% Δ^{11}-docosenoic acid, cetolic acid; no eicosenoic or hexadecenoic acids) for 4 weeks. At the end of the experiment the rats were killed and the bodies dissected and divided into skin, head, organs, and remaining carcass. Methyl esters of monoenoic fatty acids were prepared and isolated by silver nitrate–silicic acid chromatography (Bhatty and Craig[29]). The monoene fraction isolated was subdivided into monoene esters according to chain length by preparative gas-chromatography. Each monoene fraction was then oxidized with permanganate–periodate reagent (Mallard and Craig[98]) for subsequent gas-chromatography of the resultant dicarboxylic acids to determine the composition of the isomeric fatty acids and thus distinguish isomeric intermediates of the β-oxidation of the various acids. When rapeseed oil was fed, erucic acid was found in each lipid sample. Δ^{11}-Eicosenoic acid occurred in the body fat in somewhat higher concentrations than erucic acid although the dietary fat contained only half as much of the former. Dietary certolic acid, Δ^{11}-docosenoic acid, was incorporated into the body fats to the same extent as erucic acid. Thus, the rats did not distinguish between the two isomeric acids. Eicosenoic acid was absent from the semi-synthetic oil, but appreciable amounts of this acid occurred in the body fats.

Analyses of the dicarboxylic acids obtained by oxidative degradation of the various monoene fractions from rats on rapeseed oil revealed the presence of Δ^7- and Δ^9- hexadecenoic acids, Δ^9- and Δ^{11}-octadecenoic acids. From animals reared on the semi-synthetic oil the same two hexadecenoic acids were present as when rapeseed oil was fed. Three isomers, Δ^7-, Δ^9- and Δ^{11}-acids, constituted the octadecenoic fraction; the eicosenoic fraction contained two isomeric acids, Δ^9- and Δ^{11}-eicosenoic acids. Stepwise β-oxidation of erucic acid produces Δ^{11}-eicosenoic acid and oleic acid, both of which are normally present in animal tissues. A similar degradation of the positional isomer, certolic acid, would give rise to Δ^9-eicosenoic and Δ^7-octadecenoic acid which are not usually present in rat tissues. From a summation of the proportions of Δ^7-octadecenoic, Δ^9-eico-senoic, and Δ^{11}-eicosenoic acids the authors conclude that their data verify the role of β-oxidation in the metabolism of docosenoic acids.

The metabolic pathway of erucic acid has been studied recently by Carreau et al.[35, 36]. They kept male rats for 6 months on three different types of diets: a fat-free diet, and the same diet supplemented with 1.6% sunflower seed oil, or with 20% rapeseed oil. At the end of the experiment the animals were fasted for 24 hours. Then 4 ml of an emulsion containing 14-^{14}C-labeled erucic acid was injected through the femoral vein. The animals were killed 2, 26, and 50 hours after injection, and the fatty acids of the liver were analyzed. Two hours after injection the livers of animals from all 3 dietary groups contained large quantities of labeled erucic acid and appreciable amounts of other monoenoic acids, primarily oleic acid, and only very small amounts of palmitoleic acid. The specific activity of the total amount of monoenoic acids two hours after injection of labeled erucic acid was greatest in the animals fed on the diets supplemented with sunflower seed oil or rapeseed oil. The opposite effect was noticed regarding the content of radioactivity of the saturated, di- and polyenoic acids. The ratio of the specific activity of erucic acid to other monoenoic acids was of the order of 0.5. The authors conclude that oleic acid appears to be the principal metabolic product of erucic acid. Twenty-six hours after the injection very little labeled erucic acid occurred in the livers. Furthermore, the small amounts of non-labeled erucic acid in the livers are also lost, except in the animals fed on rapeseed oil. The other monoenoic acids still contain some radioactivity even after 50 hours. As erucic acid is metabolized to oleic acid, the former thus loses its character of an atypical fatty acid.

The rate of oxidation of erucic acid compared to that of oleic acid was studied by Bach *et al.*[10]. A difference in oxidation rate may possibly explain, *e.g.*, the poor growth rate of animals fed on diets containing appreciable amounts of rapeseed oil. They used a procedure by Bach *et al.*[9] for measuring the specific radioactivity of expired carbon dioxide during successive periods of 5 minutes and the percentages of the administered radioactivity collected from the expiration with time. They used 1-^{14}C-oleic acid, 10-^{14}C-oleic acid, and 14-^{14}C-erucic acid. The labeled acids were mixed with olive oil and given *per os* as a single dose. Very soon after the ingestion of the labeled samples, the amount of specific activity in the expired carbon dioxide was highest with oleic acid-1-^{14}C, which was higher than that of oleic-10-^{14}C, which again was higher than that of erucic acid-14-[14]C. This difference is ascribed to a slower absorption rate of erucic acid than of oleic acid. Over a period of 8 hours the ability to form radioactive carbon dioxide decreased in the order: 1-^{14}C-oleic > 10-^{14}C-oleic > 14-^{14}C-erucic acid. The cumulative data as well as those depicted as a function of time are much higher for the 1-^{14}C-oleic acid than for the 10-^{14}C-oleic acid, which are higher than those for 14-^{14}C-erucic acid. However, calculations based on the fact that the labeled carbon atom has a different location in the three acid molecules, *i.e.* different length of the carbon chain must be metabolized before a molecule of carbon dioxide is formed *via* β-oxidation, indicate the same rate of oxidation of the three acids. On the other hand, differences in the yield of the oxidation of the acids may be ascribed to formation of less active coenzyme A thioesters of erucic than of oleic acid (Pande and Mead[113, 144]).

The apparent discrepancies between the present results of Bach *et al.*[10] and those of Carroll[44] obtained by studies on expired radioactive carbon dioxide after administration of labeled fatty acids were found to be non-existent after recalculation. Bach *et al.*[10] also studied the rate of β-oxidation of erucic acid. Rats were kept for one month on a low-fat diet, the same diet supplemented with 1.6% sunflower seed oil or 20% rapeseed oil. Erucic acid-14-^{14}C was administered as 0.20 μCi in 0.12 ml of olive oil per 100 g of body weight. They found the rate of total oxidation of erucic acid to be faster in the rats on the fat-rich diet than in those on the low-fat diet. They suggest that the C2-units from β-oxidation in the rats deprived of dietary fat are immediately reutilized to a great extent in the synthesis of other fatty acids because of a serious demand for fatty acid synthesis, leaving only a small amount for total oxidation, and leading to low radioactivity of the expired carbon dioxide. On the other hand, rats fed on a diet rich in fat can carry out a fast total oxidation of the labeled fatty acid because synthesis is less urgent under these circumstances.

A classical general parameter of a diet being sufficient and well balanced is its effect on growth rate. It should be pointed out that parameters such as growth and weight gain must be used with forethought in nutritional experiments. In principle weight gain can take place throughout the whole life span. Such a situation is not desirable though, and especially not if good health and longevity are considered simultaneously.

Longevity, or expected longevity, has been studied in relation to various food components, *e.g.* proteins of animal or vegetable origin, dietary fat, the dietary ratio of protein and carbohydrates, and *ad libitum* versus restricted access to food. A number of valuable references to some of the early work on longevity in relation to various food constituents is given by French *et al.*[73]. The present discussion is confined to dietary fat and longevity.

French *et al.*[73] fed male and female Wistar rats on a stock diet alone, or 80% stock diet plus 20% sucrose, or 20% corn oil. The male and female rats grew well on the diet high in fat, but the life span of the males decreased markedly, and there was also a less pronounced but significant decrease in the longevity of the females. Increased efficiency of utilization of the diet was correlated with decrease in life span, but increased caloric intake per se was not so correlated. The histopathological data and the frequency of disease and abnormalities, however, revealed nothing that could be interpreted as a fat-induced cause of premature death.

Food restriction may have a bearing on life span. The fact that diets high in rapeseed oil result in a reduced intake of the experimental diet prompted Thomasson[136] to compare the longevity of male rats fed on diets containing 50 Cal% of rapeseed oil or butterfat. He found that the animals eating the rapeseed oil diet lived 20 to 25% longer than those on the corresponding butterfat diet. The pathological findings did not give distinct information regarding the cause of death. Growth rate and daily food consumption were lowest in the animals on the rapeseed oil diet. However, analysis of the experimental data showed that the longer life span of the rapeseed oil animals cannot be explained simply by a slower growth or a smaller body weight. The author suggests that if the difference in longevity of the butterfat and rapeseed oil fed animals has to do with the decreased food intake of the latter group, the active substance is probably erucic acid. In a repetition of this experiment, Thomasson *et al.*[140] fed rats on diets containing 10, 20, 30, 40, 50, 60 and 70 Cal% butterfat or rapeseed oil. However, something apparently went wrong in this experiment, and the authors state that their data only justify the conclusion that on feeding rapeseed oil the life span of rats is certainly not shortened. These unsatisfactory results led the authors to set-up a third long-term experiment. Provisional data (after 100 weeks) of this experiment, which was carried out with SPF-rats (specific pathogen-free rats), confirm the results of the first experiment in which rapeseed oil induced a longer life span than butterfat. This experiment also included diets containing corn oil, soybean oil, coconut oil and whale oil. The authors do not feel convinced that growth and mortality are invariably correlated.

Roine and Uksila[120] mention that unpublished data from rat experiments in their laboratory have shown that rapeseed oil at the 30 Cal% level did not cause any pathological changes in long-term experiments. It may also be mentioned that Murray et al.[102] found that weight gain in vitamin A deficient rats was reduced somewhat earlier when rapeseed oil was fed, but the survival time of rats fed corn oil or rapeseed oil was similar.

Thomasson et al.[140] studied longevity in mice which were fed on diets containing 10, 20, 30, 40, 50, 60 and 72 Cal% rapeseed oil or butterfat over a period of 89 weeks. For animals in all groups receiving 40 Cal% or more fat in the diet, the 50% mortality was reached significantly later with rapeseed oil than with butterfat. At the end of the experiment less than 50% of the mice fed on a diet with 10 Cal% rapeseed oil had died.

In general it may be stated that the factors determining life span of man and experimental animals are multifarious, and the type and amount of dietary fat may be two of them.

14.10 RECENT REPORTS ON THE PATHOLOGICAL EFFECTS OF DIETARY RAPESEED OIL

While the present review was in proof, the author learned of a series of very recent experiments on the pathological effects of dietary rapeseed oils. The author is grateful for having had access to the manuscripts, which will be reviewed briefly.

Abdellatif and Vles[1a] investigated the pathological effects of dietary rapeseed oil in rats. Fatty acid infiltration was found in heart, skeletal muscle and adrenals after feeding 60 Cal% rapeseed oil for only two weeks. Studies of the evolution of the pathological changes over 2, 4, 8 and 16 weeks in rats given 60 Cal% of rapeseed oil revealed a regression of the fatty infiltration of the skeletal muscles, adrenals and hearts with increase of the dietary period. However, in the heart necrotic foci, aggregations of mononuclear cells and an increase in the connective tissue elements ensued. Dose–response studies in rats over two weeks comprised diets containing 10, 20, 30, 40, 50 and 60 Cal% rapeseed oil which had an erucic acid content of 46%. The minimum level of this rapeseed oil causing fatty infiltration of heart and skeletal muscles was found to be 20 Cal%. In another experiment, rats were fed 36 Cal% of a rapeseed oil containing 48% erucic acid, or an equivalent amount of erucic acid as glyceryl trierucate for 3 weeks. In both cases the total dietary fat was made to equal 60 Cal% by addition of soybean oil. The pathological changes of the heart, adrenals and skeletal muscles were observed as described above. The authors conclude that erucic acid is apparently responsible for the pathological effects of rapeseed oil. Studies of the progression of the fatty acid infiltration of the heart and skeletal muscles showed that it occurs after only one day on a diet containing 50 Cal% rapeseed oil and becomes most severe after 3–6 days on the diet. The myocardial fatty infiltration decreased after discontinuation of the rapeseed oil diet and replacement with a fat-free diet or a diet containing 50 Cal% sunflower seed oil. None of the described pathological findings caused by dietary rapeseed oil was observed in the various control groups given sunflower seed oil. The authors suggest that the fatty acid infiltration of the heart and skeletal muscles reflects a slow oxidation rate of erucic acid in these tissues.

The same authors[1b] studied the pathological effects of dietary rapeseed oil (10–60 Cal%) in Peking ducklings. It was found that 30 Cal%, or more, caused growth retardation, mortality, increased hematocrite values and reticulocyte counts, severe hyperpericardium, hypertrophy of the heart, cirrhotic changes of the liver, lipoidosis of the spleen, and severe fatty infiltration associated with cell infiltration of the heart and skeletal muscles. The pathogenicity of rapeseed oil is found to be due to its content of erucic acid.

These studies were followed up by another experiment by Abdellatif and Vles[1c] in ducklings, where the dietary rapeseed oil was supplemented with olive oil (rich in oleic acid), safflower oil (rich in linoleic acid) or tallow (comparatively rich in palmitic acid). Unsupplemented rapeseed oil and mixtures of rapeseed oil with olive oil, or safflower oil resulted in growth retardation, mortality, lipoidosis of the spleen, cirrhotic changes of the liver and vacuolar changes of the heart and skeletal muscles. The diet containing a mixture of rapeseed oil and tallow caused no mortality, better growth, improved the liver and spleen morphology, decreased the incidences of hydropericardium, but did not markedly alter the incidence and severity of the vacuolar changes of the myocardium and skeletal muscles. The effects of supplementing the diets isocalorically in fats and in erucic acid with increasing amounts of hydrogenated palm oil or glyceryl trilaurate in place of carbohydrates (corn starch) were also studied. Increasing amounts of hydrogenated palm oil increased growth, decreased the incidence and severity of liver cirrhosis, splenic lipoidosis and hydropericardium. The vacuolar changes of the skeletal muscles were not affected, and those of the myocardium were aggravated. It is concluded that the excess of erucic acid and the deficiency of palmitic acid in rapeseed oil are responsible for the pathological changes observed in ducklings.

The effect of hydrogenated palm oil on rapeseed oil-induced pathological changes in ducklings and guinea pigs was studied by Vles and Abdellatif[142a]. In these experiments isocalorific fat mixtures containing the same amount of erucic acid (17.6 Cal%) and varying levels of palmitic acid (3.6–17.6 Cal%) for ducklings, and 11 Cal% erucic acid and 3–15 Cal% palmitic acid were used for guinea pigs. In ducklings the diet rich in palmitic acid improved growth, decreased mortality, the hydropericardium condition, splenic and liver changes, but did not affect the pathology of heart and skeletal muscles. Feeding of rapeseed oil to guinea pigs induces hemolytic anemia, hypertrophic spleen and hydropericardium/pale myocardium. The minimum effective dose is 25 Cal% rapeseed oil. In guinea pigs the increase of dietary palmitic acid in the rapeseed oil diet improved the morphology of the spleen and liver as well as the values for hemoglobin content, packed cell volume and non-electrolyte hemolysis.

Abdellatif and Vles[1d] compared the pathological effects of rapeseed oil and canbra oil in rats. The canbra oil contained 2% erucic acid. The controls were fed on a diet containing sunflower seed oil. In one experiment, the dietary fats were fed at a 60 Cal% level for two weeks. No difference was observed in food intake and growth response between canbra oil and sunflower oil. Neither did any of these groups show pathological changes, in contrast with the animals in the rapeseed oil group which ate less, grew less and developed the characteristic lesions described above. Feeding of 50 Cal% of canbra oil to rats for three days revealed no pathological changes. The authors also repeated the experiments of Rocquelin and Cluzan[119] who found that canbra oil causes myocardial fibrosis. However, Abdellatif and Vles[1d] found that 30 Cal% canbra oil or sunflower seed oil fed to rats for 24 weeks did not retard growth, cause increase of the weight of heart and kidneys, nor did it possess pathogenic properties comparable to those of rapeseed oil. Mild pathological changes were observed in both the animals fed canbra oil and those fed sunflower seed oil. The authors conclude that these changes may be designated as spontaneous and in no way comparable to the effect of rapeseed oil. Further, canbra oil produced no pathological abnormalities in rats. These studies apparently support the idea that the pathological effects caused by conventional rapeseed oil are primarily a result of its content of erucic acid.

Houstmüller et al.[87a] have investigated the amount and composition of the lipid classes in the heart of rats fed a diet containing 50 Cal% of rapeseed oil for periods varying from 1 day to 6 weeks. A sharp increase in lipid content was observed after 3 days on the diet, which is mainly due to an increase in triglycerides. An increase in the content of free fatty acids was also observed. Of the total lipids, 27% of the fatty acids was erucic acid. During the following 3 days the lipid content fell about 4% and was normal in about 5 weeks. A similar pattern was found for the triglycerides and the free fatty acids but both classes remained elevated throughout the experiment. Studies of heart mitochondria *in vitro* revealed that the rate of ATP synthesis is lower for an erucic acid-containing diet than for a diet containing sunflower seed oil. Further, erucic acid is apparently responsible for the malfunction of the mitochondria. The effect of dietary erucic acid on the rate of mitochondrial oxidation of glutamate, x-oxoglutarate, caprinate, succinate and malate was also substantially reduced. From a dose–response curve, a linear relationship was found between the amount of erucic acid ingested and reduced ATP production of the heart mitochondria. The authors suggest that the fatty infiltration of the rat heart and the malfunction of the isolated mitochondria may be related to the observed increase in free fatty acids. This suggestion is supported by the findings that erucic acid, like oleic acid, completely inhibited the oxidation of glutamate in isolated mitochondria.

Beare-Rogers[23a] has found erucic acid in rapeseed oil or other docosenoic acids in partially hydrogenated herring oil to increase the fat deposition in heart myofibrils of young Sprague-Dawley rats. More docosenoic acid was deposited from liquid rapeseed oil than from partially hydrogenated rapeseed oil or partially hydrogenated herring oil during the first week, when the oils supplied 40% of the calories in the diet. During the following two weeks the accumulated fatty acids decreased to about half of their former concentration. Less erucic acid was deposited by older rats than in experiments with very young rats (one week after weaning). A zero-effect level of abnormal fat deposition from liquid rapeseed oil fed to young rats was found to be 10 Cal% and was associated with a concentration of less than 5% of erucic acid in the cardiac fatty acids. It is concluded that cardiac tissue in young animals is unable to metabolize the long-chain fatty acids readily.

14.11 SUMMARY

Rapeseed oil in quantities of 10–20 weight % or more as the sole dietary fat to experimental animals produces growth retardation compared to practically all other vegetable oils. Rapeseed oil contains very appreciable amounts of erucic acid which is not normally present in the diet nor in the tissues of experimental animals. Much effort has been expended on the unravelling of the nutritive value and the biological effects of rapeseed oil and its particular fatty acids. However, the presence of erucic acid in rapeseed oil is apparently not the only factor responsible for the dietary effects of rapeseed oil. Within certain limits the balance between saturated and monoenoic fatty acids in the oil may be of importance. Rapeseed oil in experimental diets is found to reduce food consumption in most of the animal species studied. Heat treated or partially hydrogenated rapeseed oil also reduces the growth response of young animals. Partial hydrogenation interferes only to a minor degree with the content of the monoenoic acids. The decrease in food consumption has been related to appetite, taste, flavour, low digestibility and a slow absorption rate of rapeseed oil from the digestive tract, and especially of erucic acid.

It has been shown that the content of other dietary compounds such as calcium, phosphate and protein also influence digestibility and absorption of rapeseed oil and erucic acid. Similarly, dietary rapeseed oil may influence the excretion of endogenous lipids into the gut, but it is difficult to estimate this effect. Absorbed erucic acid is primarily deposited in the triglycerides and is present only to a small extent in cholesterol esters and phospholipids. Further, erucic acid is deposited in the organism in much smaller quantities than would be expected from the dietary content of this acid, irrespective of the fact that rapeseed oil diets reduce food consumption. In this connection it is interesting that several experiments have shown that eicosenoic acid, *i.e.* the next-lower homologue of erucic acid, which is also a characteristic fatty acid of rapeseed oil, is deposited almost to the same extent as the ordinary dietary fatty acids.

Dietary rapeseed oil has been shown to increase the content of adrenal cholesterol in Sprague-Dawley rats and in mice. The accumulation of adrenal cholesterol apparently cannot be correlated with altered adrenal cortical function. Rapeseed oil diets have been reported to interfere with reproduction in rats. Some experiments indicate that fewer young are born of rats fed on rapeseed oil, fewer of the young survive until weaning, and at weaning their average weight is lower than that of young from mothers fed on other dietary fats, *e.g.* corn oil or butterfat. These findings may be related to faulty development of mammary gland tissue resulting in reduced milk production. Further, milk from mothers fed on a rapeseed oil diet contains erucic acid and eicosenoic acid which may influence the appetite of the young and the fat absorption from the digestive tract. Rapeseed oil in the diet of experimental rats can induce hypothyroidism, increase liver weight and liver cholesterol, whereas the influence on blood cholesterol appears negligible. A possible relationship between dietary rapeseed oil and vitamins A, B_6 and ascorbic acid has not been established. Some experiments with dietary rapeseed oil have indicated a tendency to interstitial inflammatory changes of the myocardium. Studies on the nutritive value of rapeseed oil have included several species besides the rat, *e.g.* pigs, mice, guinea pigs, dogs, chickens, hamsters, ducklings, turkeys and geese. Further, in the experiments with rats possible differences in the response to dietary rapeseed oil have been considered in relation to strain, sex and age of the experimental animals.

Much work has been done to elucidate whether the metabolism of erucic acid can provide a clue to an explanation of the nutritive effects of rapeseed oil. The gross caloric value of rapeseed oil as heat in a bomb calorimeter, and the calculated caloric content from animal experiments show no significant differences from that of *e.g.* lard. A number of experiments, including studies with labeled erucic acid, have proven that erucic acid is catabolized according to the β-oxidation pathway. Its oxidation rate *in vivo* appears to be the same as for *e.g.* oleic acid. A possible difference in the yield of the oxidation of erucic acid and oleic acid has tentatively been ascribed to the formation of less active coenzyme A thioesters of erucic acid than of oleic acid. Further, the oxidation rate apparently also depends on the general dietary regimen of the animals. Thus β-oxidation of labeled erucic acid has been found to be much faster in rats fed on a diet rich in fat than in rats deprived of dietary fat. A possible effect of rapeseed oil diets on life span has been discussed with the background that although this oil causes a reduced growth rate some experiments have shown a marked increase in longevity. However, a longer or shorter life span is not just a matter of the type and amount of dietary fat but involves a multiplicity of factors.

To draw parallels from animal experiments to man is always a very delicate subject. Until very recently experience seemed to indicate that *moderate* amounts of conventional rapeseed oil in the diet of experimental animals was without specific risks. However, recent research in rats, ducklings and guinea pigs has shown serious pathological changes, *e.g.* in heart and skeletal muscles, thus indicating that conventional rapeseed oil still presents a number of intricate nutritive and biochemical problems to be solved. On the other hand, the production of canbra oil, *i.e.* rapeseed oil with very little, or zero erucic acid, proves that it is possible to control the content of a specific fatty acid in seed oils.

REFERENCES

1   E. AAES-JØRGENSEN, *Physiol. Rev.*, 41 (1961) 1.
1a A. M. M. ABDELLATIF AND R. O. VLES, *Nutrit. Metab.*, 12 (1970) 285.
1b A. M. M. ABDELLATIF AND R. O. VLES, *Nutrit. Metab.*, 12 (1970) 296.
1c A. M. M. ABDELLATIF AND R. O. VLES, *Nutrit. Metab.*, 13 (1971) 65.
1d A. M. M. ABDELLATIF AND R. O. VLES, *Proc. Intern. Conf. Sci., Technol., Marketing Rapeseed and Rapeseed Products, Canada, 1970.*
2   I. ABELIN, *Helv. Physiol. Acta*, 6 (1948) 879.
3   T. AKIYA, S. ISHII, H. SAKAI AND K. ARAI, *Eyo To Shokuryo*, 14 (1961) 397; *C.A.* 59 (1963) 11962d.
4   J. C. ALEXANDER AND F. H. MATTSON, *Can. J. Biochem. Physiol.*, 44 (1966) 35.
5   R. B. ALFIN-SLATER AND L. AFTERGOOD, *Physiol. Rev.*, 48 (1968) 758.
6   E. B. ASTWOOD, M. A. GREER AND M. G. ETTLINGER, *Science*, 109 (1949) 631.
7   E. B. ASTWOOD, M. A. GREER AND M. G. ETTLINGER, *J. Biol. Chem.*, 181 (1949) 121.
8   V. AUGUR, H. S. ROLLMAN AND H. J. DEUEL, *J. Nutrit.*, 33 (1947) 177.
9   A. BACH, P. METAIS AND P. HABEREY, *Compt. Rend. Soc. Biol.*, 161 (1967) 1361.
10 A. BACH, P. METAIS, J. RAULIN AND R. JACQUOT, *Bull. Soc. Chim. Biol.*, 51 (1969) 167.
11 V. H. BARKI, R. H. COLLINS, C. A. ELVEHJEM AND E. B. HART, *J. Nutrit.*, 40 (1950) 383.
12 R. H. BARNES, *New York State Journal of Med.*, 59 (1959) 2931.
13 J. L. BEARE, *Food Manufacture*, 32 (1957) 378.
14 J. L. BEARE, *Can. J. Biochem. Physiol.*, 39 (1961) 1855.
15 J. L. BEARE AND J. R. BEATON, *Can. J. Physiol. and Pharmacol.*, 45 (1967) 1093.
16 J. L. BEARE, J. A. CAMPBELL, C. G. YOUNG AND B. M. CRAIG, *Can. J. Biochem. Physiol.*, 41 (1963) 605.
17 J. L. BEARE, B. M. CRAIG AND J. A. CAMPBELL, *J. Am. Oil Chem. Soc.*, 38 (1961) 310.
18 J. L. BEARE, E. R. W. GREGORY AND J. A. CAMPBELL, *Can. J. Biochem. Physiol.*, 37 (1959) 1191.
19 J. L. BEARE, E. R. W. GREGORY, D. M. SMITH AND J. A. CAMPBELL, *Can. J. Biochem. Physiol.*, 39 (1961) 195.
20 J. L. BEARE, T. K. MURRAY AND J. A. CAMPBELL, *Can. J. Biochem. Physiol.*, 35 (1957) 1225.
21 J. L. BEARE, T. K. MURRAY AND J. A. CAMPBELL, *Can. J. Biochem. Physiol.*, 38 (1960) 187.
22 J. L. BEARE, T. K. MURRAY, H. C. GRICE AND J. A. CAMPBELL, *Can. J. Biochem. Physiol.*, 37 (1959) 613.
23 J. L. BEARE, T. K. MURRAY, J. M. MCLAUGHLAN AND J. A. CAMPBELL, *J. Nutrit.*, 80 (1963) 157.

23a J. L. Beare-Rogers, *Proc. Intern. Conf. Sci., Technol., Marketing Rapeseed and Rapeseed Products, Canada, 1970.*

24 K. Bernhard, F. Lindlar and H. Wagner, *Z. Ernährungsw.*, 1 (1960) 48.

25 K. Bernhard, E. Seelig and H. Wagner, *Z. Physiol. Chem.*, 304 (1956) 138.

26 K. Bernhard and E. Vischer, *Helv. Chim. Acta*, 29 (1946) 929.

27 A. von Beznak, M. von Beznak and I. Hajdu, *Ernährung*, 8 (1943) 209.

28 A. von Beznak, M. von Beznak and I. Hajdu, *Ernährung*, 8 (1943) 236.

29 M. K. Bhatty and B. M. Craig, *J. Am. Oil Chem. Soc.*, 41 (1964) 508.

30 J. Boer, *Acta Brevia Neerl. Physiol. Pharmacol. Microbiol.*, 11 (1941) 180.

31 J. Boer, B. C. P. Jansen and A. Kentie, *J. Nutrit.*, 33 (1947) 339.

32 J. Boer, B. C. P. Jansen, A. Kentie and H. W. Knol, *J. Nutrit.*, 33 (1947) 359.

33 G. Bourdel and R. Jacquot, *Révue franç. Corps Gras*; Numéro special 1967 (*cf.* Jacquot et al.[89]).

34 M. Buensod and P. Favanger, *Helv. Physiol. Pharmacol. Acta*, 14 (1956) 299.

35 J. P. Carreau, A. Thoron, D. Lapous and J. Raulin, *Bull. Soc. Chim. Biol.*, 50 (1968) 1973.

36 J. P. Carreau, A. Thoron and J. Raulin, *Compt. Rend. Acad. Sci.* 266 (1968) 417.

37 K. K. Carroll, *Proc. Soc. Exp. Biol. Med.*, 71 (1949) 622.

38 K. K. Carroll, *Endocrinology*, 48 (1951) 101.

39 K. K. Carroll, *J. Biol. Chem.*, 200 (1953) 287.

40 K. K. Carroll, *Proc. Soc. Exp. Biol. Med.*, 94 (1957) 202.

41 K. K. Carroll, *J. Nutrit.*, 64 (1958) 399.

42 K. K. Carroll, *Can. J. Biochem. Physiol.*, 37 (1959) 731.

43 K. K. Carroll, *Can. J. Biochem. Physiol.*, 37 (1959) 803.

44 K. K. Carroll, *Can. J. Biochem. Physiol.*, 40 (1962) 1115.

45 K. K. Carroll, *Can. J. Biochem. Physiol.*, 40 (1962) 1229.

46 K. K. Carroll, *Lipids*, 1 (1966) 171.

47 K. K. Carroll and R. L. Noble, *Fed. Proc.*, 8 (1949) 22.

48 K. K. Carroll and R. L. Noble, *Endocrinology*, 51 (1952) 476.

49 K. K. Carroll and R. L. Noble, *Can. J. Biochem. Physiol.*, 34 (1956) 981.

50 K. K. Carroll and R. L. Noble, *J. Can. Biochem. Physiol.*, 35 (1957) 1093.

51 K. K. Carroll and J. F. Richards, *J. Nutrit.*, 64 (1958) 411.

52 A. L. S. Cheng, M. G. Morehouse and H. J. Deuel, *J. Nutrit.*, 37 (1949) 237.

53 T. Cheniti, G. Bourdel and R. Jacquot, *Rev. Franç. Corps Gras*, 14 (1967) 151.

54 M. Chomyszyn, *Rocz. Mauk. Rol.* (B), 69 (1954) 91; *Nutr. Abstr. Rev.*, 25 (1955) 855.

55 J. Chudy, *Zesc. Mauk, WSR. Olsztyn*, 21 (1966) 37.

56 B. M. Craig and J. L. Beare, *Can. J. Biochem.*, 45 (1967) 1075.

57 B. M. Craig and J. L. Beare, *J. Can. Inst. Food Technol. Aliment.*, 1 (1968) 64.

58 B. M. Craig, C. G. Young, J. L. Beare and J. A. Campbell, *Can. J. Biochem. Physiol.*, 41 (1963) 43.

59 B. M. Craig, C. G. Young, J. L. Beare and J. A. Campbell, *Can. J. Biochem. Physiol.*, 41 (1963) 51.

60 E. W. Crampton, F. A. Farmer and F. M. Berryhill, *J. Nutrit.*, 43 (1951) 431.

61 E. W. Crampton, R. K. Shaw, V. G. Mackay and D. C. Schad, *J. Nutrit.*, 70 (1960) 81.

62 M. E. Crockett and H. J. Deuel, *J. Nutrit.*, 33 (1947) 187.

63 S. Darby, *Liebigs Ann. Chemie*, 69 (1849) 1.

64 H. J. Deuel, A. L. S. Cheng and M. G. Morehouse, *J. Nutrit.*, 35 (1948) 295.

65 H. J. Deuel, *The Lipids: Their Chemistry and Biochemistry*, Interscience Publishers Inc., New York. II, 195 (1955).

66 H. J. Deuel, S. M. Greenberg, E. E. Straub, D. Jue, C. M. Gooding and C. F. Brown, *J. Nutrit.*, 35 (1948) 301.

67 H. J. Deuel, L. Hallman and A. Leonard, *J. Nutrit.*, 20 (1940) 215.

68 H. J. Deuel, R. M. Johnson, C. E. Calbert, J. Gardner and B. Thomas, *J. Nutrit.*, 38 (1949) 369.

69 R. K. Downey and B. M. Craig, *J. Am. Oil Chem. Soc.*, 41 (1964) 475.

70 B. von Euler, H. von Euler and G. Linderman, *Arkiv Kemi, Mineral. Geol.*, B26 (1948) 1.

71 B. von Euler, H. von Euler and I. Rönnenstam-Saberg, *Arkiv Kemi, Mineralogi, Geologi*, 22A (1946) 8; 24A (1946) 15; 24A (1947) 20.

72 J. Fernandes, J. H. van de Kramer and H. A. Weijers, *J. Clin. Invest.*, 34 (1955) 1026.

73 C. H. Fench, R. H. Ingram, J. A. Uram, G. P. Barron and R. W. Swift, *J. Nutrit.*, 51 (1953) 329.

74 J. Gadamer, *Ber. Deut. Chem. Ges.*, 30 (1897) 2322.

75 J. J. Gottenbos and G. Hornstra (*cf.* Thomassen *et al.*[140]).

76 T. P. Hilditch and H. M. Thompson, *Biochem. J.*, 30 (1936) 677.

77 E. Hill and W. R. Bloor, *J. Biol. Chem.*, 53 (1922) 171.

78 R. Hill, J. M. Linazarro, F. Chevalier and I. L. Chaikoff, *J. Biol. Chem.*, 233 (1958) 305.

79 A. Holasek, *Z. Physiol. Chem.*, 298 (1954) 55.

80 A. Holasek, *Z. Physiol. Chem.*, 298 (1954) 219.

81 A. Holasek, *Z. Physiol. Chem.*, 298 (1954) 224.

82 R. T. Holman, *Progress Chem. Fats and Other Lipids*, 9 (1968) 275.

83 A. D. Holmes, *U.S. Dept. Agr. Bull.*, No. 687 (1918).

84 S. Homrowski, *J. Nutrit. and Dietet.*, 6 (1969) 22.

85 C. Y. Hopkins, M. J. Chisholm, T. K. Murray and J. A. Campbell, *J. Am. Oil Chem. Soc.*, 34 (1957) 505.

86 C. Y. Hopkins, T. K. Murray and J. A. Campbell, *Can. J. Biochem. Physiol.*, 33 (1955) 1047.

87 A. Horst, I. Kaganowicz, I. Zagorska and D. Rozynkowa, *Pathologia Polska*, 13 (1962) 139.

87a U. M. T. Houstmüller, C. B. Struijk and A. van der Beek, *Biochem. Biophys. Acta*, 218 (1970) 564.

88 R. Jacquot, J. Raulin and A. Thoron, *Rev. Franç. Corps Gras*, 14 (1967) 441.

89 R. Jacquot, G. Rocquelin and B. Potteau, *Les Corps Gras Alimentaire*, 4 (1969) 51, (Suppl. Fasc. 2).

90 S. K. Joshi and J. A. Sell, *Can. J. Anim. Sci.*, 44 (1964) 34.

91 G. Kabelitz, *Arch. Ges. Physiol.*, 247 (1944) 593.

92 T. H. Kennedy and H. D. Purves, *Brit. J. Exper. Path.*, 22 (1941) 241.

93 L. D. Kirk, G. C. Mustakas and E. L. Griffin Jr., *J. Am. Oil Chem. Soc.*, 43 (1966) 550.

94 J. M. Linazarro, R. Hill, F. Chevalier and I. L. Chaikoff, *J. Exptl. Med.*, 107 (1958) 813.

95 F. Lindlar, *Z. Vitaminforsch.*, 28 (1957) 195.

96 E. Linko, *Acta Med. Scand.*, 159 (1957) 475.

97 H. J. Lips, *Can. J. Technol.*, 39 (1962) 61.

98 T. M. Mallard and B. M. Craig, *J. Am. Oil Chem. Soc.*, 43 (1966) 1.

99 N. Matsuo, *Eyo To Shokuryo*, 12 (1959/60) 118; *C.A.*, 55 (1961) 2825b.

100 E. J. Middleton and J. A. Campbell, *Can. J. Biochem. Physiol.*, 36 (1958) 203.

101 S. Mukherjee and R. B. Alfin-Slater, *Arch. Biochem. Biophys.*, 73 (1958) 359.

102 T. K. Murray, J. L. Beare and J. A. Campbell, *Can. J. Biochem. Physiol.*, 38 (1960) 663.

103 T. K. Murray, J. L. Beare, J. A. Campbell and C. Y. Hopkins, *Can. J. Biochem. Physiol.*, 36 (1958) 653.

104 T. K. Murray, J. A. Campbell, C. Y. Hopkins and M. J. Chisholm, *J. Am. Oil Chem. Soc.*, 35 (1958) 156.

105 F. Müller, *Z. Biol.*, 20 (1884) 327.

106 U.-M. Niemi and P. Roine, *Z. Ernährungsw.*, 1 (1960) 164.

107 M. Nikonorow and S. Homrowski, *Roczn. Wojsk. Inst. Hig. Epid.*, 4 (1965) 39.

108 R. L. Noble and K. K. Carroll, *Recent Progress in Hormone Research*, 17 (1961) 97.

109 L. N. Norcia and W. O. Lundberg, *J. Nutrit.*, 54 (1954) 491.

110 L. Paloheimo and B. Jahkola, *J. Scient. Agric. Soc., Finland*, 31 (1959) 212.

111 L. Paloheimo, P. Roine and E. Uksila, *J. Sci. Agric. Soc., Finland*, 31 (1959) 251.

112 J. Paluszak, *Endocrinol. Polska*, 16 (1965) 535.

113 S. V. Pande and J. F. Mead, *J. Biol. Chem.*, 243 (1968) 352.
114 S. V. Pande and J. F. Mead, *Biochim. Biophys. Acta*, 152 (1968) 636.
115 B. Potteau, Inst. Nat. Rech. Agron., Lab. d'Etude Alim. l'Hommes, Paris, France, 1965.
116 S. Radziejewski, *Virchows Arch. Pathol. Anatom. Physiol. Klin. Med.*, 43 (1868) 286.
117 J. Raulin, C. Richir, P. Escribano and R. Jacquot, *Compt. Rend. Acad. Sci.*, 248 (1959) 1229.
118 J. F. Richards and K. K. Carroll, *Can. J. Biochem. Physiol.*, 37 (1959) 725.
119 G. Rocquelin and R. Cluzan, *Ann. Biol. Anim. Biochim. Biophys.*, 8 (1967) 395.
120 P. Roine and E. Uksila, *Acta Agralia Fennica*, 94 (1958) 11.
121 P. Roine, E. Uksila, H. Teir and J. Rapola, *Z. Ernährungsw.*, 1 (1960) 118.
122 D. Rozynkowa, W. Koziolowna and A. Horst, *Acta Physiol. Polonica*, 13 (1962) 591.
123 P. Savary and M. J. Constantin, *Biochim. Biophys. Acta*, 125 (1966) 118.
124 H. Schmid and P. Karrer, *Helv. Chim. Acta*, 31 (1948) 1017.
125 W. Schneider, *Chem. Abstr.*, 4 (1910) 3064.
126 R. Schönheimer and G. Hilgetag, *J. Biol. Chem.*, 105 (73) 1934.
127 J. L. Sell and G. C. Hodgson, *J. Nutrit.*, 76 (1962) 113.
128 J. L. Sell and J. A. McKirdy, *Poultry Science*, 42 (1963) 380.
129 W. M. Sperry, *J. Biol. Chem.*, 68 (1926) 357.
130 W. M. Sperry, *J. Biol. Chem.*, 71 (1926/27) 351.
131 W. M. Sperry and W. R. Bloor, *J. Biol. Chem.*, 60 (1924) 261.
132 B. R. Stefanson, F. W. Hougen and R. K. Downey, *Can. J. Plant Sci.*, 41 (1961) 218.
133 L. Swell, E. C. Trout, H. Field and C. R. Treadwell, *Proc. Soc. Exp. Biol. Med.*, 92 (1956) 613.
134 L. Swell, E. C. Trout, H. Field and C. R. Treadwell, *J. Biol. Chem.*, 223 (1956) 743.
135 H. J. Thomasson, *J. Nutrit.*, 56 (1955) 455.
136 H. J. Thomasson, *J. Nutrit.*, 57 (1955) 17.
137 H. J. Thomasson, *J. Nutrit.*, 59 (1956) 343.
138 H. J. Thomasson and J. Boldingh, *J. Nutrit.*, 56 (1955) 469.
139 H. J. Thomasson and J. J. Gottenbos, *Verhandel. Koninkl. Vlaam. Acad. Geneeskunde Belg.*, 19 (1957) 369.
140 H. J. Thomasson, J. J. Gottenbos, P. L. van Pijpen and R. O. Vles, *Int. Symp. Chem. Techn. Rapeseed Oil and Other Cruciferae Oils*, Gdansk, Poland 1967 .
141 A. Thoron, *Comportement nutritionel et biochimique de l'huile de colza. Métabolisme de l'acide érucique*, Thèse de 3e cycle, Physiologie de la Nutrition, Paris, 1968.
142 C. W. Turner, *Poultry Sci.*, 27 (1948) 118.
142a R. O. Vles and A. M. M. Abdellatif, *Proc. Intern. Conf. Sci., Technol., Marketing Rapeseed and Rapeseed Products, Canada, 1970.*
143 H. Wagner, E. Seelig and K. Bernhard, *Z. Physiol. Chem.*, 306 (1956) 96.
144 H. Wagner, E. Seelig and K. Bernhard, *Z. Physiol. Chem.*, 312 (1958) 104.
145 G. Wigand, *Acta Medica Scand.*, 166 Suppl. 1959 (Thesis).
146 J. D. Wood and B. B. Migicovsky, *Can. J. Biochem. Physiol.*, 36 (1958) 433.
147 E. Wysocki and H. Ziomski, *Rocz. Panstwowega Zakl. Hig.*, 14 (1963) 299 (*cf.* Potteau, [115]).

CHAPTER 15

Nutritional Value and Use of Rapeseed Meal

E. JOSEFSSON

Swedish Seed Association, Svalöf (Sweden)

CONTENTS

15.1 INTRODUCTION

Seed of rape *(Brassica napus)*, turnip rape *(Brassica campestris)* and white mustard

(Sinapis alba) contain *ca.* 47, 44, and 35 per cent oil, respectively (see Chapter 7). After removal of the oil from the seed by a prepress and solvent extraction procedure a protein-rich meal is obtained. Since world production of rapeseed (the sum of the three species mentioned above and *Brassica juncea*) amounts to *ca.* six million metric tons annually (*cf.* Chapter 2), meal production will be close to three million metric tons each year. The use made of this meal is of course of great economic importance.

Detailed reviews about rapeseed meal were published in 1965 and 1967[1, 2], to which references are sometimes given in this chapter instead of referring to early original papers.

The nutritional value of a food or feed is dependent on several factors: protein content, amino acid composition, crude fibre content, content of minerals, vitamins, and the possible presence of toxic substances. In the following sections each of these factors will be considered.

15.2 PROTEIN CONTENT

Some data on the chemical composition of rapeseed meals are given in Tables 15:1 and 15:2. Although the protein content of rapeseed meal is less than that in soya

TABLE 15:1

COMPOSITION OF RAPESEED MEAL AND SOYA MEAL[3]

Feedstuff	Dry matter %	Chemical composition, % of dry matter						
		Protein (N×6.25)	Fat	N-free extract	Crude fibre	Ash	Ca	P
Rapeseed meal:								
B. napus	89	43.1	2.3	36.9	10.7	7.0	0.8	1.2
B. campestris	89	41.1	2.0	37.8	11.6	7.5	0.6	1.3
Sinapis alba	89	45.5	1.2	36.2	9.8	7.3	0.8	1.2
Soya meal	87	50.4	0.5	35.4	6.9	6.8	0.3	0.7

meal, it represents the most concentrated vegetable protein feed produced in the cool temperate zones. Turnip rape appears to be a little lower in protein content than rape. Clandinin and Bayly[4] obtained similar results in a study of samples from various varieties grown in different areas. This may be due to the fact that the seeds of turnip rape are generally smaller than those of rape. The seed coat fraction of small seeds is usually larger than that of large ones, and the seed coat is low in protein content.

TABLE 15:2

Feedstuff	Niacin	Pantothenic acid	Ribo-flavin	Thia-mine
Rapeseed meal:				
Expeller	168	17	3	9
Soya meal:				
Expeller	23	15	4	4
Solvent	24	20	4	9

It should be noted that protein content can be greatly influenced by environmental factors. A large amount of nitrogen fertilizers increases the protein content[5]. Wetter[6] reports that the protein content of meals is usually higher in the brown soils of western Canada than in the black soil zones, and he suggests that this is due to the higher rainfall in the black soil zones.

TABLE 15:3

AMINO ACID COMPOSITION OF RAPESEED MEALS, PRODUCED BY DIFFERENT METHODS AND IN DIFFERENT PERIODS, PER CENT OF PROTEIN[7]

Amino acid	Rapeseed meals produced during 1956–1961		Rapeseed meals produced during 1965–1967	
	Expeller-type (Mean of five samples)	Prepress-Solvent type (Mean of fifteen samples)	Prepress-Solvent type (Mean of eight samples)	Solvent-type (Mean of seven samples)
Arginine	5.09	5.48	5.59	5.65
Histidine	2.40	2.66	2.66	2.63
Lysine	4.39	5.31	5.94	5.85
Tyrosine	2.16	2.10	2.30	2.28
Tryptophan	0.94	1.21	1.23	1.29
Phenylalanine	3.74	3.78	3.83	3.89
Cystine	0.73	0.80	1.19	1.22
Methionine	1.88	1.91	1.87	1.83
Serine	4.03	4.19	4.40	4.37
Threonine	4.08	4.19	4.41	4.35
Leucine	6.45	6.67	6.95	7.04
Isoleucine	3.71	3.63	3.81	3.87
Valine	4.76	4.82	4.94	4.98
Glutamic acid	16.16	16.84	18.71	18.99
Aspartic acid	6.58	6.72	7.02	7.04
Glycine	4.68	4.78	5.08	5.09
Alanine	4.21	4.29	4.44	4.46
Proline	5.71	6.13	6.13	6.32

15.2.1 Amino acid composition

The amino acid composition of the commercial meal is largely influenced by the way in which the meal is produced from the seed. Previously the oil was pressed from the seed in an expeller pressing mill, and in order to obtain a maximum oil yield the crushed seed was exposed to high pressures and temperatures. This resulted in extensive destruction of lysine, often too low in the diet. In modern processing the meal is usually produced by a combination of prepressing and solvent extraction or by solvent extraction alone (see Chapter 9). Clandinin[7] compared meals of expeller type with prepress-solvent type and solvent type produced in 1965–1967. Table 15:3 shows that the content of basic amino acids and especially of lysine is greater in the meals produced by the 'prepress-solvent extraction', or by the solvent extraction alone compared to that in the meals produced by the expeller procedure. Since lysine is often the first limiting amino acid in feed of vegetable origin, this represents an important improvement in the nutritional value of rapeseed meal.

TABLE 15:4

AMINO ACID COMPOSITION OF MEALS OF RAPE, TURNIP RAPE, WHITE MUSTARD AND SOYA BEAN
(GRAMS AMINO ACID PER 16 G N)

	Rape	Turnip rape	White mustard	Soya bean
Reference:	8	8	9	3
Number of samples:	4	4	1	27
Amino acid:				
Arginine	5.6	5.9	6.1	6.9
Histidine	2.6	2.7	2.9	2.7
Lysine	5.8	6.1	5.9	6.1
Tyrosine	2.6	2.7	3.3	2.9
Tryptophan	—	—	—	1.0
Phenylalanine	3.5	3.8	4.1	5.0
Cystine	2.4	2.4	—	1.7
Methionine	1.8	1.9	1.7	1.5
Serine	3.7	3.8	4.3	5.2
Threonine	3.8	4.0	4.6	4.0
Leucine	6.3	6.5	7.3	7.6
Isoleucine	3.7	3.8	4.1	4.8
Valine	4.8	4.9	5.6	5.0
Glutamic acid	16.6	16.8	18.0	20.6
Aspartic acid	6.2	6.6	7.9	11.9
Glycine	4.3	4.6	6.1	4.4
Alanine	3.9	4.0	4.5	4.5
Proline	6.4	6.7	6.4	—

In Table 15:4 the amino acid compositions of rape, turnip rape, white mustard and soya meals are shown. The contents of most of the amino acids in the first three meals are similar to those in soya meal. Besides lysine, methionine is the amino acid most often lacking in vegetable feed, and therefore the contents of these two amino acids are of great importance. The lysine content of rapeseed meal is slightly lower than that in soya meal, whereas the methionine content is higher in rapeseed meal. Since cystine can in part replace methionine, the higher content of this amino acid in rapeseed meal is also of importance. According to the provisional pattern of FAO[10], isoleucine and methionine are the limiting amino acids of rapeseed.

15.3 CARBOHYDRATES

Only a few studies have been made concerning the carbohydrates of rapeseed meals. When the content of protein, fat, crude fiber, and ash in the meal have been determined the rest is designated as 'N-free extracts' in the conventional feed analysis. This is a poorly defined fraction mostly containing carbohydrates, and 'crude fiber' is also an incompletely defined fraction. According to the analysis method of A.A.C.C.[11], crude fibre is composed of the substances which are insoluble upon boiling in 1.25 per cent sulphuric acid and in 1.25 per cent potassium hydroxide. The main components are cellulose and lignin. Figures for crude fibre and N-free extracts of meals of some oil crops are given in Table 15:1. The crude fibre content is higher in rapeseed meal than in meals of most other oil crops, which may cause a lowering of the metabolizable energy.

Hrdlička et al.[12] studied the saccharides in rapeseed meals as affected by different processing methods: Laboratory extraction according to Soxhlet, or extraction under industrial conditions according to a battery method or according to the methods of Bollmann, Lurgi or DeSmet. They found no appreciable differences in crude fiber, N-free extract or pentosan content caused by extraction methods, but some differences in starch content were observed. Four samples extracted by the Soxhlet method gave the mean values 10.6 per cent crude fibre, 9.9 pentosans and 6.4 starch. The corresponding values for commercial methods were on the average, 10.7 per cent, 10.0 and 5.8 respectively. The lower starch contents found in meals produced by the methods of Lurgi and Bollmann were assumed to be caused by conversion of starch to dextrins. The mean value for reducing sugars was 4.1 per cent in the meals produced according to Soxhlet, while it was only 3.2 per cent in the meals processed by the commercial procedures.

When the meal was processed according to the method of DeSmet, the content of reducing sugars was not appreciably different from that of the Soxhlet-extracted meal. On the other hand, when the meal was extracted by the battery method the content was only 2.5 per cent. This was explained as a result of the Maillard

358

reaction, in which condensation products are formed by reaction between basic amino acids, especially lysine, and reducing sugars. Besides a lower content of reducing sugars it results in lower contents of basic amino acids and a browning of the meal. Since the content of available lysine will be considerably lowered by this reaction it is detrimental to the nutritional value of the meal, and thus it is important to process the meal under conditions (time and temperature) that do not favour the Maillard reaction. When the rapeseed meal had been hydrolysed before the determination of reducing sugars, the meals produced according to Soxhlet were found to contain 11.8 per cent of these compounds, whereas the meals extracted by commercial methods contained 9.5 per cent. A great part of this sugar was derived from the glucosinolates, and since a high content of reducing sugars favours the Maillard reaction it is important to inactivate the glucosinolate-splitting enzyme, myrosinase, at an early stage of the extraction process.

In experiments with oriental mustard *(Brassica juncea)* McGhee et al.[13] obtained similar results, although the content of reducing sugars seemed to be lower in this species than in *B. napus*.

15.4 CONTENT OF MINERALS AND VITAMINS

As shown in Table 15:1 the ash content of rapeseed meal is similar to that of soya meal, while the contents of calcium and phosphorus are higher in rapeseed meal.

Our knowledge of the vitamin content of rapeseed meal is relatively incomplete. The figures in Table 15:2 indicate that rapeseed meal contains more niacin than soya meal, while the contents of pantothenic acid, riboflavin and thiamine are similar. According to Klain et al.[13a], rapeseed meal contains about three times as much choline as soya meal (6730 compared with 2170 mg/kg).

15.5 GLUCOSINOLATES

15.5.1 Chemical properties

The most limiting factor for the use of rapeseed meal as feed seems to be its content of glucosinolates. In the earlier literature these compounds were most often named mustard oil glucosides, but later they were called thioglucosides, an expression which is sometimes still in use. Their general formula (see Chapter 7) was clarified in 1956 by Ettlinger and Lundeen[14].

According to Guignard[15] the glucosinolates are distributed diffusely in the parenchymal tissues of the seed. An enzyme system, myrosinase, with the ability to split glucosinolates, seems to be separated from the glucosinolates in the intact tissue by being located in special cells[15]. If the cells are disintegrated and the water content is sufficiently high the glucosinolates will be split by the myrosinase, giving rise to glucose, sulphate and isothiocyanates (cf. Chapter 7). Fig.15:1

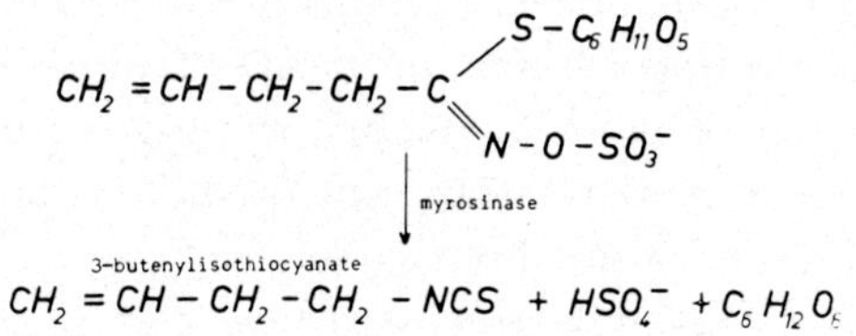

$$CH_2 = CH - CH_2 - CH_2 - C \underset{N-O-SO_3^-}{\overset{S-C_6H_{11}O_5}{<}}$$

$$\downarrow \text{myrosinase}$$

3-butenylisothiocyanate

$$CH_2 = CH - CH_2 - CH_2 - NCS + HSO_4^- + C_6H_{12}O_6$$

Fig.15:1. Enzymic hydrolysis of 3-butenylglucosinolate.

TABLE 15:5

GLUCOSINOLATES PRESENT IN SEED OF RAPE (BRASSICA NAPUS) OR TURNIP RAPE (B. CAMPESTRIS)

Glucosinolate	Reference
3-Butenylglucosinolate	16, 17
4-Pentenylglucosinolate	18, 17
4-Methylthiobutylglucosinolate	19
5-Methylthiopentylglucosinolate	19
4-Methylsulphinylbutylglucosinolate	20
5-Methylsulphinylpentylglucosinolate	20
2-Phenylethylglucosinolate	18, 17
2-Hydroxy-3-butenylglucosinolate	21, 17
2-Hydroxy-4-pentenylglucosinolate	22, 17

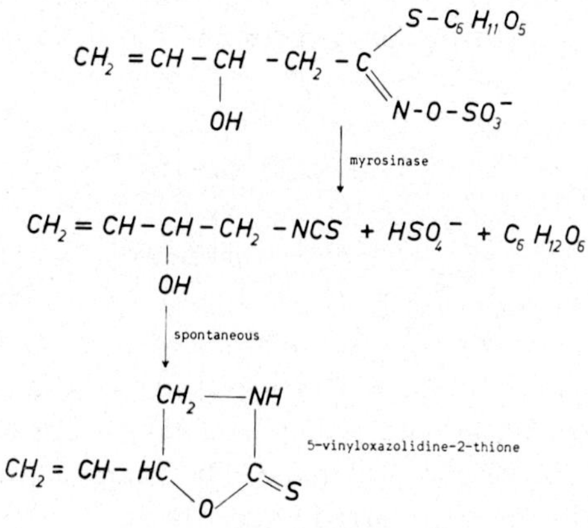

$$CH_2 = CH - \underset{OH}{CH} - CH_2 - C \underset{N-O-SO_3^-}{\overset{S-C_6H_{11}O_5}{<}}$$

$$\downarrow \text{myrosinase}$$

$$CH_2 = CH - \underset{OH}{CH} - CH_2 - NCS + HSO_4^- + C_6H_{12}O_6$$

$$\downarrow \text{spontaneous}$$

5-vinyloxazolidine-2-thione

$$CH_2 = CH - HC \underset{O}{\overset{CH_2 - NH}{<}} C = S$$

Fig.15:2. Enzymic hydrolysis of 2-hydroxy-3-butenylglucosinolate and the conversion of the aglucone to oxazolidinethione.

shows this reaction for 3-butenyl glucosinolate, one of the most predominant glucosinolates in rapeseed. The glucosinolates which have been reported to occur in meals of rape and turnip rape are enumerated in Table 15:5. Two of these glucosinolates contain an OH-group connected to the second carbon atom of the aglucone. After the enzymic splitting of the glucosinolate molecule these aglucones cyclize spontaneously and rapidly, giving rise to 5-vinyloxazolidine-2-thione (Fig.15:2) and 5-allyloxazolidine-2-thione respectively. In *Brassica napus* 2-hydroxy-3-butenylglucosinolate (progoitrin) is the predominant glucosinolate, with 3-butenylglucosinolate (gluconapin) in second place[20, 23, 17]. In *B. campestris* 3-butenylglucosinolate predominates, while glucosinolates yielding oxazolidine-thiones are most often low in amount, or lacking, in this species[20, 23]. Figures

TABLE 15:6

RANGE IN GLUCOSINOLATE CONTENT IN SEED MEALS OF SOME CRUCIFERAE OIL CROPS CULTIVATED IN NORTH OR MIDDLE EUROPE

The formula weight of the potassium salts of the glucosinolates has been used in the calculations with the exception of *Sinapis alba* for which the formula weight of the *p*-hydroxybenzylgluco-sinolate ion was used.

Species and type	Number of varieties	Number of samples	Range in gluco-sinolate content (% in dry matter)	Reference
Brassica napus:				
Winter type	16	43	3.39–7.77	25
Summer type	5	15	3.72–5.75	25
Brassica campestris:				
Winter type	4	8	3.09–4.30	25
Summer type	2	7	3.01–3.59	25
Sinapis alba	5	23	7.91–10.12	26

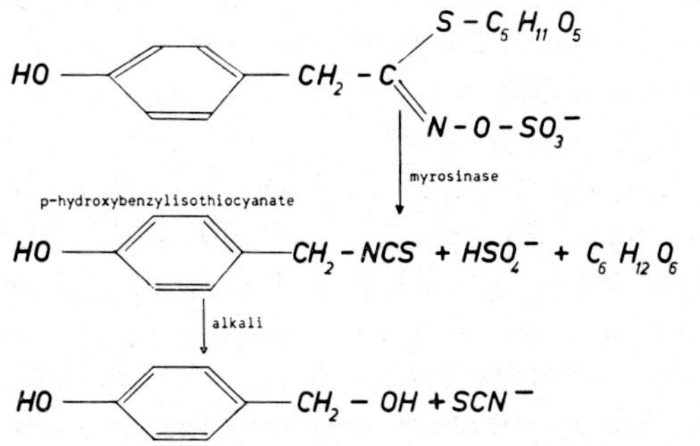

Fig.15:3. Enzymic hydrolysis of *p*-hydroxybenzylglucosinolate, showing products formed and further conversion in alkaline solution.

for the range in glucosinolate content of the seed meals of some varieties cultivated in Europe are given in Table 15:6.

Only p-hydroxybenzylglucosinolate has been shown to occur in white mustard seed[24]. The hydrolysis product p-hydroxybenzyl isothiocyanate is relatively labile in neutral and still more in alkaline solutions, giving rise to p-hydroxybenzyl alcohol and thiocyanate ion[27-29] (Fig.15:3).

3-Butenyl-, 4-pentenyl- and phenylethyl-isothiocyanate are volatile with steam. This may be of practical importance for meals containing mainly these gluco-sinolates, since the cleavage products after hydrolysis may be removed by steaming.

The glucosinolates are easily soluble in water, and in ethanol, methanol and acetone. They are more soluble in 10–20 per cent water mixtures of these solvents than in the pure solvents. The isothiocyanates and oxazolidinethiones are soluble in ethyl ether and in chloroform. Steam volatile isothiocyanates are lipophilic and soluble, for instance in hexane, while the non-volatile isothiocyanates are less lipophilic, and the oxazolidinethiones relatively hydrophilic.

In studies on the variation in enzymic degradation products from epi-progoitrin after autolysis of meal from Crambe abyssinica, VanEtten et al.[30] found that nitriles could be produced in large quantities instead of oxazolidinethione. The relative amounts of nitriles and oxazolidinethiones varied greatly depending on the auto-lysis conditions. If the seed had not been stored or heat-treated, and if hydrolysis was performed at pH ca. 5 and with relatively small quantities of water, nitrile formation predominated. The formation of oxazolidinethione was favoured by storing the seed for instance one year at room temperature, by heat-treatment of the seed at 100–120 °C, by carrying out the hydrolysis at an increased temperature, by dilution with large quantities of water and by raising the pH to above 5. Studies of the hydrolysis products of progoitrin from B. napus, which is a steric isomer of the epi-progoitrin of crambe, gave similar results.

15.5.2 The toxicity of various cleavage products

According to Belzile et al.[31] presence of myrosinase is important for the undesirable properties of rapeseed meal. The products formed by the enzymic reaction, isothiocyanates, oxazolidinethiones, nitriles and thiocyanate ions show harmful effects. The effect of the oxazolidinethiones seems to be that most studied.

In 1941 Kennedy and Purves[32] reported that Brassica seed fed to rats produced enlarged thyroids in these animals, and they also showed that this effect could not be wholly counteracted by feeding iodine. When the rats were hypophy-sectomized their thyroid glands weighed only ca. 6 mg as compared to ca. 44 mg for the intact controls[33]. These results indicated that rapeseed causes thyroid enlargement by interfering with the synthesis of thyroxin, whereupon the anterior pituitary is stimulated to produce TSH (thyroid-stimulating hormone) which acts on the thyroid glands causing hypertrophy and hyperplasia. Recent results by

Matsumoto *et al.*[33a] suggest that oxazolidinethione also interferes with the secretion of the thyroid hormone in the thyroid gland into the blood. In 1952 Frölich[34] reported that more than 5 per cent rapeseed meal fed to growing chicks caused thyroid hypertrophy and depressed growth. Nordfeldt *et al.*[35] found similar results when rapeseed meal was used at levels of 10–20 per cent in swine diets. No evidence has been reported for goitrogenic effects in cattle.

In 1949 Astwood *et al.*[36] isolated a goitrogenic substance from rapeseed and other *Brassica* seeds. The substance was shown to be (−)-5-vinyl-2-oxazolidinethione.

Subsequent to the work of Kennedy and Purves, several authors have shown by experiments on various animal species that the goitrogenic effects of rapeseed meal can only be partly corrected by feeding iodine. Use of iodinated casein or thyroxine counteracted the goitrogenicity of rapeseed when added to the diet of rats, poultry, or pigs, but the growth-depressing effect of the rapeseed meal was not always prevented. However, we now have reason to believe that when growth depression was noted, it was probably due to heat damage to the protein of the meal during processing or to failure to adjust the metabolizable energy value of the experimental feeds.

The thiocyanate ion which might be split off from the *p*-hydroxybenzyl isothiocyanate produced in white mustard seed meal is also goitrogenic. It exercises its effect by competition with the iodine uptake in the thyroid gland[37] and can be counteracted by feeding iodine.

Several times palatability problems have been reported from feeding with rapeseed meals, which may be caused by several of the cleavage products of the meal as well as by other substances. Since the isothiocyanates have pungent tastes and odours they may be greatly responsible for the palatability problems. In high concentrations they damage skin and mucous membranes, and by injuring the surface of the alimentary canal they may depress the weight gain. Bell[38] has found that the reduction in weight gain of mice fed with rapeseed meal is approximately correlated with the sum of contents of glucosinolates yielding oxazolidinethiones and isothiocyanates. According to Langer and Štolc[39] the isothiocyanates also show goitrogenic properties on rats. In a recent study VanEtten *et al.*[40] fed rats on a diet containing 10 per cent crambe seed meals that had been treated in various ways. A meal containing *epi*-progoitrin, however, with the enzymes destroyed by heat treatment, gave a weight gain that was 77 per cent of that of the control group. When the enzymes had not been destroyed the growth was poorer or the rats died. Meals containing active enzymes and autolyzed under such conditions that mainly oxazolidinethione was formed yielded 85 per cent growth, while meals autolyzed to form mainly nitriles caused death of the rats. Rats fed meals containing active enzymes or meals autolyzed to contain nitriles showed enlargement and microscopic lesions in the liver and kidneys. Since it seems probable that nitriles formed from the glucosinolates of rapeseed behave similarly[41], more attention should be paid in the future to the production of nitriles in rapeseed meal.

Bell[42] fed mice with soya and linseed meal to which was added oxazolidine-thione to a level of 0.15 per cent of the diet. This quantity is less than what may be released from a diet with rapeseed meal containing average quantities of glucosinolates as the only protein supplement*. The weight gain of the mice was only *ca.* 50 per cent of the weight gain which is typical of mice fed a good diet. In another experiment he added allyl isothiocyanate to soya meal and meat meal in a quantity to provide 0.05 per cent in the diet[43]. In this case the weight gain, adjusted to a comparable feed intake basis, was reduced by 12%. The results of these experiments may seem contradictory to the result mentioned above that the reduction of weight gain is correlated with the sum of contents of different glucosinolates. It should be noted, however, that nothing is known about which cleavage products were formed by feeding rapeseed meal, whether oxazolidine-thiones and isothiocyanates or nitriles, or other split products, while pure oxazo-lidinethione or isothiocyanate was added in the other experiments.

In 1956, Clements and Wishart[44] described some experiments indicating that goiter in man could be caused by a goitrogenic factor in milk from cows fed on marrow stem kale (*Brassica oleracea* var. *moelleria*). This induced Virtanen[45] to perform experiments to investigate the transfer of oxazolidinethione and thio-cyanate ion from fodder to milk. He found that maximally 0.05 per cent of the oxazolidinethione fed to the cows was transferred to the milk. When the milk from cows fed oxazolidinethione, or given oxazolidinethione-producing feed, was consumed by rats or men no inhibition of the radio-iodine uptake nor any thyroid enlargement was found. Furthermore, no harmful quantities of thio-cynate ion were found in the milk. On the other hand, Krusius and Peltola[46] reported that milk from areas of endemic goiter in Finland contained oxazolidine-thione in quantities which caused thyroid enlargement in rats. This may indicate that measurement of the radio-iodine uptake is not a suitable method of detecting goitrogenic effects caused by small amounts of a goitrogen ingested over a long period of time.

15.6 COUNTERACTING THE TOXICITY

15.6.1 Enzyme inactivation

Since the glucosinolates do not seem to be toxic a meal with the enzymes com-pletely destroyed should not be toxic if it is not exposed to enzymes from other sources. In modern procedures of extracting rapeseed oil a heat treatment is included, which causes inactivation of glucosinolate splitting enzymes. After

---

* This is true if the myrosinase activity is sufficient for the complete cleavage of the gluco-sinolates.

the seed has been crushed it is heated at approximately 80 °C. This enzyme in-activation step is important not only for the quality of the meal but also for the quality of the oil. If the myrosinase is not destroyed in an initial stage of the extraction process extensive enzymic splitting of the glucosinolates would occur. Hydrolysis products, for instance isothiocyanates, would dissolve in the oil and poison the catalyst used for its hydrogenation (cf. Chapter 9).

The effect of the heat treatment on the myrosinase inactivation depends on temperature, humidity and time of treatment. Appelqvist and Josefsson[47] found in laboratory studies that the myrosinase of seed with 8 per cent moisture content was completely inactivated by heating at 90 °C for 15 minutes. It is, however, possible that the animals may get myrosinase from other sources than the meal, for instance by eating fresh parts of Cruciferae plants.

In 1959 Greer and Deeney[48] reported that progoitrin was hydrolysed to goitrin (oxazolidinethione) in the gastro-intestinal tracts of rats and men although no myrosinase was included in the diet. Later Oginsky et al.[49] demonstrated myrosinase activity in a variety of bacteria, particularly Paracolobactrum, which is common in the intestinal tract of man.

Although myrosinase inactivation does not completely inhibit the hydrolysis of glucosinolates, it greatly improves the meal. This was demonstrated by Bell[50] in a swine feeding experiment, in which he added a myrosinase source to an in-activated rapeseed meal and showed that the growth inhibition was three times larger than without addition of myrosinase.

15.6.2 Technical methods of removal or destruction of glucosinolates or cleavage products

(a) Autolysis and distillation

Goering et al.[51] added water to meal of *Brassica juncea* (brown or yellow mustard) together with crushed seed as a source of myrosinase. The glucosinolates were split during this treatment and steam volatile isothiocyanates were removed by distillation. When the product obtained replaced 50 per cent of the soya meal in a diet for rats, the nutritional value of the diet was found to be at least as good as with soya meal alone. *B. juncea*, however, contains only glucosinolates that yield steam-volatile isothiocyanates, while only ca. 30 per cent of the glucosinolates of rapeseed *(B. napus)* meal yields volatile products. In the winter turnip rape variety most cultivated in Sweden approximately 90 per cent of the products are volatile. If it is possible to distinguish between *B. napus* and *B. campestris* seed in the trade, an improvement of the *B. campestris* meal might be obtained by using the distillation technique.

(b) Autoclaving and steam stripping

Bell and Belzile[52] reported that most of the glucosinolates were destroyed in a meal that had been autoclaved for 60 minutes at a pressure of 1.2 kg/cm². When the meal was fed to rats, the animals showed normal weight gains, demonstrating that the main part of the toxicity had been removed.

In another study Bell and Belzile[52] passed steam through rapeseed meal which at the same time was maintained at a temperature of 110 °C. Meal that had been treated in this way for at least 1 hour caused good growth responses when tested in diets for mice.

Bell and Belzile[52] also point out, however, that there was a gradual deterioration in protein quality as the time of autoclaving or steam stripping was extended. Meal that was autoclaved for 2 hours at a pressure of 1.2 kg/cm² did not support growth when used as the only protein source in an otherwise adequate diet. It seems probable that this was an effect of lysine destruction, but this was not proved.

(c) Treatment with ammonia and heat

Kirk et al.[53] reported that treatment of meal of *Crambe abyssinica* with ammonia and heat destroyed the glucosinolates effectively. Preliminary feeding tests indicated a great improvement of the nutritional value. In recent experiments, however, there have been palatability problems on feeding ammonia-treated meal to cattle[54].

(d) Soda cooking

Mustakas et al.[54] treated crushed seed, or defatted meal of crambe, by cooking with sodium carbonate under pressure. The glucosinolates were destroyed completely by this treatment, but the reaction products formed remain to be elucidated. Feeding tests with cows demonstrated that the palatability of the soda-treated meal was better than that of untreated or ammonia-treated meal. In feeding experiments with poultry the soda-treated as well as the ammonia-treated meal was more palatable than meal which was only toasted. The palatability was, however, not quite satisfactory when the ratio of rapeseed meal in the feed was as high as 20 per cent. Since the soda treatment is inexpensive it may be of practical value.

(e) Treatment with ferrous salts

Youngs and Perlin[55] found that when certain metallic salts, especially ferrous sulphate, were added to Brassica seed meals, the glucosinolates decomposed to yield nitriles. When this treatment was followed by steam treatment the nitriles from the glucosinolates of *B. juncea* were removed, while nitriles formed from the glucosinolates of *B. napus* and *B. campestris* were not completely removed. Since the toxicity of the nitriles is still higher than that of isothiocyanates or oxazolidine-thiones the feeding tests did not show any improvement in the nutritional value as a result of the ferrous sulphate treatment[41].

366

(f) Extraction of the toxic factors

Extractions aimed at removing glucosinolates have been made partly with water or with buffers in water solution, and partly with various organic solvents.

(A) Rapeseed meal has been extracted with cold or warm water[56, 57, 31] or with buffers[58]. By extraction of myrosinase-free meal with buffer, Belzile and Bell[58] could remove more than 80 per cent of the glucosinolates. When the meal contained active myrosinase the water extraction was less effective since the cleavage products are less water-soluble than the intact glucosinolates. In bio-assays which involved feeding the myrosinase-active extracted meals to mice the weight gains of the animals were less than expected according to the assays of isothiocyanates and oxazolidinethiones. This may indicate that these meals contained other split products from glucosinolates which were harmful to the animals.

(B) Several authors report that extraction of rapeseed meal with 70–95 per cent ethanol results in a lowered glucosinolate content and an improved nutritional value. Most of these studies were, however, made many years ago and the chemical analyses were performed with methods that are unreliable according to our present knowledge.

In a recent study[59] myrosinase-free meals of rape and white mustard were extracted with 80 per cent ethanol. The results of these extractions on the bio-assays with mice are shown in Table 15:7. The weight gain per g protein of the

TABLE 15:7

EFFECT OF MEALS OF RAPE (BRASSICA NAPUS) AND WHITE MUSTARD (SINAPIS ALBA), EXTRACTED WITH 80% ETHANOL, ON MICE GROWTH AND PROTEIN EFFICIENCY[59]

Protein	Gain 20 days g	Consumed feed g	Gain per g protein
Egg albumin	10.0	53	1.90
Casein	9.5	57	1.65
Brassica napus seed meal	9.1	49	1.87
Brassica napus seed meal, supplemented with methionine and isoleucine	9.5	50	1.89
Sinapis alba seed meal	7.6	50	1.58
Sinapis alba seed meal, supplemented with methionine and isoleucine	9.0	50	1.81

mice fed on rapeseed meal extracted with 80 per cent ethanol exceeded that of the mice fed on casein.

Attempts at removing glucosinolates, combined with chemical analyses and bio-assays of the products with rats, have also been performed on seed meals of *Crambe abyssinica*[60]. The glucosinolate content of dehulled seed of crambe is on a level similar to that of winter rape or somewhat higher. More than 90 per cent of the crambe seed glucosinolates are composed of *epi*-progoitrin that yields oxazolidinethione on enzymic hydrolysis under suitable conditions. When acetone was used as solvent the water content had to be at least 20 or 25 per cent, if the glucosinolates were to be removed effectively. An increase in the water content of the acetone, however, resulted in a greater extraction of dry matter from the meal. When crambe meal extracted with 80 per cent acetone was fed to rats at a 23 per cent level the gains in weight were close to normal. On the other hand, when untreated crambe meal was fed to rats at a 25 per cent level all the rats died within 21 days.

In other experiments a small volume of water was added to crambe meal, or crushed seed, containing active myrosinase[61]. After hydrolysis of the glucosinolates the cleavage products were extracted with acetone. Since these products are more lipophilic than the glucosinolates the water content of the acetone could be kept on a lower level than in the former experiment. The loss in dry matter decreased to *ca*. 10 per cent, and these losses involved protein only to a small extent. According to the chemical analyses no glucosinolates or oxazolidinethiones remained in the meal after the extraction. It was also possible to completely remove the nitriles from the hydrolysed meal. In a feeding test with rats at a 28 per cent level of hydrolysed and extracted crambe meal, the weight gains of the animals were *ca*. 88 per cent of that of the control group fed on casein and linseed cake meal.

Extraction procedures of crambe meal or rapeseed meal with water or water solutions of ethanol or acetone have generally resulted in a loss of approximately 20 per cent of the dry matter of the meal. Approximately one fourth of this represents glucosinolates. As indicated above the losses of dry matter in crambe seed meal were less when the meal was extracted after hydrolysis. This procedure has, however, been shown to be not very successful when reproduced on rapeseed meal[59]. The losses of dry matter from meal in combination with the costs for evaporation of the solvent used in extraction methods seem to cause these methods to be too expensive for preparing meals for animal feeding.

If the detoxified meal product is intended to be used for human consumption it may be sold at a price which can bear higher production costs. Eapen *et al*.[62] have studied the possibilities of making a meal that could be used for this purpose. In their procedure they started with an enzyme inactivation by cooking the seed in water for 2 min. Glucosinolates are removed by aqueous extraction prior to or after the defatting process. The crude fiber content of the meal is lowered by removing the seed coat by air classification, a procedure that is facilitated by the previous hot water treatment of the seed. A white glucosinolate-free meal, with an increased content of protein and a decreased content of crude fiber is obtained according to this method.

15.6.3 Isolation of the protein

It should be possible to produce protein isolates of rapeseed meal according to procedures similar to those used for soya meal today, in which the proteins are extracted into a buffered water solution. The extracted proteins are then precipitated by adjustment of the pH to the point at which most of the proteins in question have their lowest solubility. After filtration or centrifugation the precipitate is dried. The losses of protein in these procedures amount to 30–40 per cent, and the protein content of the protein isolate obtained is *ca.* 90 per cent. Although the price per kg protein obtained as isolate is considerably higher than the price per kg protein in meal because of losses and manufacturing costs, it is still low compared to the price of animal proteins.

15.6.4 Removal of glucosinolates by plant breeding methods

There are two requirements that must be fulfilled if the chemical composition of a plant material is to be changed by plant breeding methods. First, analytical methods must be available which are sufficiently accurate, precise, sensitive and suitable for serial analysis so that a large number of individual plants can be analysed. Secondly, there must exist a plant material that deviates genetically in the direction of the desired property.

Up to 1965 little progress had been made regarding these two requirements. In that year an accurate analytical method used for glucosinolates in *Crambe abyssinica* was published by VanEtten et al.[60]. The method was based on photometric titration of enzymatically released sulphate and consequently was a determination of total glucosinolate content. It was combined with the determination of steam-volatile isothiocyanates according to Wetter[63]. In 1965 and 1967 Appelqvist and Josefsson[64, 47] published another method including a separation of isothiocyanates and oxazolidinethiones by selective extraction and the determination of these two components by spectrophotometric measurements. A method that included gas chromatographic determinations of individual isothiocyanates and spectrophotometric determination of oxazolidinethiones was published by Youngs and Wetter in 1967[19].* By the use of these methods it was possible to study a large number of samples including different species, varieties, lines, and individual plants. In 1967 it was found independently at the Prairie Regional Laboratory of the National Research Council at Saskatoon, Canada[81], and at the Swedish Seed Association, Svalöf, Sweden[25] that the Polish summer rape variety Bronowski had a genetically determined glucosinolate content that was much lower than the average. Although this variety is the one which contains the lowest

---

* In 1969 Lein and Schön[65] published a sensitive method for assay of total glucosinolate content based on estimation of enzymatically released glucose.

amounts of glucosinolates known hitherto in rapeseed, other varieties have also been found that are considerably lower in glucosinolate content than the average. Some results on material low in glucosinolates found at the Swedish Seed Association are shown in Table 15:8. Kondra[66] has found that there exist qualitative as

TABLE 15:8

EXAMPLES OF LOW CONTENT OF GLUCOSINOLATES IN SEED MEAL OF RAPE (BRASSICA NAPUS) AND TURNIP RAPE (B. CAMPESTRIS)

Species and variety	Glucosinolate content % in dry matter
Brassica napus, summer rape, cv. Ukraina	3.03
Brassica napus, summer rape, cv. Bronowski	0.64
Brassica napus, summer rape, selected plant from Bronowski	0.11
Brassica campestris, summer turnip rape, Sv 67/1683, pl 27	1.76

well as quantitative differences in glucosinolate patterns between varieties. This may open possibilities for obtaining varieties without glucosinolates or low in such compounds by a plant breeding programme (see Chapter 6).

The seed meal of Bronowski has been compared with meal from normal varieties in mice tests. Preliminary results of such tests are shown in Table 15:9. Similar results have been obtained by Bell[67] who only used heat-treated meals. The reason for the relatively large growth inhibition due to the meal of Bronowski that was not derived from heat-treated seed still remains to be elucidated. It does

TABLE 15:9

EFFECT OF DIFFERENT RAPESEED MEALS ON MICE GROWTH

Exp. No.	Protein	Gain 20 days g
1	Casein	11.2
	Meal from summer rape, cv. Rigo, no heat treatment	−1.4
	Meal from summer rape, cv. Bronowski, no heat treatment	5.6
2	Egg albumin	10.4
	Meal from summer rape, cv. Rigo. The seed was heat treated at 90 °C for 15 minutes before defatting	8.9
	Meal from summer rape, cv. Bronowski. The seed was heat treated at 90 °C for 15 minutes before defatting	10.6

not seem probable that isothiocyanates and oxazolidinethiones released from the glucosinolates of the Bronowski meal would have resulted in such a growth inhibition since Bell did not obtain any significant difference in weight gains when he fed mice on heat-treated meal with and without added myrosinase. Since it has been found that glucosinolates in meals that have not been heat-treated often give rise to large amounts of nitriles[40] these highly toxic products might have been released in the digestive tract of the animals.

The seed yield of Bronowski is considerably lower than that of the modern Swedish summer rape varieties. It is also important to obtain winter varieties with a low glucosinolate content. For these reasons Bronowski is now used in plant breeding programmes at several European institutes, being crossed with high yielding summer rape varieties as well as with winter varieties which are high yielding and winter resistent. For more details of this work the reader is referred to Chapter 6 in this volume.

15.7 DIGESTIBILITY

Rutkowski and Kozłowska[2] reported that the digestibility coefficients of rapeseed meal protein obtained in experiments *in vitro* were similar to those of other oil seed meals. They also found that processing methods favouring formation of melanoids caused a lower digestibility coefficient.

Jarl[68] reported a digestibility coefficient *in vivo* (ruminants) for the organic matter of rapeseed meal as 76 per cent and for crude protein 83 per cent. Nordfeldt[69] found corresponding values of 76 and 85 per cent.

When Bell[50] replaced equal parts of linseed and soya meal with *Brassica campestris* meal in a diet for swine he found that the energy and protein components of this meal were at least as digestible as those in the soya-linseed meal mixture. In studies with chicks, Lodhi *et al.*[82] obtained nitrogen absorbability values of 80 and 85 per cent for rapeseed meal and soya meal, respectively.

Rutkowski *et al.*[70] found a coefficient of digestibility of the rapeseed meal protein that was *ca.* 10 per cent lower in experiments *in vivo* (poultry) than *in vitro*. This might have been due to the glucosinolate content.

15.8 FEEDING VALUE

15.8.1 Ruminants

Fewer studies have been made concerning the feeding value of rapeseed meal for ruminants than the number of studies made on swine and poultry. The reasons for this may be that studies with cattle are relatively expensive to perform and that sheep are not a large animal group in the rapeseed growing areas. The fact that the problems of feeding rapeseed are less for ruminants than for non-ruminant animals may be another reason. In contrast to swine, poultry, and other non-ruminants, the ruminants do not develop goiter when fed on rapeseed meal, nor do there seem to be any toxic effects produced on these animals. The reason for this was clarified in an experiment by Virtanen et al.[71], in which 500 mg vinyl oxazolidinethione was fed to a cow, and it was found that only 0.05 per cent of this quantity was transferred to the milk. Approximately two-thirds of this amount was transferred within two hours. Therefore, the substance had apparently passed through the rumen wall into the blood. When the cow was fed crystalline progoitrin only traces of oxazolidinethione were found in the milk, thus demonstrating that the progoitrin was not hydrolysed in the rumen, or the oxazolidinethione formed was destroyed there. It has recently been reported that feeding rapeseed meal to sheep affected the thyroid gland[83].

TABLE 15:10

CONSUMPTION OF RAPESEED MEAL IN SWEDEN AND USE OF THIS MEAL IN VARIOUS CONCENTRATE MIXES. 1000 TONS[72]

Year	Consumption of rapeseed meal	Rapeseed meal in concentrate mixes for		
		cattle	poultry	swine
1962/63	71.1	52.6	9.0	2.9
1963/64	56.8	40.7	9.7	2.9
1964/65	75.8	57.9	9.6	3.4
1965/66	93.2	74.5	7.9	6.4
1966/67	82.0	61.3	7.2	0.4

Since the cleavage products of glucosinolates do not exercise toxic effects on ruminants, rapeseed meal is mainly used for these animals. The consumption of rapeseed meal by different domestic animals in Sweden is shown in Table 15:10.

15.8.2 Young ruminants (below 6 months)

Few studies have been made on feeding rapeseed meal to young calves and lambs. In the studies made, rapeseed meal was compared to linseed meal. Generally rapeseed meal was less palatable initially than linseed meal, but after the initial stage there were no differences in feed consumption between the rapeseed groups and the linseed groups and the weight gains were equal[73].

15.8.3 Growing and fattening ruminant animals

The results of the studies made on growing and fattening ruminant animals are similar to those for the youngest ruminants: there seem to be some palatability problems in the beginning of the feeding period, but the animals consume their rations and gain in weight at approximately the same rate as on other oilseed meals[73].

15.8.4 Milk cows

In experiments with milk cows Jarl[68] daily fed the animals *ca.* 2.5 kg of an oilcake mixture containing 0, 25, 50, and 60 per cent rapeseed meal. There were some palatability problems in the beginning of the feeding period but not when the cows had become accustomed to the rapeseed. The milk production was less when 50 or 60 per cent rapeseed was used compared to 0 or 25 per cent. There was a small decrease in the fat content of the milk in all the combinations with rapeseed meal. The greater part of this decrease was explained, however, by the fact that the feed without rapeseed meal contained coconut oil. Jarl concluded that 2 kg of rapeseed meal could be given per cow and day, and suggested that the meal should be given in a dry state.

Nordfeldt[69] compared meals of rapeseed and white mustard seed with two different fat levels (*ca.* 1.5 and 7.0 per cent) with soya meal. The quantities of rapeseed and white mustard meals amounted to *ca.* 1.5 kg per cow per day. 1.2 kg soya meal was given to the cows in the control group. The meals of rapeseed and white mustard seed were less palatable than soya meal, especially in the beginning of the experimental period. Rapeseed meal gave a higher milk production than soya meal, but also a slightly lower fat content of the milk. Therefore, there were no differences between groups in fat production, nor were there any differences in the taste of the milk produced.

In a review in 1965 Whiting[73] concluded that rapeseed meal can be considered to be equivalent in nutritional value (on an equivalent protein basis) to linseed and soya meals when it makes up to 10 per cent of the total dry matter of the ration.

15.8.5 Swine

For non-ruminant animals such as swine the amino acid composition of the protein fed is more important than for ruminants. When the amino acid composition of rapeseed meal is compared to the amino acid requirements of the young pig it is found to be quite favourable. According to results of feeding experiments, swine, and especially young ones, are relatively sensitive to the cleavage products of glucosinolates, which cause goiter and growth inhibition. Therefore, in Sweden the ratio of rapeseed meal in concentrated feed for swine has most often been below one per cent. It seems possible that this figure is less than necessary. In a review in 1965 Bowland[74] concludes that up to 4 per cent of the total ration may be composed of rapeseed meal for young pigs up to 25 kg in weight. Market pigs from 25 to 90 kg liveweight could efficiently use up to 10 per cent rapeseed meal in the ration. For mother-sows during pre-gestation and lactation rapeseed meal should not be fed at a level above 3 per cent of the total ration. Recently Schuld and Bowland[75] reported that when 8 per cent rapeseed meal, substituted on an isonitrogenous basis for soya meal and wheat, was fed to young sows, there was a decreased number of pigs weaned per litter. In the second reproductive cycle, however, no differences were found that were caused by differences in diet. The same authors studied the progenies from these sows and found that the feed given to the sows did not influence the progeny[76]. Eight per cent rapeseed meal given to the progeny depressed the rate of gain and efficiency of feed utilization in one experiment, but not significantly in another one.

Results of experiments with market pigs indicate that the levels of rapeseed meal recommended will have no adverse effects on carcass characteristics[74].

15.8.6 Poultry

It is known from several reports that feeding of rapeseed or rapeseed meal to poultry causes thyroid enlargement[34, 56]. In a review in 1965 Clandinin[77] concluded that the initial effects of this feed on the thyroid gland of poultry includes a decrease in the uptake of iodine by the gland and an increase in the release of iodine from the gland. After the poultry have become accustomed to the rapeseed meal (3–4 weeks), the uptake of iodine by the enlarged thyroid gland is increased while the secretion rate from the glands becomes normal. Although the thyroid gland will still be enlarged, a physiological equilibrium seems to have been reached.

Extensive studies of the effect of rapeseed feeding on poultry have been performed by Clandinin et al.[78]. When comparing expeller-processed rapeseed meals with soya meal, they found that the former resulted in decreased growth rate and depressed feed efficiency, even when only small amounts of the supplementary protein in the ration was composed of rapeseed meal. When the meal was processed at lower temperatures the nutritional value increased. Amino acid analyses revealed that the poor results of feeding expeller-processed rapeseed meal largely reflected lysine destruction. In a series of experiments Clandinin[77] established that the growth rate and feed conversion of chicken were the same when 15 per cent prepress-solvent rapeseed meal was used in the diet or when a diet isonitrogenous and isocaloric with soya meal was used. Therefore he concluded that 10–15 per cent rapeseed meal could be used in chick starter rations.

With regard to laying and breeding chicks and turkeys Clandinin concluded that 10 per cent rapeseed meal yields just as satisfactory production, feed conversion, fertility and hatchability as corresponding amounts of protein from soya meals. These results have been confirmed for turkeys in a recent paper by Robblee and Clandinin[79]. Later studies[84], however, suggest that rations for laying and breeding chickens should not contain more than 5 per cent rapeseed meal. Summers et al.[80], however, reported recently that when rations containing 10, 20, and 30 per cent of rapeseed meal were given to turkeys starting at 1 day of age there was a significant linear decrease in weight gain and feed utilization when compared with an isocaloric and isonitrogenous basal diet including corn and soya bean. Steam-pelletting the meal reduced these differences. Most of the adverse effects of rapeseed meal were obtained before the birds were 56 days of age. Lodhi et al.[85] showed that the ability of the chick to utilize the energy in rapeseed meal was relatively low but that it increased with age.

Since it has not often been reported whether the rapeseed meal fed was derived from *Brassica napus* or *B. campestris*, and no glucosinolate analyses were made of the meals used in experiments some years ago (*cf.* Table 15:6) it is not possible to draw definite conclusions about the levels of rapeseed meal that may be used successfully as feed for swine or poultry. Most of the feeding experiments have been performed in Canada where the rapeseed grown is mostly composed of *B. campestris*. On the other hand, mainly *B. napus* is grown in European countries. Generally this species is higher in glucosinolates, and especially in those yielding oxazolidinethiones, than *B. campestris*. Thus it is possible that seed meal of *B. napus* should not be fed to swine or poultry at so high a level as meal of *B. campestris*. Work is in progress in order to decide if there are any differences in nutritional value between *B. napus* and *B. campestris* meal for swine or poultry as well as to decide the upper limit for favourable feeding with these meals.

REFERENCES

1 *Rapeseed Meal for Livestock and Poultry – A Review*, The Canada Department of Agriculture Publication 1257, Ottawa, 1965.
2 A. RUTKOWSKI AND H. KOZŁOWSKA, *Rapeseed Meal*, Wydawnictwo Ministerstwa Przemysłu Lekkiego i Spozywczego, Warszawa, Poland, 1967.
3 *NJF:s Fodermiddeltabel*, Mariendals Boktrykkeri A/S, Gjøvik, 1969.
4 D. R. CLANDININ AND L. BAYLY, *Can. J. Animal Sci.*, 43 (1963) 65.
5 E. JOSEFSSON, *J. Sci. Food Agr.*, 21 (1970) 98.
6 L. R. WETTER, in *Rapeseed Meal for Livestock and Poultry – A Review*, The Canada Department of Agriculture, Publication 1257, 1965, p.32.
7 D. R. CLANDININ, *Poultry Sci.*, 46 (1967) 1596.
8 R. W. MILLER, C. H. VANETTEN, C. MCGREW, I. A. WOLFF AND Q. JONES, *J. Agr. Food Chem.*, 10 (1962) 426.
9 C. H. VANETTEN, W. F. KWOLEK, J. E. PETERS AND A. S. BARCLAY, *J. Agr. Food Chem.*, 15 (1967) 1077.
10 Food and Agr. Organization, U.N., *FAO Nutritional Studies*, No. 16, Rome, Italy, 1957.
11 *A.A.C.C. Cereal Laboratory Methods*, 7th edn., Method 32-15, 1962.
12 J. HRDLIČKA, H. KOZTOWSKA, J. POKORNÝ AND A. RUTKOWSKI, *Nahrung*, 9 (1965) 71.
13 J. E. MCGHEE, L. D. KIRK AND G. C. MUSTAKAS, *J. Amer. Oil Chem. Soc.*, 41 (1964) 359.
13a G. J. KLAIN, D. C. HILL, H. D. BRANION AND J. A. GRAY, *Poultry Sci.*, 35 (1956) 1315.
14 M. G. ETTLINGER AND A. J. LUNDEEN, *J. Amer. Chem. Soc.*, 78 (1956) 4172.
15 L. GUIGNARD, *J. Botanique*, 4 (1890) 385.
16 A. KJAER, J. CONTI AND K. A. JENSEN, *Acta Chem. Scand.*, 7 (1953) 1271.
17 E. JOSEFSSON AND C. MÜHLENBERG, *Acta Agr. Scand.*, 18 (1968) 97.
18 A. KJAER AND R. B. JENSEN, *Acta Chem. Scand.*, 10 (1956) 1365.

19 C. G. YOUNGS AND L. R. WETTER, *J. Amer. Oil Chem. Soc.*, 44 (1967) 551.
20 M. G. ETTLINGER AND C. P. THOMPSON, *Final Report to Quartermaster Research and Engineering Command*, Contract DA 19-129-QM-1689 (1962). AD-290 747, Office of Technical Services, U.S. Dept. of Commerce.
21 M. A. GREER, *J. Amer. Chem. Soc.*, 78 (1956) 1260.
22 B. A. TAPPER AND D. B. MACGIBBON, *Phytochemistry*, 6 (1967) 749.
23 M. E. DAXENBICHLER, C. H. VANETTEN, F. S. BROWN AND Q. JONES, *J. Agr. Food Chem.*, 12 (1964) 127.
24 A. KJAER AND K. RUBINSTEIN, *Acta Chem. Scand.*, 8 (1954) 598.
25 E. JOSEFSSON AND L.-Å. APPELQVIST, *J. Sci. Food Agr.*, 19 (1968) 564.
26 E. JOSEFSSON, *J. Sci. Food Agr.*, 21 (1970) 94.
27 H. WILL AND A. LAUBENHEIMER, *Justus Liebigs Ann. Chem.*, 199 (1879) 150.
28 R. GMELIN AND A. I. VIRTANEN, *Acta Chem. Scand.*, 14 (1960) 507.
29 J. BAROTHY, *Über p-Hydroxybenzyl-isothiocyanat, das Senföl aus Samen von Sinapis alba L.*, Prom. Nr. 3537, Eidg. Tech. Hochsch., Zürich, 1964.
30 C. H. VANETTEN, M. E. DAXENBICHLER, J. E. PETERS AND H. L. TOOKEY, *J. Agr. Food Chem.*, 14 (1966) 426.
31 R. BELZILE, J. M. BELL AND L. R. WETTER, *Can. J. Animal Sci.*, 43 (1963) 169.
32 T. H. KENNEDY AND H. D. PURVES, *Brit. J. Exp. Pathol.*, 22 (1941) 241.
33 W. E. GRIESBACH, T. H. KENNEDY AND H. D. PURVES, *Brit. J. Exp. Pathol.*, 22 (1941) 249.
33a T. MATSUMOTO, H. ITOH AND Y. AKIBA, *Poultry Sci.*, 48 (1969) 1061.
34 A. FRÖLICH, *Kgl. Lantbruks-Högskol. Ann.*, 19 (1952) 205.
35 S. NORDFELDT, N. GELLERSTEDT AND S. FALKMER, *Acta Pathol. Microbiol. Scand.*, 35 (1954) 217.
36 E. B. ASTWOOD, M. A. GREER AND M. G. ETTLINGER, *J. Biol. Chem.*, 181 (1949) 121.
37 J. E. VANDERLAAN AND W. P. VANDERLAAN, *Endocrinology*, 40 (1947) 403.
38 J. M. BELL, Personal communication.
39 P. LANGER AND V. ŠTOLC, *Endocrinology*, 76 (1965) 151.
40 C. H. VANETTEN, W. E. GAGNE, D. J. ROBBINS, A. N. BOOTH, M. E. DAXENBICHLER AND I. A. WOLFF, *Cereal Chem.*, 46 (1969) 145.
41 *Fats and Oils in Canada*, Department of Industry, Ottawa, 2:2 (1967) 37.
42 J. M. BELL, *Can. J. Animal Sci.*, 37 (1957) 43.
43 J. M. BELL, *Can. J. Animal Sci.*, 37 (1957) 31.
44 F. W. CLEMENTS AND J. W. WISHART, *Metabolism*, 5 (1956) 623.
45 A. I. VIRTANEN (Editor), *Investigations on the Alleged Goitrogenic Properties of Milk*, Helsinki, 1963.
46 F.-E. KRUSIUS AND P. PELTOLA, *Acta Endocrinologica*, 53 (1966) 342.
47 L.-Å. APPELQVIST AND E. JOSEFSSON, *J. Sci. Food Agr.*, 18 (1967) 510.
48 M. A. GREER AND J. M. DEENEY, *J. Clin. Invest.*, 38 (1959) 1465.
49 E. L. OGINSKY, A. E. STEIN AND M. A. GREER, *Proc. Soc. Exp. Biol. Med.*, 119 (1965) 360.
50 J. M. BELL, *J. Animal Sci.*, 24 (1965) 1147.
51 K. J. GOERING, O. O. THOMAS, D. R. BEARDSLEY AND W. A. CURRAN, Jr., *J. Nutr.*, 72 (1960) 210.
52 J. M. BELL AND R. J. BELZILE, in *Rapeseed Meal for Livestock and Poultry – A Review*, The Canada Department of Agriculture, Publication 1257, 1965, p.45.
53 L. D. KIRK, G. C. MUSTAKAS AND E. L. GRIFFIN, JR., *J. Amer. Oil Chem. Soc.*, 43 (1966) 550.
54 G. C. MUSTAKAS, L. D. KIRK, E. L. GRIFFIN, JR. AND D. C. CLANTON, *J. Amer. Oil Chem. Soc.*, 45 (1968) 53.
55 C. G. YOUNGS AND A. S. PERLIN, *Can. J. Chem.*, 45 (1967) 1801.
56 C. E. ALLEN AND D. S. DOW, *Sci. Agr.*, 32 (1952) 403.
57 A. FRÖLICH, *Kgl. Lantbruks-Högskol. Ann.*, 20 (1953) 105.
58 R. J. BELZILE AND J. M. BELL, *Can. J. Animal Sci.*, 46 (1966) 165.
59 L.-Å. APPELQVIST AND L. MUNCK, to be published.
60 C. H. VANETTEN, M. E. DAXENBICHLER, J. E. PETERS, I. A. WOLFF AND A. N. BOOTH, *J. Agr. Food Chem.*, 13 (1965) 24.

61 H. L. TOOKEY, C. H. VANETTEN, J. E. PETERS AND I. A. WOLFF, *Cereal Chem.*, 42 (1965) 507.
62 K. E. EAPEN, N. W. TAPE AND R. P. A. SIMS, *J. Amer. Oil Chem. Soc.*, 46 (1969) 52.
63 L. R. WETTER, *Can. J. Biochem. Physiol.*, 33 (1955) 980.
64 L.-Å. APPELQVIST AND E. JOSEFSSON, *Acta Chem. Scand.*, 19 (1965) 1242.
65 K.-A. LEIN AND W. J. SCHÖN, *Angew. Botanik*, 43 (1969) 87.
66 Z. KONDRA, *Some environmental effects on and genetics of the isothiocyanate and oxazolidinethione content of seed meal of Brassica napus L. and B. campestris L.*, Order No. 68-5910, University Microfilms, 300 North Zeeb Road, Ann Arbor, Michigan, 1967.
67 J. M. BELL, personal communication.
68 F. JARL, *Kungl. Lantbrukshögskolan och Statens Lantbruksförsök, Statens Husdjursförsök*, Meddelande Nr 45, Tierp, Sweden, 1951.
69 S. NORDFELDT, *Kungl. Lantbrukshögskolan och Statens Lantbruksförsök, Statens Husdjursförsök*, Meddelande Nr 66, Uppsala, 1958.
70 A. RUTKOWSKI, H. KOZOWSKA, J. CHUDY AND J. POKORNÝ, *Nahrung*, 9 (1965) 207.
71 A. I. VIRTANEN, M. KREULA AND M. KIESVAARA, *Z. Ernährungswiss.*, (Suppl. 3) (1963) 23.
72 T. ROBERTSSON, The National Swedish Agricultural Marketing Board, personal communication.
73 F. WHITING, in *Rapeseed Meal for Livestock and Poultry – A Review*, The Canada Department of Agriculture, Publication 1257, 1965, p.61.
74 J. P. BOWLAND, in *Rapeseed Meal for Livestock and Poultry – A Review*, The Canada Department of Agriculture, Publication 1257, 1965, p.69.
75 F. W. SCHULD AND J. P. BOWLAND, *Can. J. Animal Sci.*, 48 (1968) 57.
76 J. P. BOWLAND AND F. W. SCHULD, *Can. J. Animal Sci.*, 48 (1968) 189.
77 D. R. CLANDININ, in *Rapeseed Meal for Livestock and Poultry – A Review*, The Canada Department of Agriculture, Publication 1257, 1965, p.81.
78 D. R. CLANDININ, R. RENNER AND A. R. ROBBLEE, *Poultry Sci.*, 38 (1959) 1367.
79 A. R. ROBBLEE AND D. R. CLANDININ, *Can. J. Animal Sci.*, 47 (1967) 127.
80 J. D. SUMMERS, W. F. PEPPER, E. T. MORAN, JR. AND H. S. BAYLEY, *Can. J. Animal Sci.*, 47 (1967) 131.
81 R. K. DOWNEY, B. M. CRAIG AND C. G. YOUNGS, *J. Am. Oil Chem. Soc.*, 46 (1969) 121.
82 G. N. LODHI, R. RENNER AND D. R. CLANDININ, *Poultry Sci.*, 49 (1970) 991.
83 N. GRENET, *International Rape Conference, Paris, Abstracts*, C.E.T.I.O.M., 174 Avenue Victor-Hugo, Paris, 1970.
84 D. R. CLANDININ, *International Conference on the Science, Technology and Marketing of Rapeseed and Rapeseed Products*, Ste. Adele, Quebec, Canada, 1970, *Abstracts*.
85 G. N. LODHI, R. RENNER AND D. R. CLANDININ, *Poultry Sci.*, 48 (1969) 964.

APPENDICES

378

Important Varieties of Brassica Oil Crops

Name of variety	Species	Type	Breeder	Country	Licenced or released in year	Special properties
Argus	B. napus	Winter	Weibull	Sweden		
Arlo	B. campestris	Summer	Svalöf	Sweden	1957	
Bele	B. campestris	Summer	Svalöf	Sweden	1960	
Bronowski	B. napus	Summer		Poland	1955	Low content of glucosinolates
Crésus	B. napus	Summer	Ringot	France	1964	
Diamant	B. napus	Winter	N.P.Z.	W. Germany		Tolerant to late sowing
Dippes	B. napus	Winter	Crebrüder	W. Germany	1943	
Donoslaski	B. napus	Winter	P.B.S.P.A.	Poland	1969	
Duro	B. campestris	Winter	Svalöf	Sweden	1956	
Echo	B. campestris	Summer	Indian Head Farm	Canada	1964	
Etampes	B. campestris	Summer	Manchina	France	1965	
Golden				Canada		
Górczański	B. napus	Winter	P.B.S.P.A.	Poland	1955	
Gruber	B. campestris	Winter	Gruber	W. Germany	1942	
Gylle	B. napus	Summer	Svalöf	Sweden	1969	
Hambourg	B. napus	Winter	Mansholt	Holland		
Hector	B. napus	Winter	Hammenhög	Sweden	1969	
Heiner	B. napus	Winter				
Janetzki	B. napus	Summer	Janetzki	Germany	1942	
Janus	B. napus	Summer	Svalöf-Ringot	France	1969	
Lembkes	B. campestris	Winter	N.P.Z.	W. Germany		
Lembkes	B. napus	Winter	N.P.Z.	W. Germany		
Ludowy	B. campestris	Winter	I.B.P.A.	Poland	1964	
Marcus	B. napus	Winter	Ringot	France	1969	Certain field resistance to Phoma
Matador	B. napus	Winter	Svalöf	Sweden	1949	
Mazowiecki	B. napus	Summer	P.B.S.P.A.	Poland	1955	
Mlochowski	B. napus	Summer	I.B.P.A.	Poland	1969	
Nilla	B. napus	Summer	Hammenhög	Sweden		
Norde	B. napus	Winter	Svalöf	Sweden	1969	Winter hardy
Nugget	B. napus	Summer	Can. Agr. Res. Station	Canada	1961	
Oléor	B. napus	Winter	I.N.R.A.	France	1964	
Oro	B. napus	Summer	Can. Agr. Res. Station	Canada	1968	Free of erucic acid
Panter	B. napus	Winter	Svalöf	Sweden	1968	Tolerant to late sowing
Polar			Svalöf	Sweden		
Pollux	B. napus	Winter	Gross Lüsewitz	E. Germany		Tolerant to late sowing

Name of variety	Species	Type	Breeder	Country	Licenced or released in year	Special properties
Rapido III	B. campestris	Winter	Svalöf	Sweden	1969	
Rapol	B. napus	Winter	N.P.Z.	W. Germany		
Regina II	B. napus	Summer	Svalöf	Sweden	1953	
Rigo	B. napus	Summer	Svalöf	Sweden	1966	
Sarepta	B. napus	Winter	I.N.R.A.	France	1962	
Sinus	B. napus	Winter	Svalöf	Sweden	1971	
Skrzesowicki	B. napus	Winter	P.B.S.P.A.	Poland	1955	
Slapska	B. napus	Winter		Czecho-slovakia		Tolerant to late sowing
Span	B. campestris	Summer	Can. Agr. Res. Station	Canada	1971	
Sylvi	B. campestris	Winter	Kumla	Sweden	1969	
Tanka	B. napus	Summer	Univ. of Manitoba	Canada	1963	
Target	B. napus	Summer	Univ. of Manitoba	Canada	1966	
Titus	B. napus	Winter	Ringot	France	1966	
Tonus	B. napus	Winter	Ringot	France	1963	
Torpe Vestal	B. campestris	Summer	Svalöf	Sweden	1969	
Victor	B. napus	Winter	Svalöf	Sweden	1964	
Zephyr	B. napus	Summer	Can. Agr. Res. Station	Canada	1971	
Zollerngold	B. napus	Summer	Späth	W. Germany	1952	

Characteristics for Rapeseed Oil

Compressibility, $\times 10^{-6}$	60.9
Dielectric constant, cgs units	3.06
Flash point, open cup, °C	282
Iodine number	97–108
Pour point, °C	−12
Refractive index at 25 °C	1.470–1.474
Refractive index at 40 °C	1.465–1.469
Saponification number	170–180
Specific gravity at 25 °/25 °C	0.906–0.914
Specific heat at 20 °C	0.488
Surface tension, dynes per cm at 12 °C	36.6
Thermal conductivity,	
cals/sq.cm.sec for 1 °C and 1 cm thick	41
Titer, °C	11.5–15.0
Viscosity (Seybolt) 38 °C	260
100 °C	54

Physical properties of erucic acid

Specific gravity, d_4^{20}	0.85321
Specific gravity, d_4^{55}	0.860
Melting point, °C	34.7
Boiling point, °C	281 (30 mm)
Refractive index n_D^{20}	1.44438
Iodine value	74.98

Physical properties of methyl erucate

Density, d_4^{20}	0.8706
d_4^{40}	0.8565
Molar volume, V_m^{20}	404.96
	411.63

Recommended International Standard for Edible Rapeseed Oil

1.   DESCRIPTION

Rapeseed Oil (synonyms: Turnip Rape Oil; Ravison Oil; Sarson Oil and Toria Oil) is derived from the seeds of *Brassica campestris* L., *Brassica napus* L. and *Brassica tournefortii* Gouan.

2.   ESSENTIAL COMPOSITION AND QUALITY FACTORS

2.1   *Identity Characteristics*

2.1.1 Relative Density (20 °C/water at 20 °C)	0.910–0.920
2.1.2 Refractive Index (n_D 40 °C)	1.465–1.469
2.1.3 Saponification Value (mg KOH/g)	168–181
2.1.4 Iodine Value (Wijs)	94–120
2.1.5 Crismer Value	80–85
2.1.6 Unsaponifiable Matter	not more than 20 g/kg

2.2   *Quality Characteristics*

2.2.1 Colour

Characteristic of the designated product.

2.2.2 Odour and Taste

Characteristic of the designated product, and free from foreign and rancid odour and taste.

2.2.3 Acid Value	
(a) Virgin oil	not more than 4.0 mg KOH/g
(b) Non-virgin oil	not more than 0.6 mg KOH/g
2.2.4 Peroxide Value	Not more than 10 milliequivalents peroxide oxygen/kg

3.   FOOD ADDITIVES

3.1   *Colours*

The following colours are permitted for the purpose of restoring natural colour lost in processing or for the purpose of standardizing colour, as long as the added colour does not deceive or mislead the consumer by concealing damage or inferiority or by making the product appear to be of greater than actual value:

	Maximum level of use
3.1.1 Beta-carotene	Not limited
3.1.2 Annatto*	Not limited
3.1.3 Curcumin*	Not limited
3.1.4 Canthaxanthine	Not limited
3.1.5 Beta-apo-8'-carotenal	Not limited
3.1.6 Methyl and ethyl esters of Beta-apo-8'-carotenoic acid	Not limited

3.2   *Flavours*

Natural flavours and their identical synthetic equivalents, except those which are known to represent a tonic hazard, and other synthetic flavours approved by the Codex Alimentarius Commission are permitted for the purpose of restoring natural flavour lost in processing or for the purpose of standardizing flavour, as long as the added flavour does not deceive or mis-

---

* Temporarily endorsed.

lead the consumer by concealing damage or inferiority or by making the product appear to be of greater than actual value.

3.3	*Antioxidants*	*Maximum level of use*
3.3.1	Propyl, octyl, and dodecyl gallates	100 mg/kg individually or in combination
3.3.2	BHA, BHT	200 mg/kg individually or in combination
3.3.3	Any combination of gallates with BHA or BHT, or both	200 mg/kg, but gallates not to exceed 100 mg/kg
3.3.4	Ascorbyl palmitate Ascorbyl stearate	200 mg/kg individually or in combination
3.3.5	Natural and synthetic tocopherols	Not limited
3.3.6	Dilauryl thiodipropionate	200 mg/kg

3.4	*Synergists*	*Maximum level of use*
3.4.1	Citric acid	Not limited
3.4.2	Sodium citrate	Not limited
3.4.3	Isopropyl citrate mixture	
3.4.4	Monoglyceride citrate	100 mg/kg individually or in combination
3.4.5	Phosphoric acid*	

3.5 *Anti-foaming Agents*
Dimethyl polysiloxane (syn. Dimethyl silicone) singly or in combination with silicon dioxide*   10 mg/kg

3.6 *Crystallization Inhibitor*
Oxystearin*   1250 mg/kg

4.	CONTAMINANTS	*Maximum level*
4.1	*Matter volatile at 105°C*	2000 mg/kg
4.2	*Insoluble impurities*	500 mg/kg
4.3	*Soap content*	50 mg/kg
4.4	*Iron* (Fe)	
(a)	Virgin Oil	5.0 mg/kg
(b)	Refined Oil	1.5 mg/kg
4.5	*Copper* (Cu)	
(a)	Virgin Oil	0.4 mg/kg
(b)	Refined Oil	0.1 mg/kg
4.6	*Lead* (Pb)	0.1 mg/kg
4.7	*Arsenic* (As)	0.1 mg/kg

5.   HYGIENE
It is recommended that the product covered by the provisions of this standard be prepared in accordance with the appropriate Sections of the General Principles of Food Hygiene recommended by the Codex Alimentarius Commission (Ref. No. CAC/RCP. 1-1969)

6.   LABELLING
In addition to Sections 1, 2, 3, 4, 5 and 6.1 of the General Standard for the Labelling of Prepackaged Foods (Ref. CAC/RS 1-1969) the following specific provisions apply:
6.1   All products designated as *rapeseed oil, turnip rape oil, ravison oil, sarson oil* or *toria oil* must conform to this standard.
6.2   Oil produced from the seeds of *Eruca sativa* Mill. and conforming to the standard may be designated *jamba rape* oil.

---

* Temporarily endorsed.

6.3   Where *rapeseed oil* has been subject to any process of esterification or to processing which alters its fatty acid composition or its consistency the name *rapeseed oil* or any synonym shall not be used unless qualified to indicate the nature of the process.

7.   METHODS OF ANALYSIS

The methods of analysis described hereunder are international referee methods.

7.1   *Determination of relative density*

According to the method of the British Standards Institute (BS. 684:1958, p.10 Method 1, with reference temperature 20 °C in place of 15.5 °C).

7.2   *Determination of refractive index*

According to the IUPAC method (1966, II.B.2, *Refractive index*)

7.3   *Determination of saponification value*

According to the IUPAC method (1966, II.D.2, *Saponification value (Is)*)

7.4   *Determination of iodine value*

According to the IUPAC method (1966, II.D.7.1, II.D.7.2 and II.D.7.3, *The Wijs method*)

7.5   *Determination of crismer value*

According to the AOCS official method. Cb. 4-35.

7.6   *Determination of unsaponifiable matter*

According to the IUPAC method (1966), II.D.5.1 and II.D.5.3, *Diethyl ether method*).

7.7   *Determination of acid value*

According to the IUPAC method (1966, II.D.1.2, *Acid value*)

7.8   *Determination of peroxide value*

According to the IUPAC method (1966, II.D.13, *Peroxide value*)

7.9   *Determination of matter volatile at 105 °C*

According to the IUPAC method (1966, II.C.1.1, *Moisture and volatile matter*)

7.10   *Determination of insoluble impurities*

According to the IUPAC method (1966, II.C.2, *Impurities*)

7.11   *Determination of soap content*

According to the method of the British Standard Institute (BS.684:1958, p.49)

7.12   *Determination of iron*

According to the method of the British Standards Institute (B.S. 864:1958, p.92)

7.13   *Determination of copper*

According to the AOAC method (1965, 24.023-24.028, *IUPAC Carbamate Method*)

7.14   *Determination of lead*

According to the AOAC method (1965, 24.053, colorimetric dithizone determination)

7.15   *Determination of arsenic*

According to the AOAC method (1965 Diethyldithiocarbonate method, 24.011, 24.014-24.017, 24.006-24.008).

Useful Conversion Factors

acres	to	hectares	0.40468
bushels	to	hectoliters	0.3637
bushels (Brit.)	to	liters	36.367
bushels (U.S.)	to	liters	35.238
bushels (of rapeseed)	to	kilograms (of rapeseed)	22.68
bushels (of rapeseed)	to	pounds (of rapeseed)	50
bushels (of rapeseed)	to	tons (metric)	0.02268
bushels/acre	to	kg/hectare	67.25
gallons	to	liters	3.785
hectares	to	acres	2.4710
kilograms	to	pounds	2.2046
liters	to	gallons	0.2642
pounds	to	grams	453.59
pounds	to	kilograms	0.4536
pounds/bushel	to	kilograms/hectoliter	1.247
tons (metric)	to	tons (short)	1.10231
tons (metric)	to	pounds	2204.6
tons (metric)	to	bushels (US)	36.743
tons (short)	to	tons (metric)	0.90718
tons (short)	to	pounds	2000
tons (short)	to	bushels	37.33

Centigrade - Fahrenheit Conversion

C ← F	C → F		C ← F	C → F	
−90	−130	−202	1.7	35	95.0
−84	−120	−184	2.2	36	96.8
−79	−110	−166	2.8	37	98.6
−73	−100	−148	3.3	38	100.4
−68	− 90	−130	3.9	39	102.2
−62	− 80	−112	4.4	40	104.0
−57	− 70	− 94	5.0	41	105.8
−51	− 60	− 76	5.6	42	107.6
−46	− 50	− 58	6.1	43	109.4
−40	− 40	− 40	6.7	44	111.2
−34	− 30	− 22	7.2	45	113.0
−29	− 20	− 4	7.8	46	114.8
−23	− 10	14	8.3	47	116.6
−17.8	0	32	8.9	48	118.4
−17.2	1	33.8	9.4	49	120.2
−16.7	2	35.6	10.0	50	122.0
−16.1	3	37.4	10.6	51	123.8
−15.6	4	39.2	11.1	52	125.6
−15.0	5	41.0	11.7	53	127.4
−14.4	6	42.8	12.2	54	129.2
−13.9	7	44.6	12.8	55	131.0
−13.3	8	46.4	13.3	56	132.8
−12.8	9	48.2	13.9	57	134.6
−12.2	10	50.0	14.4	58	136.4
−11.7	11	51.8	15.0	59	138.2
−11.1	12	53.6	15.6	60	140.0
−10.6	13	55.4	16.1	61	141.8
−10.0	14	57.2	16.7	62	143.6
− 9.4	15	59.0	17.2	63	145.4
− 8.9	16	60.8	17.8	64	147.2
− 8.3	17	62.6	18.3	65	149.0
− 7.8	18	64.4	18.9	66	150.8
− 7.2	19	66.2	19.4	67	152.6
− 6.7	20	68.0	20.0	68	154.4
− 6.1	21	69.8	20.6	69	156.2
− 5.6	22	71.6	21.1	70	158.0
− 5.0	23	73.4	21.7	71	159.8
− 4.4	24	75.2	22.2	72	161.6
− 3.9	25	77.0	22.8	73	163.4
− 3.3	26	78.8	23.3	74	165.2
− 2.8	27	80.6	23.9	75	167.0
− 2.2	28	82.4	24.4	76	168.8
− 1.7	29	84.2	25.0	77	170.6
− 1.1	30	86.0	25.6	78	172.4
− 0.6	31	87.8	26.1	79	174.2
0.0	32	89.6	26.7	80	176.0
0.6	33	91.4	27.2	81	177.8
1.1	34	93.2	27.8	82	179.6

C ← F	C → F		C ← F	C → F
28.3	83	181.4	127	260	500
28.9	84	183.2	132	270	518
29.4	85	185.0	138	280	536
30.0	86	186.8	143	290	554
30.6	87	188.6	149	300	572
31.1	88	190.4	154	310	590
31.7	89	192.2	160	320	608
32.2	90	194.0	166	330	626
32.8	91	195.8	171	340	644
33.3	92	197.6	177	350	662
33.9	93	199.4	182	360	680
34.4	94	201.2	188	370	698
35.0	95	203.0	193	380	716
35.6	96	204.8	199	390	734
36.1	97	206.6	204	400	752
36.7	98	208.4	210	410	770
37.2	99	210.2	216	420	788
37.8	100	212.0	221	430	806
38	100	212	227	440	824
43	110	230	232	450	842
49	120	248	238	460	860
54	130	266	243	470	878
60	140	284	249	480	896
66	150	302	254	490	914
71	160	320	260	500	932
77	170	338	266	510	950
82	180	356	271	520	968
88	190	374	277	530	986
93	200	392	282	540	1004
99	210	410	288	550	1022
100	212	413.6	293	560	1040
104	220	428	299	570	1058
110	230	446	304	580	1076
116	240	464	310	590	1094
121	250	482	316	600	1112

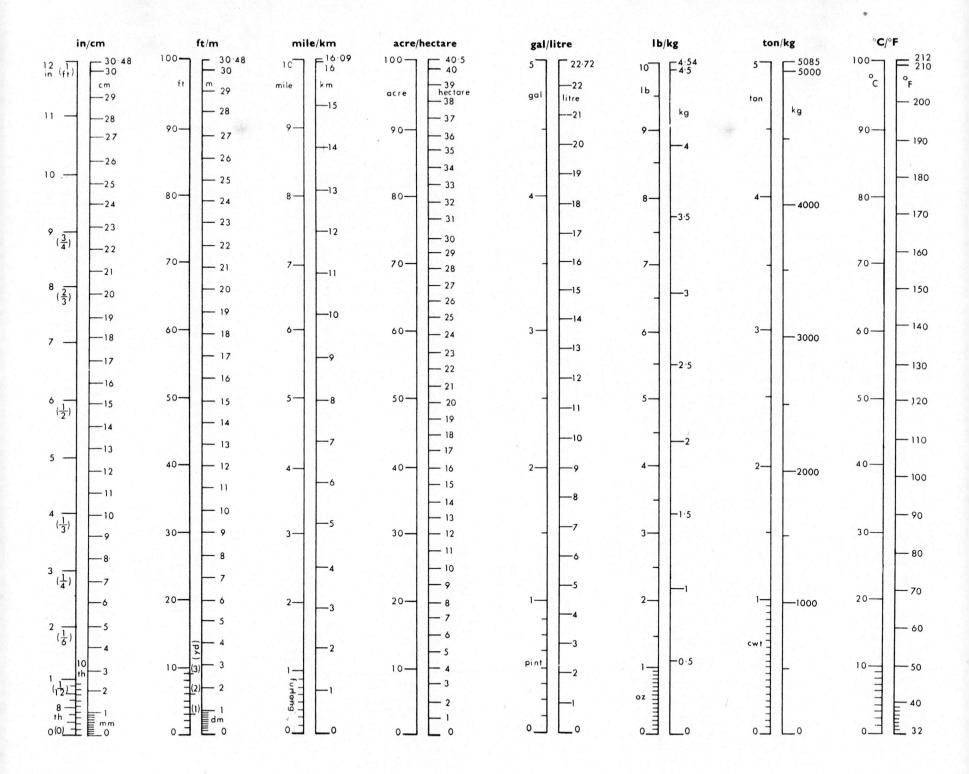

Index

391

Date Due

UML 735